Russia's Agriculture in Transition

RURAL ECONOMIES IN TRANSITION

Series Editors

Zvi Lerman
The Hebrew University of Jerusalem
Csaba Csaki
The World Bank and the Budapest Economics University
Johan Swinnen
Katholieke Universiteit Leuven

Much of the world's rural population lives in transition countries that are attempting to transform their economies from the centrally planned socialist system to a market-oriented system. The Rural Economies in Transition series evaluates the deep impact these transformations are having in Central and Eastern Europe, the former Soviet Union, and Asia. Based on original empirical work by international teams at the cutting edge of research, each volume presents the results of one complete empirical study devoted to a single country or region. The editors have designed the series to provide agricultural, development, and political economists, as well as scholars in other disciplines, with authoritative and up-to-date factual information about rural transition. The editors welcome manuscript submissions from scholars engaged in original empirical research in the field.

Private Agriculture in Armenia, by Zvi Lerman and Astghik Mirzakhanian
Romanian Agriculture and Transition toward the EU, edited by Sophia Davidova and Kenneth J. Thomson
Transition, Institutions, and the Rural Sector, edited by Max Spoor
Greening Industrialization in Asian Transitional Economies: China and Vietnam, edited by Arthur P. J. Mol and Joost C. L. van Buuren
Agriculture in Transition: Land Policies and Evolving Farm Structures in Post-Soviet Countries, by Zvi Lerman, Csaba Csaki, and Gershon Feder
Building Market Institutions in Post-Communist Agriculture: Land, Credit, and Assistance, edited by David A. J. Macey, William Pyle, and Stephen K. Wegren
Russia's Food Policies and Globalization, by Stephen K. Wegren

Cooperation in the Romanian Countryside: An Insight into Post-Soviet Agriculture, by Rachel Sabates-Wheeler

Reinventing the Cuban Sugar Agroindustry, edited by Jorge F. Pérez-López and José Alvarez

Integrated Development of Agriculture and Rural Areas in Central European Countries, edited by Sophia Davidova, Kai Bauer, and Michael Cuddy

Measuring Social and Economic Change in Rural Russia: Surveys from 1991 to 2003, by David J. O'Brien and Valery V. Patsiorkovsky

Russia's Agriculture in Transition: Factor Markets and Constraints on Growth, edited by Zvi Lerman

Russia's Agriculture in Transition

Factor Markets and Constraints on Growth

EDITED BY
ZVI LERMAN

LEXINGTON BOOKS

A division of
ROWMAN & LITTLEFIELD PUBLISHERS, INC.
Lanham • Boulder • New York • Toronto • Plymouth, UK

LEXINGTON BOOKS

A division of Rowman & Littlefield Publishers, Inc.
A wholly owned subsidiary of The Rowman & Littlefield Publishing Group, Inc.
4501 Forbes Boulevard, Suite 200
Lanham, MD 20706

Estover Road
Plymouth PL6 7PY
United Kingdom

British Library Cataloguing in Publication Information Available

Library of Congress Cataloging-in-Publication Data

Russia's agriculture in transition : factor markets and constraints on growth / edited by Zvi Lerman.
p. cm. — (Rural economies in transition)
Includes index.
ISBN-13: 978-0-7391-2009-5 (cloth : alk. paper)
ISBN-10: 0-7391-2009-3 (cloth : alk. paper)
1. Agriculture—Economic aspects—Russia (Federation) 2. Agriculture and state—Russia (Federation) I. Lerman, Zvi, 1941–
HD1995.15.Z8R89 2008
338.10947'090511—dc22 2007033268

Printed in the United States of America

™ The paper used in this publication meets the minimum requirements of American National Standard for Information Sciences—Permanence of Paper for Printed Library Materials, ANSI/NISO Z39.48–1992.

Contents

Chapter 1
Introduction

Zvi Lerman

This volume grew out of a three-year research project supported by the U.S. Agency for International Development as part of its BASIS/CRSP collaborative research support program. BASIS is the acronym for Broadening Access and Strengthening Input Market Systems and the program "seeks to improve rural prosperity by making markets work for all" (see the BASIS/CRSP web site www.basis.wisc.edu). BASIS had supported numerous policy-oriented research projects in developing countries all over the world, and the "emerging market" economies (such as Russia) naturally appeared to fit its focus on global market constraints. Bruce Gardner from the University of Maryland at College Park and Evgeniya Serova from the Analytical Centre on Agri-Food Economics (AFE) in Moscow accordingly organized an international group of Russian and Western researchers, all with proven experience in studying agricultural transition. The group embarked on a study of the development of factor markets in Russian agriculture and the impact of existing constraints on agricultural growth. The study was officially launched in July 2001 in Golitsino, a pleasant training and conference facility in a small village some 40 km west of Moscow, and wrapped up in September 2004, at a summary conference held at the Academy of National Economy in Moscow.

The study focused specifically on the development of factor markets in Russian agriculture—markets for labor, purchased inputs, land, and credit. This distinguishes the proposed volume from many other publications on agricultural transition to market in former socialist countries, which pay more attention to markets for farm products and not markets for farm inputs.

Another distinctive feature of the study is the integration of official statistical data with the findings of fairly a large questionnaire-based survey.

This survey, referred to as the 2003 BASIS survey throughout the volume, was specifically designed to cover issues of agricultural land, labor, supply and use of purchased inputs, access to credit, and ultimately farm production with a view to efficiency estimations.

Structure of the Book

The volume chapters are organized in three parts: changing farm structure (Chapters 2-4), developing input markets, including markets for land, labor, farm inputs, and credit (Chapters 5-9), and production efficiency (Chapters 10-14).

Farm Structure (Chapters 2-4)

The Soviet farm structure was notoriously inefficient, and one of the goals of transition was to achieve higher productivity and efficiency through market-oriented restructuring of farms. Vasilii Uzun (Chapter 2) provides an overview of the changes in Russia's farm sector during the transition, covering both large agricultural business (corporate farms) and small individual farms. In his analysis, Uzun uses a unique database of all 24,000 corporate farms in Russia, supplementing the data with national and regional statistics. Sergei Sazonov and Damira Sazonova (Chapter 3) report the results of a unique survey of peasant farms in Tambov Oblast in Central Russia, which has been going on since 1992. Their paper provides an in-depth look into the functioning of this new sector of small-scale Russian agriculture. In Chapter 4, Dmitri Rylko and Robert Jolly, a Russian-American team supported by contributions from Irina Khramova and Vasilii Uzun, discuss the other extreme of the farm structure spectrum: this chapter uses case studies and interviews to describe the development of agroholdings, those super-large mega-farms that began to emerge in the mid-1990s with investments from non-agricultural interests. This unique Russian phenomenon has not been sufficiently explored in Western literature, and its impact on the rural scene justifies much greater attention.

Land (Chapters 5-6)

The economic function of land markets is, first, to put each plot of land in the hands of the person who values it most and, second, to enable agricultural land to flow from less efficient to more efficient users. Natalya Shagaida and Zvi Lerman, another international team, discuss the constraints on the development of agricultural land markets in Russia (Chapter 5). Although the full legal framework for transactions in land is now in place, lack of market information and bureaucratic complexity of registration procedures are major obstacles to the development of land markets. The analysis in this article exploits national and regional data on land ownership and transactions in land, as well as the results of the 2003 BASIS survey. Leonard Rolfes (Chapter 6) supplements this analysis with a Western view of the changes in Russian land policies and legislation.

Labor (Chapter 7)

Vladimir Bogdanovskii provides in this chapter a sweeping overview of the changes in rural and agricultural employment since 1990. Using national statistics, the 2003 BASIS survey, and a number of smaller original surveys, Bogdanovskii shows how rural labor shifted from corporate farms to the individual sector. He demonstrates that registered rural unemployment in Russia is kept in check by the dual mechanism of reducing the effective working time (i.e., allowing hidden unemployment) and paying labor less than the value of marginal product. Underutilization of labor is particularly pronounced in the small individual farms, which act as a "labor sink" for rural workers who have been dismissed by corporate farms.

Farm Inputs (Chapter 8)

Evgeniya Serova and Olga Shick discuss the supply and demand for five groups of farm inputs: machinery, fuel, fertilizer, seeds, and animal feed. They conclude that the state no longer has a role as a direct supplier of inputs to agricultural producers and that this function has shifted to wholesalers, traders, and manufacturers, who sell mainly for cash and bank transfers, not barter. The strong imperfections that still prevail in input markets have encouraged vertical integration, with fertilizers, fuel, and machinery delivered in substantial quantities through internal channels of large holding structures (these conclusions echo the discussion on agroholdings in Chapter 4).

Finance (Chapter 9)

Because of persistent data limitations, financial questions can be analyzed only for corporate farms. Chapter 9 co-authored by three Russian researchers (Olga Yastrebova, David Epshtein, and Aleksandr Subbotin) paints a broad canvas of financial issues in Russian agriculture. Using national statistics to address the pressing issues of non-payments and bankruptcy, the chapter concludes that Russian agriculture does not need more subsidies; it needs better functioning market channels and more qualified managers to be successful and profitable. An econometric analysis of access to credit based on the 2003 BASIS survey data shows in that efficiency and profitability are the main factors enabling farms to borrow, whereas asset endowments (land and machinery) are less important, probably because of underdeveloped collateralization mechanisms. A unique database of Russian corporate farms in Leningrad Oblast is used to develop a measure of financial solvency and to show that management quality is the main factor reponsible for financial health.

Production Efficiency (Chapters 10-14)

Since the main objective of transition is to improve productivity and efficiency, the volume appropriately concludes with this subject. The discussion is divided into two topics, echoing the methodological dichotomy between estimation of allocative efficiency (which uses production function methods) and technical efficiency (which uses production frontier methods drawing on Data Envelopment Analysis and Stochastic Frontier Analysis).

The first topic, focusing on allocative efficiency of Russian agriculture, starts with an overview by William Liefert, who summarizes the various analyses carried out by the BASIS researchers (Chapter 10). This is followed by three chapters (Margarita Grazhdaninova and Zvi Lerman in Chapter 11, Vasilii Uzun in Chapter 12, and David Epshtein in Chapter 13) that give more technical detail of the respective analyses. The main conclusions of the allocative efficiency chapters are that material inputs in general are overused, while fertilizer in particular is underused. A surprising result is that labor is underused in the sense that the value of labor's marginal product is higher than the cost of labor.

The second topic, focusing on technical efficiency, is covered in Chapter 14, which integrates an overview by Gregory Brock with detailed technical results from Margarita Grazhdaninova, Vasilii Uzun, David Epshtein, and Zvi Lerman. Roughly half of Russia's corporate farms are at the efficiency frontier while the other half produce about 50% of the output that they could produce with their inputs if they were managed as well as the upper half. If reforms could get the lower half of farms into managerial hands of the quality of the upper half, Russian agricultural output would rise by 25% at no additional resource cost. Even if done over a decade, a 2.5% annual increase in total factor productivity growth would outstrip the performance of the Western agricultural economies during their periods of rapid technological change.

Because of difficulties with data the main thrust of the allocative efficiency analysis is on corporate farms—the reorganized successors of Soviet kolkhozes and sovkhozes. The technical efficiency analysis, on the other hand, presents comparative results for corporate and individual farms. Some allocative efficiency results for peasant farms in Central Russia are presented in Chapter 3 by Sergei Sazonov and Damira Sazonova.

The 14 core contributions are summarized in Chapter 15, where Bruce Gardner pulls the various chapters together by drawing lessons and suggesting policy implications that may be relevant to Russian decision makers.

Survey Design

One of the main data source for this study is a questionnaire-based farm survey covering broad issues of agricultural land, labor, supply and use of purchased inputs, access to credit, and ultimately farm production with a view to efficiency estimations. The survey was implemented between November 2002 and February 2003 in three Russian provinces (oblasts) with different agro-climatic and political characteristics—Rostov, Nizhnii Novgorod, and Ivanovo. Rostov is a grain-producing region in the fertile chernozem belt, located some 1,200 km south-east of Moscow, on the eastern border of Ukraine. Nizhnii Novgorod is located some 600 km east of Moscow, on the Volga River; because of its relatively poor non-chernozem soils Nizhnii Novgorod emphasizes livestock production, but its climate is still suitable for growing grain. Ivanovo, located 300 km north-east of Moscow, has mixed crop-livestock agriculture, but its northern climate is generally unsuitable for grain, and crop production includes mainly potatoes, root vegetables, and grasses. The three provinces are characterized by different levels of political and economic reform: Rostov follows the most interventionist agricultural policies, while Nizhnii Novgorod is the least interventionist of the three (see Chapter 8 for the corresponding ranking).

Table 1.1. Mean characteristics of sample farms by type

	Corporate farms	Peasant farms	Household plots
Sales revenue ('000 rubles)	6,113	401	38
Agricultural land (ha)	3,541	267	1.2
Agricultural workers (persons)	114	3.9	3.2
Input costs ('000 rubles)	2,968	144	15
Machinery (pieces adjusted for quality)	21.6	2.1	0.13
Livestock (st. head)	417.5	6.4	5.8

Source: 2003 BASIS survey.

The survey covered the basic organizational forms of agricultural producers in Russia: corporate farms (or "farm enterprises," the successors of former kolkhozes and sovkhozes) and individual farms (including both peasant farms and household plots). The corporate farm sample was obtained by randomly selecting three districts in each province and then surveying all the corporate farms in these districts. This approach produced a sample of 142 corporate farms. Individual farms were sampled in the second stage by selecting three individual farms from the village associated with each corporate farm. The individual farms were selected at random from lists provided by local departments of agriculture, which by request included only farms with significant commercial orientation (i.e., individual farms with sales equal to 50% of output or more). After the fact, the total sample of 425 individual farms was found to consist of 223 units with substantially more than 10 hectares of land (classified as peasant farms for the purposes of our analysis) and 202 units smaller than 10 hectares (classified as household plots).[1] The mean characteristics of the farms surveyed are summarized in Table 1.1 by farm type—corporate farms, peasant farms, and household plots. The numbers in the table clearly show the size differences between corporate farms (largest), peasant farms (mid-sized), and household plots (smallest).

The survey database created at the Analytical Centre for Agri-Food Economics in Moscow was used by all group members for analysis, and their joint results are comprehensively reflected in this volume. The survey data are supplemented, when appropriate, with official statistics from the Russian Statistical Office (Goskomstat) and the Russian State Cadastre (Roszemkadastr, formerly known as Goskomzem).

Terminology and Abbreviations

Throughout this volume the term "corporate farms" denotes those large-scale farms that succeeded the former collective and state farms, generally through legally mandated reorganization. The creation and development of these entities is described in Chapter 2. Alternative terms often used in the literature are "farm enterprises" or "agricultural enterprises." We prefer the term "corporate farms" as a counterpoint to "individual farms."

"Individual farms" are subdivided into "household plots" and "peasant farms" (the latter often called "private farms" in the literature—a misnomer, as corporate farms are also private in the sense of "not state-owned"). Although household plots and peasant farms are two distinct organizational

---

[1] The two sets of survey instruments used in face-to-face interviews with corporate farm managers and individual farmers can be downloaded from the volume editor's web site at http://departments.agri.huji.ac.il/economics/lerman-main.html.

categories in Russia, our sampling methodology randomly selected individual farms without regard to their legal classification. The respondents were classified into "peasant farms" and "household plots" after the fact, based on differences in land holdings (see above in *Survey Design*). The size differences between household plots and peasant farms in the survey are discussed in Chapter 5, and the national differences between the two categories are examined in Chapter 2.

The Russian acronym Goskomstat stands for the State Committee for Statistics of the Russian Federation—the official source for all Russian statistics. Goskomstat regularly publishes the Statistical Yearbook for Russia and a number of specialized yearbooks (Agriculture in Russia, Labor, Finances), all of which are cited through this volume. In addition to comprehensive regular yearbooks, Goskomstat has a range of periodic publications (e.g., results of quarterly income and employment surveys) and some of its unpublished databases can be accesses for a fee.

Another common acronym is Roszemkadastr—Russia's State Cadastre. This organization has gone through a number of name changes since 1992, which are listed in a special note in Chapter 5.

Other abbreviations used in this volume (in alphabetical order):

AE	Allocative Efficiency
ALP	Agricultural Labor Productivity
CRS	Constant Returns to Scale
DEA	Data Envelopment Analysis
GAO	Gross Agricultural Output
GDP	Gross Domestic Product
HH	household (as in HH plot for household plot)
Kolkhoz	collective farm
MinAg	Russia's Ministry of Agriculture
Sovkhoz	state farm
RSFSR	Russian Soviet Federated Socialist Republic—the official name Russia (or the Russian Federation) had until the dissolution of the Soviet Union in December 1991
SFA	Stochastic Frontier Analysis
TE	Technical Efficiency
TFP	Total Factor Productivity
VMP	Value of Marginal Product
VRS	Variable Returns to Scale

The full literature references as a rule are given in notes. Separate lists of references have been assembled only for three "citation-intensive" chapters (Chapters 6, 10, and 14). In these chapters, the references in the body of the text follow the author-year convention.

Part I

Changing Farm Structure

Chapter 2

Large and Small Business in Russian Agriculture

Vasilii Uzun

Russian agriculture relies on two distinct groups of producers: large and superlarge corporate farms on the one hand; small and ultrasmall family farms on the other. This chapter describes how the farms in these two categories evolve and adapt to market conditions. The term "small agricultural business" is used in this chapter to denote individual and family farms (both peasant farms and household plots), as well as small agricultural enterprises, which are legal persons employing up to 60 permanent workers. All other agricultural enterprises (corporate farms) are included in the rubric of "large agricultural business."

In market economies, most of the agricultural output is produced by small business. The predominance of small business in market agriculture has been widely discussed in the literature. Economic theory explains the sustainability and efficiency of family farms by the higher motivation of the owner compared with that of a hired worker, by the orientation toward family consumption instead of profit maximization, by the psychological value of working on one's own farm, and by the unity of work and family life. Moreover, lack of economies of scale in crop production does not justify a large body of hired workers, and limited family labor is sufficient.

Large agricultural business develops in response to economies of scale. In this respect, capitalist economic theory is in agreement with socialist ideology, which justified the emphasis on large farms by considerations of efficiency improvement achievable through economies of scale. However, the inevitable increase of management difficulties, transaction costs, and ecological hazards with the increase of farm size limits the growth of large business in agriculture. Modern institutional economics describes these limits to size in terms of moral hazard or opportunism. Oliver Williamson defined opportunism as "self-interest seeking with guile," when the agent practices lying, stealing, and cheating.[1] These considerations are highly relevant in modern Russia, where opportunistic behavior has been inherited from the socialist system and most agricultural workers still regard it as the norm.

Moral hazard can be counteracted by provision of appropriate safeguards in labor contracts and by increased monitoring of the workforce. This works very well in manufacturing industries, where the workplace is spatially restricted, each worker has precisely defined tasks and responsibilities (as on an assembly line), the quantity and quality of labor input is easily measured. The situation is different in agriculture. Farming is a spatially dispersed activity, and it is difficult, or virtually impossible, to monitor all contract workers, to safeguard all machinery and equipment, to protect all crops and all livestock everywhere. Moreover, it is generally impossible to define the precise tasks of each agricultural worker in the operations-research sense: the workers are required to exercise initiative and improvise in response to the vagaries of weather and diversity of activities prescribed by the changing crop and livestock mix. A manager cannot monitor and control everything everywhere on a large farm, and opportunism carries unavoidable transaction costs that offset the benefits of scale economies. The unavoidability of these transaction costs distinguishes corporate farms from family farms, where the owner is simultaneously the manager, the laborer, the watchman, and the controller. Family members do not require intensive monitoring, as they are more conscientious than hired labor. The number of hired workers in family farms is usually small, and they are always under the watchful eye of family members. Moral hazard and opportunism are therefore much less of a problem in family farms.

It may be interesting to note that in market economies large agricultural business may still develop if there is no spatial dispersion and no diversity of activities. Such industrial conditions exist in specialized poultry and pig operations, which often grow to very large size indeed. They also exist in specialized one-crop plantations (e.g., banana plantations), where strict monitoring is built into a vertically integrated industrial system to ensure timely delivery of a perishable commodity from the field to the ship and thence to the processing plant.

During the socialist era, Russian agriculture was dominated by large business. In 1990, on the eve of transition, there were 25,800 collective and state farms (kolkhozes and sovkhozes) in Russia, which on average cultivated 7,800 hectares of agricultural land with 320 permanent workers. So-

[1] O. Williamson, *The Economic Institutions of Capitalism: Firms, Markets, Relational Contracting*, New York: The Free Press (1985), p. 47.

cialist agricultural policy was based on two precepts. The first precept stated that *large farms are more efficient than small farms.* This led to continued aggregation of farms and creation of ever larger entities. In the late 1920s-early 1930s, collectivization combined hundreds of peasant households into one kolkhoz (collective farm). In the 1950s, kolkhozes were aggregated into still larger farms, and the average size of a kolkhoz increased five-fold. In the 1960s-1970s, Soviet agriculture was subjected to the policy of interfarm cooperation and integration, which increased farm sizes by creating so-called interfarm enterprises. In the 1980s, regional agroindustrial associations began to be created as a means for controlling the agricultural production of a whole administrative region as one unit.

The second precept advocated that *state property is the highest and most efficient form of ownership.* The resulting policy led to a growth of the state sector in agriculture (sovkhozes or state farms) and a reduction of the private and even the collective (kolkhoz) sector. Between 1960 and 1990, the number of state farms in Russia increased from 4,000 to 13,000, while the number of collective farms conversely declined from 18,500 to 12,800. State farms controlled about 60% of agricultural land in 1990. The share of agricultural assets in state farms increased from less than 40% in 1960 to more than 60% in 1990.

The predominance of large agricultural business during the socialist era was achieved by the following means:

- "The return of serfdom." Kolkhoz members did not have passports for domestic travel and could not leave the village without official permission. Rural people had no choice but to work in the local farm enterprise.
- Allocation of farmland by the state (state farms also received their non-land assets from the state). Land was given in perpetual use and remained nontransferable. Farm sizes were determined administratively and independently of efficiency considerations.
- The use of the legal enforcement apparatus to combat moral hazard behavior among agricultural workers by ensuring compliance with officially sanctioned behavioral norms. The socialist theory claimed that people could be re-educated and that opportunism would be ultimately eradicated. As long as this had not happened, the state directly protected farm property, imposing severe prison sentences even for stealing very small quantities of wheat.
- Simplification of the farm management system through top-down assignment of production targets, input deliveries, state prices, exclusive suppliers for each user, and so on.

- Introduction of soft budget constrains. If a farm enterprise did not have enough money, the state would usually cover the shortfall and write off debt. Bankruptcy was not practiced.
- Restrictions on small business. Large farms survived because they did not have to compete on the domestic market. The state limited the size of individual plots, the number of livestock in individual ownership, the use of farm machinery by individual farms (a rural household was not allowed to have a horse or a tractor). Family members were allowed to farm their plot only after completing a full workday at the local farm enterprise.
- Absence of external competition. Collective and state farms were protected from external competition, because the state monopolized foreign trade and did not allow importers to penetrate the domestic market. Although food imports existed, they did not affect the domestic prices and the procurement of agricultural commodities on the domestic market.

Most of these factors have not survived the process of transition to market; the influence of others has been dramatically attenuated. Peasants can no longer be forced to work on a farm enterprise, land and other assets have to be leased or purchased, the state is unable to help combat moral hazard, managers now have to worry about what to produce, how to sell, and where to buy inputs. If a farm does not meet its financial obligations, it may be forced into bankruptcy. Farm enterprises are under constant pressure from competitors—both domestic (small agricultural business) and external (importers).

This chapter is organized as follows. We start with a classification of the forms of agricultural business in Russia. We then proceed to describe the main features of the various organizational forms, their evolution and adaptation to market conditions, and the changing role of large and small business during the transition. The impact of government policies on large and small business is discussed. We conclude with a summary of findings and some policy recommendations.

The analysis is based on several statistical sources, which include official Goskomstat data; the Goskomstat database of all large farm enterprises in Russia with a time series of 250 variables for 1995-2002; a special database of small agricultural enterprises; a MinAg database of annual financial reports of large farm enterprises for 1994-2002. This information has been applied to group the farms by organizational form, various size measures, and an index of financial health.

Organizational Forms of Agricultural Producers in Russia

Figure 2.1 presents a general hierarchical classification of Russian farms as it is observed today. For the purposes of this chapter, we classify agricultural producers into "large business" and "small business." A more conventional classification is by organizational form into corporate farms and individual farms. There is unfortunately a partial overlap between the two classifications: large and medium corporate farms with more than 60 annual-average workers are included in "large business," whereas small corporate farms with up to 60 workers are included in "small business." "Large business" is thus entirely corporate farms. "Small business" is a mixture of small corporate farms and individual farms.

Fig. 2.1. Organizational forms in Russian agriculture

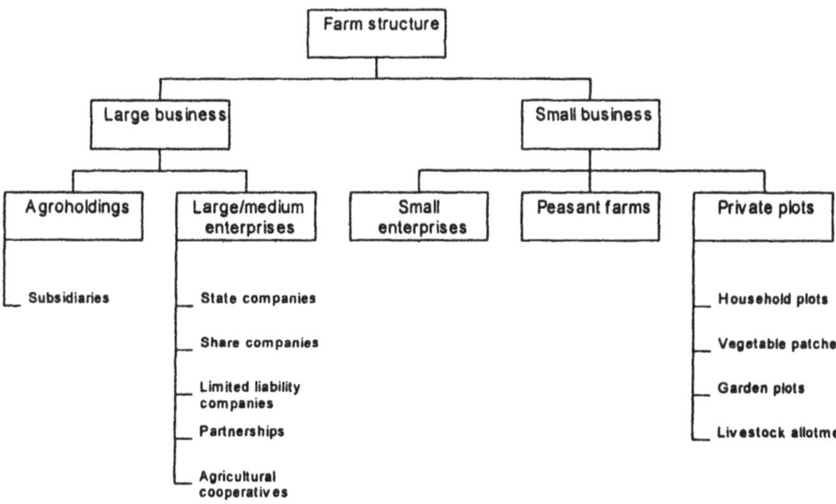

Large and medium corporate farms ("large business") may be organized in a variety of legal forms: state and municipal enterprises, shareholding companies (closed or open), limited liability companies, partnerships, agricultural cooperatives, and other forms. Small corporate farms (included in "small business") are typically organized as small enterprise—a special legal form in Russian company law. A subcategory of corporate farms includes "superlarge" farms—so-called agrofirms and agroholdings. The emergence of these "superlarge" entities is a recent phenomenon, which is discussed in Chapter 4.

Individual farms fall into two categories: peasant farms and household plots. Farms in both categories are treated as physical persons (unincorporated entities), although peasant farms have to register with regional land and tax authorities. Household plots are divided into four subgroups. The traditional household plot is owned by a rural family and it typically consists of a plot around the house in the village plus additional plots in the open fields (which may be reallocated from year to year, unlike the "permanent" plot in the village). In this subgroup, in addition to 14 million established household plots of rural families, there are more than 6 million new household plots on land allocated to rural and urban families for housing construction in villages. Vegetable patches are plots allocated to rural residents who live in apartment blocks and thus do not have a house with a traditional household plot (construction is not allowed on these plots). Garden plots are allocated to urban residents for farming and recreation (the so-called dacha plots for the construction of a small country house are included in this category). The fourth subgroup is "livestock allotments," i.e., land allocated for livestock grazing and hay mowing around villages and rural townships.

One of the main aims in the early stages of agrarian reform was the creation of a "multi-organizational" farming structure, with a level playing field for farms of all organizational forms. In the socialist era, state farms (sovkhozes) and collective farms (kolkhozes) enjoyed preferential treatment, while operating under strict central controls. Private farming was prohibited. All farms cultivated state land. Sovkhoz assets were always owned by the state, whereas kolkhoz assets, although legally collective property, were indivisible and thus practically state-owned. Agrarian reform began in the early 1990s with the privatization of land and assets in sovkhozes and kolkhozes. Everything was divided between active and retired workers of farm enterprises. Workers of rural social services also participated in land privatization. Land was divided equally among all beneficiaries, and the net assets of farm enterprises were divided proportionately to "labor participation," allowing for seniority and cumulative wages.

Privatization made it possible to move from state-controlled to private farm management. Kolkhozes and sovkhozes were reorganized into corporate farms, peasant farms began to emerge, restrictions on household plots were gradually removed or relaxed. The reforms produced a sharp reduction of the state sector in agriculture and encouraged the emergence of small rural business. The number of large farm enterprises did not change much (it fluctuated between 24,000-28,000), but their role changed significantly. Large farms lost about one-third of their agricultural land; much of their labor and assets migrated to farms of other types. The number of small agricultural enterprises increased rapidly, reaching 30,000. An entirely new private sector emerged, with 260,000 peasant farms and 17 million hectares of land. The land in household plots increased from 2 million hectares to 28 million hectares), although their number did not change much.

Large Agricultural Business

Changing Organizational Forms

As of the beginning of 2003, Russia's "large agricultural business" comprised 24,343 large and medium corporate farms. These farms were essentially created in two legal waves. First came the RSFSR Law on Enterprises and Entrepreneurial Activity, which was passed in 1991 before the dissolution of USSR (when Russia was still called RSFSR—Russian Soviet Federated Socialist Republic). This law recognized a variety of corporate organizational forms and thus provided the framework for Yeltsin's presidential decrees of December 1991, which mandated the reorganization of kolkhozes and sovkhozes into closed and open shareholding societies, limited liability partnerships, mixed (*komandit*) partnerships, associations of peasant farms, and agricultural cooperatives (created on the basis of yet another Soviet law, the 1988 USSR Law on Cooperation). The variety of organizational forms created in the early 1990s was deceptive, however, as all these corporate farms despite their different sounding names were essentially linear successors of Soviet kolkhozes and sovkhozes. They operated like agricultural production cooperatives, with each member having one vote regardless of invested capital, which was probably inevitable in a setting where each farm had 500-800 shareholding members.

Table 2.1. Organizational forms among corporate farms 1995 and 2002

Organizational forms	1995		2002	
	number	%	number	%
State and municipal enterprises, sovkhozes	3,385	13	2,728	11
Shareholding companies	5,417	20	4,197	17
Open shareholding companies	298	1	1,191	5
Limited liability companies, partnerships	7,256	27	2,739	11
Agricultural production cooperatives, kolkhozes, other forms	10,828	40	14,679	60
Total	26,896	100	24,343	100

Source: Goskomstat.

The second wave of corporate reorganizations came in 1995 with the adoption of the new Civil Code and a number of associated laws, such as the Law on Shareholding Companies, the Law on Agricultural Cooperation, and the Law on Limited Liability Companies. The new legal framework recognized closed and open shareholding companies, limited liability companies, partnerships, and agricultural production cooperatives. In addition, it introduced the category of state and municipal "unitary" enterprises. Just as the 1991 law was used to initiate a massive reorganization of kolkhozes and sovkhozes into new organizational forms, the 1995 legislation forced large-scale reorganization of farms from old organizational forms that became illegal into new forms compatible with the Civil Code.

Table 2.1 gives an inventory of various organizational forms at two points in time before and after the adoption of the Civil Code (1995 and 2003). It should be mentioned that after nearly a decade and a half of reorganization and reform, there still exist significant remnants of old Soviet farm types. As of the beginning of 2003, there were 1,568 kolkhozes and 391 sovkhozes among the 24,000 corporate farms. Thus, 8% of all corporate farms have survived the two reorganization waves and persistently kept their Soviet organizational form.

Comparison of the organizational profiles for 1995 and 2002 reveals a significant decrease in the percentage of shareholding companies, limited liability companies, and partnerships, which is accompanied by a very substantial increase in the percentage of agricultural production cooperatives. This post-1995 transformation was induced by a feature of the new legislation, which limited to 50 the maximum number of owners in shareholding companies and limited liability companies. In most farm enterprises that succeeded the Soviet kolkhozes and sovkhozes the number of shareholders was much greater than 50, and they were forced to reorganize as agricultural production cooperatives during 1995-2000. In addition to this technical reason, there was yet another economic factor that encouraged reorganization as agricultural cooperatives. The new 1998 law on agricultural cooperation prevented creditors and executors from foreclosing on certain assets in a cooperative: farm machinery, breeding livestock, and everything classified as indivisible assets were exempt from foreclosure in bankruptcy or debt-recovery proceedings. This legal provision gave a clear advantage to agricultural cooperatives compared with farms of other organizational forms, which often fell prey to creditors in the tumultuous financial environment of the late 1990s. Although open shareholding companies (i.e., companies with tradable shares) account for a small percentage of corporate farms, their number increased significantly between 1995 and 2002, probably due to the entry of outside investors, who ensure liquidity of their investment by insisting on tradability of shares (see Chapter 4).

Table 2.2 presents some partial productivity measures for farms of various organizational forms. Agricultural production cooperatives are clearly the least productive organizational form by all six measures. This is probably a direct outcome of the 1998 law on agricultural cooperation, which attracted the weakest and most vulnerable farms to the safe haven of the cooperative form of organization. For the other organizational forms, no clear performance ranking can be established, as they score differently by different productivity measures.

While recognizing the emerging "multi-organizational" structure of agriculture, Russian scholars often cite agricultural production cooperatives as the most preferred form—without any real economic substantiation. Thus, Shut'kov and Ushachev unequivocally state that "the formation of a modern 'multi-organizational' economy in agriculture requires a gradual transition from shareholding organizations, so alien to farming, to cooperative forms of organization" and "cooperative farms are the most promising organizational form in agriculture."[2] To the extent that various authors find it necessary to substantiate their position regarding cooperatives, reference is usually made to communal and collective traditions of the Russian peasant. Another argument is that all other organizational forms are strictly private, and cooperatives are the only form that retains some flavor of socialism.

Table 2.2. Partial productivity measures for different organizational forms of large farms (2001, in '000 rubles)

Organizational forms	Ratio of sales revenue to		
	labor*	land**	assets#
State and municipal enterprises, sovkhozes	83	727	411
Shareholding companies	100	694	543
Open shareholding companies	114	896	713
Limited liability companies, partnerships	69	421	511
Agricultural production cooperatives, kolkhozes, other forms	51	307	294
Total	69	449	390
	Ratio of value added to		
	labor*	land**	assets#
State and municipal enterprises, sovkhozes	34	294	166
Shareholding companies	39	272	213
Open shareholding companies	38	301	240
Limited liability companies, partnerships	26	159	192
Agricultural production cooperatives, kolkhozes, other forms	23	136	130
Total	29	185	161

* per worker; ** per 100 ha arable land; # per 1 million rubles production assets.
Source: Calculated from Goskomstat database of large and medium farm enterprises.

[2] A. A. Shut'kov and I. G. Ushachev, eds., *Mnogoukladnaya ekonomika APK Rossii. Voprosy teorii i praktiki*, Moscow: Kolos (1998), pp. 10, 13.

The "pros" and "cons" of agricultural production cooperatives have been discussed in detail by the author in a recent Russian publication.[3] A comprehensive description of the world experience with production cooperatives and an analysis of a theoretical model of a production cooperative can be found in Serova's textbook.[4] A recent discussion of this issue in Western literature is provided in Lerman et al.[5] It suffices to mention that production cooperatives are very rare in market economies; that they suffer from weak motivation and low competitiveness. Frohberg[6] points to the weak motivation of both workers and managers in cooperatives as one of the shortcomings of the cooperative form of organization. He stresses that cooperatives are unable to attract highly qualified managers, because they cannot offer competitive compensation. Russian economists often bring the example of the Israeli kibbutz as a successful production cooperative. This is a highly idealized view, as the kibbutz sector in Israel accumulated unsustainable debts and went through virtual bankruptcy in after 1986.[7] The rare examples of relatively successful cooperatives are cases of very small groups (e.g., the French HAJEK cooperatives formed by two-three families) or cooperatives whose activity is controlled by a strong leader with a small group of members, while all other members are in fact hired workers.

Development Trends among Corporate Farms

Concentration of capital

In the process of reform, the assets of former kolkhozes and sovkhozes (both their fixed assets and working capital) were divided into asset shares and distributed to active workers and retirees of farm enterprises. Most of the new owners simply invested their asset shares in the equity capital of the local corporate farm. At the time of reorganization, the entire equity of corporate farms, and therefore also 100% of the votes at shareholder meetings, belonged to physical persons—former kolkhoz and sovkhoz workers. As

[3] V. Ya. Uzun, *Sel'skokhozyaistvennyi proizvodstvennyi kooperativ*, Moscow: ERD (1999).

[4] E. V. Serova, *Agrarnaya ekonomika*, Moscow: GUVShE (1999), pp. 99-114.

[5] Z. Lerman, C. Csaki, and G. Feder, *Agriculture in Transition: Land Policies and Evolving Farm Structures in Post-Soviet Countries*, Lanham, MD: Lexington Books (2004).

[6] K. Frohberg, "Optimal size of a farm enterprise in a market economy" [in Russian], *Politekonom*, No. 1[12]: 22-23 (1999).

[7] Y. Kislev, Z. Lerman, and P. Zusman, "Recent experience with cooperative farm credit in Israel," *Economic Development and Cultural Change*, 39(4): 773-789 (July 1991). A discussion of these issues is available in Russian in T. Hanstad, "Are small farms appropriate for former Soviet republics?" RDI Report on Foreign Aid and Development #97, Seattle: Rural Development Institute (February 1998).

time went by, equity capital began to undergo significant concentration and the composition of shareholders changed radically. As of the beginning of 2003, physical persons no longer have the majority of votes in corporate farms: the control has shifted from former workers to legal bodies and outside investors (Table 2.3). Today, physical persons control only 27% of the equity in corporate farms, while the remaining 73% is controlled by outside investors and other corporations that have acquired farm shares. The situation is even more extreme in some regions. In Moscow and Belgorod oblasts, for instance, corporations control 90% of the votes in farm enterprises, while physical persons are left with 10%.

Table 2.3. Distribution of equity in corporate farms by groups of owners (percent, beginning of 2003)

	Russia	Moscow Oblast	Belgorod Oblast
Corporations and entrepreneurs	73.2	90.5	89.8
Peasant farms and small enterprises	0.3	0.1	0.0
Physical persons	26.5	9.4	10.2

Source: Goskomstat database of large and medium farm enterprises.

Unfortunately, capital concentration measures were introduced into the Goskomstat database only in 2002 and it is impossible to get a more complete picture of changes over time. If the snapshot in Table 2.3 is valid, this basically means that farm enterprises no longer belong to their workers: the original vision of agrarian reform is gone, and the majority control in farm enterprises has passed to corporate investors and outside entrepreneurs.

Concentration of output and resources

We are also witnessing concentration of output and resources in Russian agriculture as production and assets increasingly shift to the largest corporate farms. The disproportionate share of the largest corporate farms in sales revenue is shown in Table 2.4.

Table 2.4. Share of largest farms in sales (2001)

Largest corporate farms				Small corporate farms generating equivalent sales	
Number of farms	% of all corporate farms	Sales, billion rubles	% of all corporate farm sales	Number of farms,	% of all corporate farms
10	0.0	8.5	2.9	7,700	31
300*	1.2	54.7	18.5	17,000	68
1734**	6.9	142.3	47.9	23,000	92

*The Agro300 group of farms as defined in the body of the text.

** Farms with sales greater than 31.5 million rubles = US$1 million.

In 2001, the 10 largest corporate farms (out of the total 24,000) generated the same volume of sales as 7,700 of the smallest farms. The sales of the 300 very large farms from the Agro300 Club matched the sales of 17,000 small corporate farms (Table 2.4).

The number of very large farms in Russia is relatively modest, just 1,700 out of 24,000 (in 2001). Yet these 7% of farms account for 48% of sales—the same volume of sales as generated by virtually all the remaining 23,000 smaller farms. The concentration of sales becomes more pronounced with time. This is evident from the two Lorenz curves that show the concentration of sales in 1995 and 2001 (Figure 2.2): the Lorenz curve in 2001 has a larger bulge, which is a sign of greater inequality of sales across farms. We see from Table 2.5 that 50% of the smaller farms (the left-hand tail of the two curves in Figure 2.2) generated 13% of sales in 1995 and only 8% of sales in 2001. At the other extreme, the share of 10% of the largest farms (the right-hand tail) increased from 42% of sales in 1995 to 55% in 2001. The largest farms are increasing their sales share at the expense of the small farms.

Fig. 2.2. Concentration of sales revenue in corporate farms 1995-2001

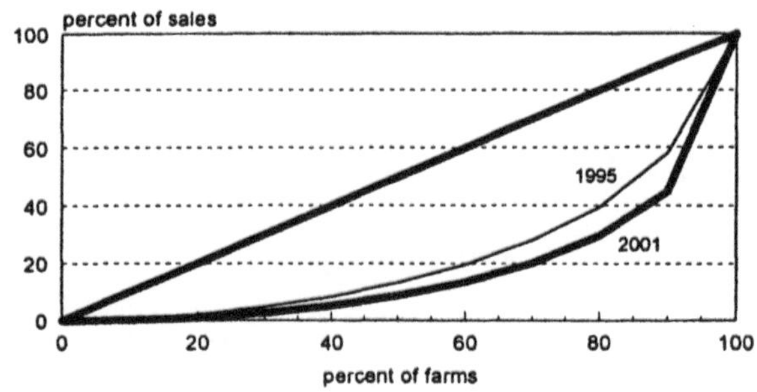

Table 2.5. Concentration of sales revenue in corporate farms 1995-2001

% of farms	10.0	20.0	30.0	40.0	50.0	60.0	70.0	80.0	90.0	100.0
% of sales										
1995	0.6	2.3	4.8	8.4	13.3	19.6	28.0	39.5	58.0	100.0
2001	0.2	1.0	2.5	4.9	8.3	13.2	20.1	29.9	45.3	100.0

Source: Calculated from Goskomstat database of large and medium farm enterprises.

An additional view of agricultural concentration is provided by the Agro300 farms, which are mentioned in Table 2.4 and presented in more detail in Table 2.6. The Agro300 Club is a group of 300 largest farms in a ranking of all corporate farms based on two measures—sales revenue and net profit.[8] In 2000-2002, this group included 1.2% of all corporate farms, but they generated 18.5% of all sales revenue. Their share in profit was close to 30%, with each farm earning on average 37.3 million rubles (compared with 1.1 million rubles on average for all other corporate farms reporting profits).

Table 2.6. Comparison of Agro300 farms with all other corporate farms

	Agro300 % of all farms	Agro300 farms (*n*=300)		All other corporate farms (*n*=25,402)	
		Per farm	Per 100 workers	Per farm	Per 100 workers
Number of farms	1.2				
Sales revenue, million rubles	18.5	172.0	18.8	8.9	5.7
Crops	13.1	46.6	5.1	3.7	2.3
Livestock	22.6	111.6	12.2	4.5	2.9
Profit, million rubles*	29.3	37.3	4.1	1.1	0.7
Agricultural land, '000 ha	1.8	9.0	1.0	5.9	3.7
Sown land, '000 ha	3.2	7.1	0.8	2.5	1.6
Number of workers	6.4	913		158	
Agricultural fixed assets, million rubles	6.1	154.7	16.9	28.0	17.7
Accounts payable, million rubles	4.7	34.7	3.8	8.4	5.3
Overdue payables	0.6	2.4	0.3	4.5	2.8
Subsidies, million rubles	8.0	4.0	0.4	0.5	0.3

* Calculated for 12,894 farms reporting profit.

Fig. 2.3. Change in selected indicators from 2000 to 2002: Agro300 and other corporate farms

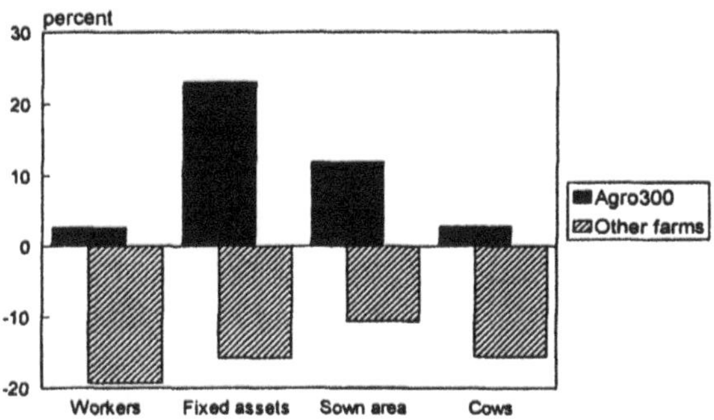

Source: Goskomstat

[8] All corporate farms in the Goskomstat database are first ranked by each measure separately, and the final ranking is based on the arithmetic average of the two ranks. The data for ranking are averaged over three years (2000-2002) to smooth out short-term fluctuations.

The degree of concentration increased over time, as Agro300 farms added both labor and land, while the rest of corporate farms shed their resources. Between 2000 and 2002, sown area increased by 11.9% in Agro300 farms and decreased by 11.5% in all other farms; labor increased by 2.7% in Agro300 farms and decreased by 19.3% in other farms. Figure 2.3 shows that Agro300 farms grew by several indicators between 2000 and 2002, while the rest of Russia's corporate farms contracted.

The Agro300 farms are larger than the average farm, but not dramatically so: the average Agro300 farm has 9,000 hectares of agricultural land compared with 6,000 hectares for all other corporate farms. Yet they utilize more of their land for production: 80% of agricultural land is sown in Agro300 farms compared with only 40% in all other farms. The size differences are more pronounced when we look at the number of workers (900 for the average Agro300 farm, 160 for all other farms) and the value of fixed assets (5 times greater for the average Agro300 farm), but the asset endowment per 100 workers is practically the same in farms of both categories. These numbers provide clear evidence of the greater efficiency of Agro300 farms: sales per 100 workers reach 18.8 million rubles in Agro300 farms compared with only 5.7 million rubles in all other corporate farms; asset utilization efficiency (the ratio of sales revenue to agricultural fixed assets) is 1.1 in Agro300 farms and 0.3 in all other farms. Partial productivity measures for Agro300 and all other corporate farms are presented in Table 2.7.

Table 2.7. Partial productivity measures for Agro300 farms and all other corporate farms

	Agro300 farms (n=300)	All other corporate farms (n=25,402)
Productivity of land: sales per ha, '000 rubles	19.1	1.5
Productivity of labor: sales per worker, '000 rubles	188.3	56.6
Asset utilization efficiency: sales to value of agricultural fixed assets	1.1	0.3

The Agro300 farms emphasize livestock production to a greater extent than other corporate farms: 65% of their sales revenue derives on average from livestock, compared with 50% for all other farms. And yet the Agro300 farms are much more profitable than all the rest, which clearly goes against the common wisdom that livestock is unprofitable. The reasons for the success of the Agro300 farms therefore lie in factors other than the product mix. Moreover, Agro300 farms occur in 52 of the 75 administrative regions spread from Yaroslavl' just north of Moscow to Kamchatka on the Pacific coast. The group includes representatives from all 11 natural-economic regions of Russia. Thus, the success of the Agro300 farms cannot be attributed to natural conditions. The organizational form is not relevant either, as group includes a wide range of farm types.[9] The main reason for the success of Agro300 farms is probably a combination of high entrepreneurial motivation of the managers, strict discipline (both labor and financial—see the low level of overdue payables in Table 2.6), and integration of primary agricultural production with processing and marketing.

Table 2.8. Performance characteristics of Agro100 farms and all other corporate farms by commodity specialization

Commodity	Production*		Yields*	
	Agro100 farms	Other farms	Agro100 farms	Other farms
Grain	32477	2602	34.6	17.1
Sunflower	2538	85	15.5	7.9
Sugar beet	22605	451	340	175
Potatoes	4337	62	184	89
Vegetables	6781	36	301	92
Milk	8825	631	5503	2529
Beef	778	51	595	342
Pork	2742	14	427	214
Poultry	7912	18	34	13
Eggs	132	0.4	296	230

Commodity	Specialization, %***		Profitability# (Agro100 farms), %
	Agro100 farms	Other farms	
Grain	40.9	20.7	80
Sunflower	14.9	2.1	123
Sugar beet	7.9	1.1	50
Potatoes	18.6	0.7	74
Vegetables	21.1	0.5	46
Milk	40.2	17.2	53
Beef	25.6	8.8	27
Pork	49.6	3	24
Poultry	84.8	3.5	15
Eggs	70.4	8.8	27

* ton per farm; eggs in million pieces; beef, pork, and poultry—weight gain in tons per year; ** centner/ha for crops; kg per cow per year for milk; g per head per day for livestock weight gain; eggs per layer for eggs; *** share of sales of main commodity in total farm sales; # calculated as the ratio of profit to costs.

Similar phenomena are observed on the commodity level. Following the same procedure as for the selection of Agro300 farms, we have selected the 100 large farms for each commodity sector, creating "sectoral" Agro100

[9] The Agro300 farms include 124 closed shareholding companies, 43 open shareholding companies, 12 partnerships and limited liability companies, 68 agricultural production cooperatives and kolhozes, 44 state enterprises and sovkhozes, 1 association of peasant farms.

groups.[10] The sectoral groups represent all the major agricultural commodities in both crop and livestock production. The sectoral Agro100 farms are larger than all other corporate farms; achieve higher yields; operate at higher levels of specialization; and are highly profitable (Table 2.8). Moreover, they increase their sales share of the respective commodities over time, which points to growing concentration in the commodity sectors as well (Figure 2.4). In 2000-2002 the sectoral Agro100 farms accounted for more than 50% of corporate farm sales for vegetables, pork, poultry, and eggs. While their market share for eggs was very high also in previous years, the increase in the market share for pork and poultry is phenomenal.

Fig. 2.4. Share of sectoral Agro100 farms in commodity sales 1996-1998 and 2000-2002

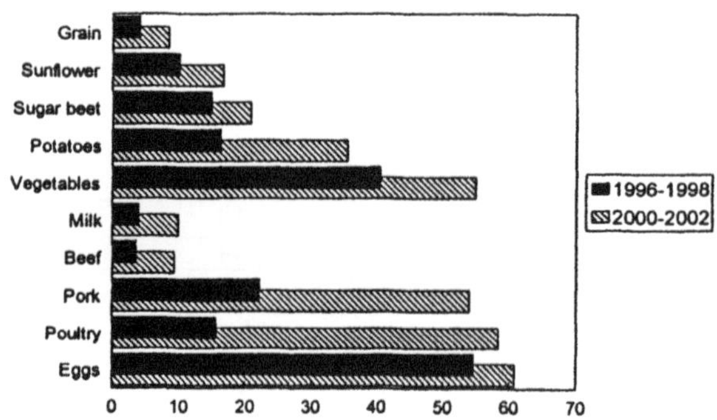

Source: Goskomstat

What are the main factors responsible for the success of sectoral Agro100 farms? In addition to the factors that have been mentioned for the entire group of Agro300 farms, we conjecture that the relatively high levels of commodity specialization, combined with the relatively large size of these farms, contribute to cost efficiency and hence profitability.

Adjustment driven by financial factors

The efficiency of resource utilization in corporate farms is largely dependent on their financial situation. A hint of this linkage is observed in Table 2.6, where the highly successful Agro300 farms are seen to have a much lower level of overdue debt, despite their larger size and volume of operations. We have classified the 24,000 corporate farms in the Goskomstat database into five solvency groups using two financial measures: the net

[10] The ranking for the sectoral Agro100 groups was based on three variables: volume of gross production, sales revenue, profit from sales.

profit and the current indebtedness ratio (net payables over sales). Group 1 (the best) contains financially healthy, solvent farms characterized by positive profit and low current indebtedness. Group 5 (the worst) contains financially ailing farms with high indebtedness and generally reporting losses. Judging by their indebtedness and lack of repayment capacity, group 5 farms are technically bankrupt. The definitions of the five solvency groups are given in Table 2.9 (for more details see Chapter 9).

Table 2.9. Classification of corporate farms into solvency groups (2002)

Solvency groups	Classification criteria	Descriptive characterization	Percent of farms	Percent of sales
1	Profit ≥ 0, Indebtedness ≤ 0.5	Solvent (financially healthy) farms	24	55
2	Profit ≥ 0, 0.5 < Indebtedness ≤ 1 or Profit < 0, Indebtedness ≤ 0.5	Farms with temporary solvency difficulties	18	19
3	Profit ≥ 0, 1 < Indebtedness ≤ 2 or Profit < 0, 0.5 < Indebtedness ≤ 1	Insolvent farms	16	12
4	Profit ≥ 0, 2 < Indebtedness ≤ 4 or Profit < 0, 1 < Indebtedness ≤ 2	Unsustainable farms	16	8
5	Profit ≥ 0, Indebtedness > 4 or Profit < 0, Indebtedness > 2	Technically bankrupt farms	27	5

The performance of financially healthy and financially weak farms is strongly polarized. The financially healthy farms in group 1 represent 24% of all corporate farms and generate 55% of sales. Roughly the same number of farms in group 5 (27% of all corporate farms) account for only 5% of sales. Instead of profits, the weakest farms generate only losses. The productivity of labor in group 5 farms is one-quarter of than in group 1 farms (31,000 rubles per worker compared with 118,000 rubles in 2002).

Analysis of changes between 1995 and 2001 shows a definite decrease in the physical endowments of all corporate farms. Yet the financially healthy, solvent farms contracted immeasurably less than the technically bankrupt farms (Figure 2.5). In financially healthy farms (group 1), the labor force decreased by 16% from 1995 to 2001, sown areas declined by about 5%, and the dairy herd contracted by 20%. The changes in the financially ailing farms (group 5) were much more dramatic: these farms lost fully 60% of their labor between 1995 and 2001, their sown areas decreased by 55%, and the dairy herd shrank by a catastrophic 70%.

Fig. 2.5. Changes in financially healthy and financially weak farms: 1995-2001

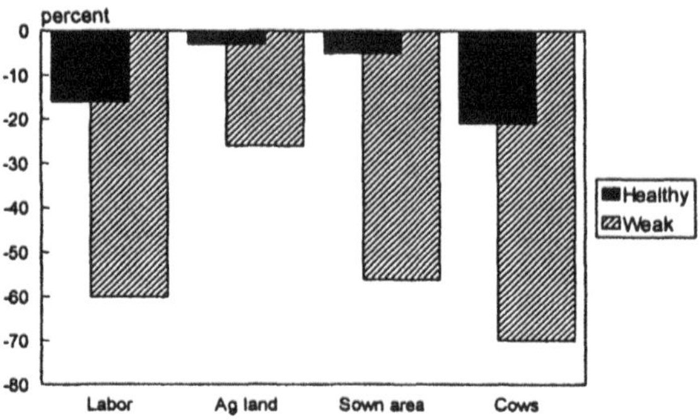

Agricultural labor is thus leaving the financially weak farms much more rapidly than it is leaving the good farms (the national trend in the 1990s reveals general decline of agricultural labor; see Chapter 7). The adjustment of land and physical assets lags behind labor force changes in all farms, and as a result the endowments per worker increase. However, the increase in asset endowments per worker is more pronounced in financially weak farms. Agricultural land per worker increased in group 5 farms by more than 150% between 1995 and 2001, while the corresponding increase in group 1 farms was less than 20%. The value of farm assets per worker increased nearly four-fold in group 5 farms and only doubled in group 1 farms. Unfortunately, the higher per worker endowments of land and assets in financially weak farms are not translated into higher output per worker: they actually remain part of a steadily growing pool of underutilized capacity. The weak farms simply do not have the required cash to make use of their land and assets.

Liquidation of insolvent farms

If a farm enterprise cannot function successfully, it should be restructured: its land and assets should be redistributed to peasant farms, household plots, or possibly more efficient new users and owners who are capable of restoring production to full capacity. This is essential not only from economic considerations. The deteriorating situation of the financially ailing farms goes beyond abandoned land and idle assets: it also has a strong negative impact on the rural society, strengthening the undesirable behavioral traits (stealing, drinking, etc.). The only way to ensure rural recovery is by providing successful and gainful employment, in part in successful and profitable farms.

Fig. 2.6. Distribution of "vanished" and new farms by solvency groups

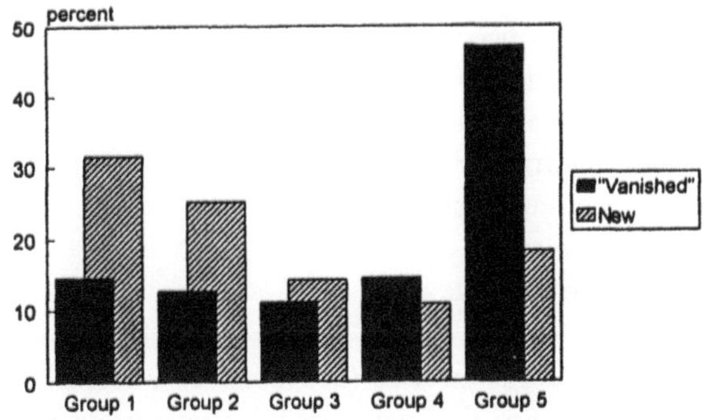

Market forces appear to be driving weak farms into liquidation, redistributing their resources among more efficient new producers. Almost half the corporate farms that ceased to exist between 2000 and 2001 (i.e., dropped out from the Goskomstat database) were financially weak group 5 farms (Figure 2.6). More than 50% of the new farms created between 2000 and 2001 were strong, financially healthy farms from groups 1 and 2. The "vanished" farms (4,704 in total) had employed 283,000 workers, controlled 11 million hectares of agricultural land, and had 70 billion rubles of fixed assets. The newly created farms (1,930 in total) employed 252,000 workers, controlled 11 million hectares of land, and had 24 billion rubles of fixed assets (Table 2.10). On balance, all the land and virtually the entire labor force of 4,704 "vanished" farms had been redistributed to 1,930 much larger new farms. As to fixed assets, only 36% of the existing stock was reabsorbed, because new investors pick and choose only the most attractive assets in the interest of financial recovery. A large stock of unusable assets is left in the empty shell of the old farm to offset the outstanding debt for bankruptcy proceedings.

Table 2.10. Liquidation and creation of corporate farms between 2000 and 2001

	"Vanished" farms	Percent	New farms	Percent
Group 1	688	14.6	608	31.5
Group 2	596	12.7	484	25.1
Group 3	523	11.1	276	14.3
Group 4	682	14.5	209	10.8
Group 5	2215	47.1	353	18.3
Total farms	4704	100.0	1930	100.0
		Per farm		Per farm
Workers	283,000	60	252,000	131
Land	11 mln ha	2,300 ha	11 mln ha	5,700 ha
Fixed assets	69.8 bln rub.	14.8 mln rub.	23.8 bln rub,	12.3 mln rub.

Small Agricultural Business

In our typology, small agricultural business includes the entire individual sector (peasant farms and household plots of various types) plus the category of small agricultural enterprises, i.e., corporate farms that are not part of the official classification of 24,000 large agricultural enterprises. Table 2.11 presents some basic characteristics of large and small agricultural business, and breaks down the small business sector into its components.

Small Agricultural Enterprises

This category includes legal entities (i.e., corporate farms) that are neither part of the individual sector (by virtue of their corporate form of organization) nor part of the established category of large farm enterprises (the successors of former kolkhozes and sovkhozes). It does not include peasant farms, even when they are registered as legal persons. Small enterprises (*malye predpriyatiya*) is a relatively new category in Russian statistics, and the available data are very fuzzy. Even the number of small agricultural enterprises is not certain, as different sources give different values of this basic statistic. General statistical data on small enterprises are available since 1998 in a special Goskomstat publication *Maloe predprinimatel'stvo v Rossii* (*Small Enterprises in Russia*). Data on small agricultural enterprises also can be inferred from Russia's statistical yearbooks, which provide information on all agricultural enterprises and large agricultural enterprises separately: the data for small agricultural enterprises are then calculated by difference.

Table 2.11. Selected characteristics of large and small agricultural business 2002

	Units	Agricultural land, mln ha	Average size, ha	Agricultural production per farm, '000 rubles
Large enterprises	24,200	142.6	5,900	15,974
Small enterprises	29,800	7.8	260	782
Peasant farms	264,000	17.7	67	145
Household plots	15.9 mln	6.9	0.43	26*
Garden plots	14.6 mln	1.3	0.09	n.a.
Vegetable patches	4.4 mln	0.4	0.10	n.a.
Other individual uses	n.a.	19.2	n.a.	n.a.
Total		194.6		

*The calculation assumes that household plots contribute 70% of total agricultural production of the individual sector excluding peasant farms (expert estimate).

Table 2.12 presents basic information for small agricultural enterprises and compares them with large corporate farms. All the agricultural data have been calculated (by difference) from the statistical yearbooks. The number of small agricultural enterprises is given from two different sources: as reported in *Small Enterprises in Russia* and as calculated by difference from the total number of legal bodies engaged in agriculture and the number of large corporate farms reported by Roszemkadastr—the Russian Land Cadastre Authority. The two sources unfortunately produce widely different numbers, but the general trend is clear: the number of small agricultural enterprises increases over time and so does the labor force employed in small agricultural enterprises. In contrast to the observed growth of small agricultural enterprises, the large corporate farms are shrinking, both in number and the labor force they employ. By average size of agricultural land, the small enterprises correspond to the partnerships and family corporations in the United States. The small enterprises are private companies, with absolutely no state ownership. They are primarily owned by Russian capital, and only 6% are in mixed ownership with foreign investors.

Table 2.12. Comparison of large and small agricultural enterprises 1999-2002

	Large corporate farms#		Small corporate farms#		Small in large, %
	1999	2002	1999	2002	2002
Number of enterprises*	27.3	24.2	25.4	29.8	
Number of enterprises**			13.5	15.6	
Number of employed, '000**	5100	3800	179.5	186.6	4.9
Agricultural land, mln ha	152.7	142.6	9.1	7.8	5.5
Sown area, mln ha	73.0	64.6	4.6	6.2	9.6
GAO, mln rubles (current prices)	229.0	386.6	15.4	23.3	6.0
Livestock, mln head					
cattle	17.3	14.9	0.2	0.1	0.7
cows	6.9	5.6	0	0.1	1.8
pigs	9.5	8.9	0.5	0.3	3.4
sheep and goats	4.8	4.4	0	0.3	6.8
Production, mln ton					
grain	47.8	69.3	2.5	6.0	8.7
sugar beet	13.9	13.6	0.4	0.8	5.9
potatoes	2.0	1.6	0.2	0.3	18.8
vegetables	2.4	1.9	0.2	0.2	10.5
milk	15.7	16.1	0	1.2	7.5

* Calculated by difference from Roszemkadastr report "O nalichii zemel' u predpriyatii . . ., zanimayushchikhsya proizvodstvom sel'skokhozyaistvennoi produktsii" (form 22-4).
** As reported in *Maloe predprinimatel'stvo v Rossii 2003*, Moscow: Goskomstat (2003).
# All agricultural variables (from agricultural land to production of milk) from Agricultural Statistical Yearbook for 1999 and from Statistical Yearbook for 2002 (large corporate farms as reported; small corporate farms calculated by difference from all agricultural enterprises and large corporate farms).

Small agricultural enterprises are generally equivalent to 5%-10% of the large corporate sector (Table 2.12). Thus, they employ about 5% of labor and 5% of agricultural land used by large agricultural enterprises. Their gross agricultural product is 6% of that in large corporate farms; they produce 9% of grain, 10% of vegetables, 19% of potatoes, and 8% of milk produced the large farms. Their livestock herd is substantially smaller, however: the number of cows and pigs is only 2%-3% of the corresponding numbers in large corporate farms. Their meat production is therefore negligible relative to the large farms.

Peasant Farms

The first peasant farms began to be formed in the early 1990s. The number of peasant farms rose rapidly, peaking out at 280,000 in 1996. After that the number of peasant farms slightly decreased, and in recent years it has stabilized at around 260,000 (Table 2.13). The total land in peasant farms, however, continued to increase steadily over time, so that the average farm size rose from about 40 hectares to 67 hectares in 2003. The additional land was acquired at a rate of about 1 million hectares per year by offering higher lease payments to owners of land shares who had previously leased their land to the local corporate farm. In parts of Saratov Oblast, for instance, these competitive pressures have actually led to total liquidation of corporate farms, culminating in the transfer of all their land and assets to peasant farms.

Table 2.13. Growth of peasant farming 1992-2003

Year	Number of registered farms, '000	Total land, '000 ha	Average farms size, ha
1992	49	2,068	42
1996	280.1	12,011	43
1997	278.6	12,237	44
1998	284.1	13,045	48
1999	270.2	13,845	51
2000	261.1	14,384	55
2001	261.7	15,292	58
2002	265.5	16,525	62
2003	264	17,662	67

Source: Statistical Yearbook 2003, p. 408.

Some peasant farms are registered as legal entities (under the 1990 law on peasant farms), others are run by entrepreneurs without being registered as legal persons (under the 1994 Civil Code). Peasant farms control 8.7% of agricultural land and produce 3.7% of gross agricultural output. Although peasant farms account for a small share of total agricultural production, their share of some products is substantial. In 2002 peasant farms produced 20% of sunflower, 12% of grain, 7% of sugar beet, and 8% of wool. Peasant farms mainly specialize in grain (64% of farms) and sunflower (4%). The availability of farm machinery in peasant farms is adequate for their crop specialization. In 1999, there were 76 tractors, 36 vehicles, and 23 combines per 100 peasant farms. The level of mechanization in peasant farms is even higher than in corporate farms. Thus, in 1999, the load per tractor was 50 hectares of arable land in peasant farms and 130 hectares in corporate farms; for grain harvester the respective loads were 67 hectares of land under grain in peasant farms and nearly 200 hectares in corporate farms.

Table 2.14 groups the peasant farms by size. The smallest peasant farms (the first group with only 2.8% of land and 4.6 hectares per average farm) are no different from household plots. The farm provides part-time employment to the members of the family, who produce mainly for own consumption. Some farms from this group have actually changed their status to household plots in recent years. The second and third groups consist of peasant farms that provide the main source of employment and the main source of income to family members, who produce for commercial sales. These farms have sufficient land for developing a commercial operation. The last group can be described as entrepreneurial peasant farms. These are relatively large peasant farms that rely mainly on hired labor. Their average size is 593 hectares and they control 55% of all land in peasant farms. Their technological level and organization of production are essentially similar to those in large corporate farms. For these reasons, many scholars and officials recommend reorganizing them as limited liability companies or other corporations.

Table 2.14. Grouping of peasant farms by size (beginning of 2003)

	Percent of farms	Percent of land	Average size, ha
Up to 10 ha	41.4	2.8	4.6
10-100 ha	45.3	26.2	39
100-200 ha	7.0	15.8	150
More than 200 ha	6.2	55.1	593
Total	100.0	100.0	67
	264,000 farms	17,662,000 ha	

Source: Statistical Yearbook 2003, p. 408.

No official data are published on the financial results of peasant farms and it is impossible to analyze their solvency along the same lines as for corporate farms. The results of the 2003 BASIS survey in three oblasts indicate that peasant farms accumulate substantially less overdue debt than corporate farms (see Chapter 9). However, this encouraging conclusion is offset by survey findings in Tambov Oblast, where peasant farms have consistently reported losses every year since 1994 and where nearly 50% of the cumulative number of peasant farms went out of business during the last decade (see Chapter 3).

Household Plots

As noted previously, the category of household farms includes several subgroups, such as the traditional household plots of rural residents, garden plots of rural people living in apartment blocks, vegetable patches of urban people, and livestock allotments. The official land use statistics show that the amount of land in household use has doubled since 1990 (Table 2.15). In the main subgroup of household plots—those cultivated by rural residents, the total number of farming families did not increase, so the average plot size more than doubled from 0.20 hectares in 1990 to 0.43 hectares in 2002. On the other hand, the number of the much smaller garden plots nearly doubled, and the average allotment per family hardly changed.

Table 2.15. Household farming 1991-2003 (beginning of year data)

	1991	1996	2000	2003
Household plots:				
number of families, millions	16.3	16.3	15.5	15.9
land, '000 ha	3250	5810	6137	6914
average per family, ha	0.2	0.36	0.4	0.43
Garden plots:				
number of families, millions	8.5	15.0	14.1	14.6
land, '000 ha	576	1242	1262	1260
average per family, ha	0.07	0.08	0.09	0.09
Vegetable patches:				
number of families, millions	5.1	7.4	5.1	4.4
land, '000 ha	379	603	437	420
average per family, ha	0.08	0.08	0.09	0.1
Total land in household farms, '000 ha	4205	7655	7836	8594

Source: Statistical Yearbook 2002, p. 407.

The data in Table 2.15, however, do not reflect the full phenomenon of household farming in Russia. Specifically, the 8.6 million hectares shown in this table do not include certain categories of land that are registered to corporate farms and village councils, but are actually used by households. In this category we have 16 million hectares of state-owned agricultural land allotted to rural households for hay mowing and livestock grazing by the municipalities, half a million hectares for rural residential construction and individual agricultural entrepreneurship, and 3 million hectares allotted to individuals by regional authorities for unspecified uses. According to Roszemkadastr records, individuals (excluding peasant farmers) used 27.8 million hectares of agricultural land as of January 2003, which amounts to 14% of land used by agricultural producers.

In addition to land that statistical organs recognize as being in individual use, rural households use extensive areas of common pastures that are owned by village councils. As of January 2002, 17.1 million hectares of village council land had been transferred in use rights to physical and legal persons and the remaining 7.8 million were in common use of rural residents.[11] Thus, in total, individual use encompasses not 8.6 million hectares as in Table 2.15, but 27.8 million of officially allotted land and another 7.8 million hectares of commons managed by village councils. The total land in identified individual uses thus reaches 35.6 million hectares, or 18% of all land in agricultural uses (Table 2.16).

Table 2.16. Land in household use 2002

	'000 hectares	Percent of land in household use	Percent of agricultural land
Household plots of rural residents	6,914	19.3	3.5
Garden plots	1,260	3.5	0.6
Vegetable patches	420	1.2	0.2
Hay meadows	15,253	42.7	7.8
Livestock allotments	532	1.5	0.3
Individual residential construction	395	1.1	0.2
Individual entrepreneurs	201	0.6	0.1
Allotted to unspecified uses	2,973	8.3	1.5
Common pastures	7,800	21.8	4.0
Total	35,748	100.0	18.4

Source: Constructed by the author from official land statistics.

Not all rural households actually use hay meadows and common pastures. If we assume that these additional allotments are used only by the 6 million households with cattle and sheep (the number of households with animals based on the 1996 livestock census), then the average size of their household plots increases to 4 hectares. In many regions, households with livestock have effective access to as much as 10 hectares on average.

There are no statistics on the number of employed in household plots. The number of people engaged in commercial agricultural production in the individual sector is estimated at 3.3 million,[12] which is comparable with the

[11] *Zemel'nyi fond Rossiiskoi Federatsii na 1 yanvarya 2002 g.*, Moscow: Roszemkadastr (2002), pp. 724, 766.

[12] This estimate is obtained by subtracting the number of employed in large corporate farms (3.8 million), small agricultural enterprises (186,600), and peasant farms (417,000) from total agricultural employment in 2002 (7.683 million).

number of employed in large corporate farms (3.8 million). Recently begun Goskomstat surveys indicate that about 10 million people (yearly average equivalents) are engaged in subsistence oriented agricultural production in household plots. Thus, the total number of employed in agriculture is estimated at 13.3 million, which is three times the number of employed in all other farm types combined. (For a more detailed discussion see Chapter 7.)

Household plots specialize in potatoes, vegetables, fruits, and wool. In recent years, the share of household plots in meat and milk increased markedly (55% of meat production and 47% of milk in 2002), but this was mainly due to the decline of corporate production, not to any physical growth. Households produce mainly for own consumption, and their level of commercialization has decreased over time (Table 2.17). This is somewhat surprising, as we were naturally expecting the household plots to become more commercially oriented as they acquired more land and increased their volumes of production.

Household farming mainly relies on manual labor and primitive technology. As a result, the productivity of labor is lower than in farms of other types. However, for many rural people farming the household plot is the only occupation: they have lost their salaried job at the local corporate farm and are unable to find alternative employment in the area. Moreover, household plots enjoy a tax advantage relative to peasant farms. Income from the household plot is exempt from personal income tax, while peasant farmers have to pay personal income tax on their farm income (after a grace period of 5 years after the establishment of the farm). As a result of these policies, we are witnessing many instances of peasant farmers who have changed their status to household plot in recent years to take advantage of the tax shelter. Other peasant farmers do not register their livestock as part of the farm and leave it in the household plot to avoid taxes on income from animal production. Federal legislators have left the decision on new principle for tax exemption of household plots to regional authorities, and these are beginning to introduce in an inconsistent manner various size limits up to which a farming unit may be regarded as a household plot enjoying the associated tax benefits.

Table 2.17. Commercialization of household plots (percent of output sold)

	1991	1995	1998	2001
Potatoes	28	12	10	10
Vegetables	17	9	10	8
Fruits	21	14	12	. . .
Meat	30	23	22	19
Milk and dairy products	25	18	18	21
Eggs	19	13	13	8

Source: *Sel'skokhozyaistvennaya deyatel'nost' khozyaistv naseleniya v Rossii*, Moscow: Goskomstat (1999), p. 38; data for 2001 from Agricultural Statistical Yearbook 2002.

The importance of the household plot in rural society clearly emerges from a rough comparison of the income from this source with wage income earned in corporate farms. According to Goskomstat household budget surveys,[13] the annual income from the household plot in 2002 was 4,192 rubles per rural resident. The annual wages paid in cash and in kind by corporate farms amounted to 900 rubles per rural resident.[14] The income from the household plot is thus 4.7 times the wage earnings received from the local corporate farms.

Table 2.18. Comparative characteristics of individual farms in Russia, United States, and West Germany (1997)

	Russia		USA[a]		West Germany	
	Peasant farms	Household plots	Primary	Subsidiary[b]	Primary	Subsidiary[b]
Number of farms, '000	274	15,984	758	886	228	324
Agricultural land per farm, ha	48	1.7[c]	143	16	41	9
Workers per farm[d]	3.2	0.2	2.2	0.3	1.4	0.7
Agricultural output, rubles[e]	29,000	9,000	801,000	10,000	736,000	72,000
Gross earnings, rubles	n.a.	n.a.	231,000	−16,000	168,000	9,000

Source: V. Ya. Uzun, "Corporate and family farms in Russia, United States, and West Germany," *Politekonom*, No. 1(12), pp. 91-107 (1999).

[a]USA statistics include family farms and family corporations.

[b]Subsidiary farms are those with less than $10,000 in sales per year. All others are primary farms.

[c]Includes hay meadows and common pastures (see Table 2.16).

[d]In Russia and West Germany, average number of workers per year. In the United States, number of permanent and seasonal hired workers.

[e]For USA and West Germany, revenue from sales of farm products. Exchange rates used for conversion into rubles: $1 = 6 rubles, DM1 = 3.7 rubles.

Household plots are not unique to Russia. Such farms are also observed in other countries. Nearly 60% of farms in the United States and Germany are part-time subsidiary operations (Table 2.18). In these farms, the owner earns most of the income from another job and devotes only a part of the working time to agriculture. The value of output produced by a household plot in Russia is comparable to the value of output from a small part-time farm in the United States (respectively 9,000 rubles and 10,000 rubles in 1997). It is not the existence of household plots, but their excep-

[13] *Sel'skokhozyaistvennaya deyatel'nost' khozyaistv naseleniya v Rossii*, Moscow: Goskomstat (2003), p. 151.

[14] Calculated from consolidated annual report of corporate farms for 2002.

tionally large role in agriculture that makes Russia's household plot sector so unique. Household plots contribute only 2% of farm output produced by family farms in the United States and 10% in Germany. In Russia, on the other hand, household plots account for more than 50% of farm output.

A Comparative View of Large and Small Business

Changing role of corporate and individual farms

The farm restructuring philosophy in the early 1990s was based on two competing premises. The first premise focused on the peasant as an owner. It was argued that once the peasant had received his shares in land and other assets, he would exercise his legal right to withdraw from the former collective farm and use the land and the assets to create an independent peasant farm. According to this premise, Russian agriculture would mainly consist of family farms, as in all developed market economies.

The second premise focused on the peasant as a collectivist or a cooperator. Having received the right to decide freely what to do with land and assets, having been freed from production and procurement targets, from constraints on input purchases, from government intervention in finances, the collective farms will fully realize their cooperative essence and the state farms will also transform into true cooperatives. These farms would become highly efficient, supplying food to both domestic and export markets. They would survive the competition with family farms in Russia and elsewhere in the world.

Neither premise has materialized. Only 5% of peasants have left the collective farms to establish their own private farms. The rest decided to avoid the risk of independent farming and preferred to entrust their land and asset shares to the restructured corporate farms. The majority of the Russian peasantry simply did not rise to the occasion of establishing sufficiently large, commercially oriented family farms.

Collective farms failed to become highly efficient producers in the new environment. Only 20% of corporate farms have managed to adapt to market conditions, while the rest rapidly contracted their production and sales volumes, dismissed many of their workers, and became technically bankrupt. As a result, small agricultural business gradually emerged to a more prominent place.

Russian peasants have chosen a third option: the land and asset shares have been entrusted to corporate farms, even though these remained highly inefficient, and the household plot has emerged as the main source of family income.

The corporate farms, which were the dominant sector in Russian agriculture during the Soviet era, have shrunk dramatically since 1990. The shrinkage is observed in their labor force, their agricultural land, and agricultural production. The number of employed in large and medium corporate farms decreased from 8.3 million in 1990 to 3.8 in 2002. On the other hand, the number of employed in commercial agricultural production in the small business sector increased from 1.4 million to 3.9 million during the same period (Figure 2.7). In addition to these 3.9 million workers in commercial small farms, another 10 million people are engaged in subsistence farming, so that small agricultural business actually employs today close to 14 million people compared with less than 4 million people in large corporate farms.

Fig. 2.7. Number of employed in large and small business 1990-2002

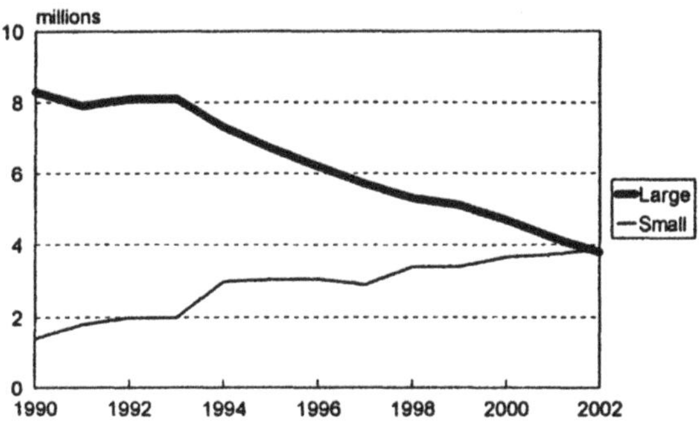

Fig. 2.8. Agricultural production in corporate and individual farms: 1990-2002

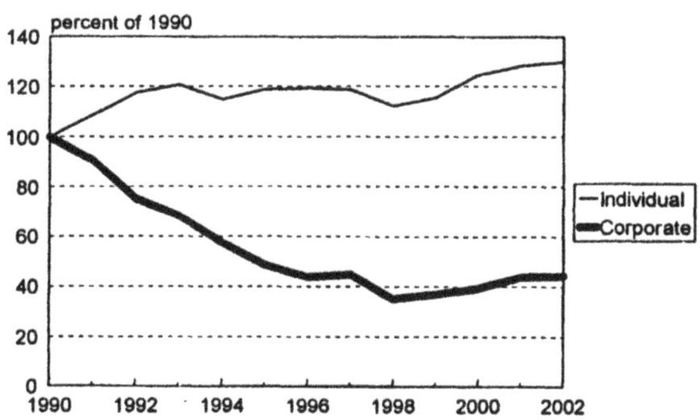

The agricultural land managed by corporate farms decreased from 209.8 million ha in 1990 to 150.4 million ha in 2002, whereas the land used by the family sector increased by a factor of almost 20 and reached 44 million hectares in the end of 2002. This land in individual tenure is based on official statistics, and it does not include some 8 million hectares allotted to households by rural councils and the land assigned by corporate farms to individual use without execution of formal documents.

The volume of agricultural production in corporate farms decreased by more than 50% between 1990 and 2002, while the production of the individual sector increased by more than 20% (Figure 2.8). As a result, the corporate and individual farms reversed their roles in Russian agriculture. The corporate farms lost their dominant place in agriculture, and their share in agricultural production dropped from 80% in 1990 to 40% in 2002. The individual sector, on the other hand, rose from a strictly subsidiary role with about 25% of agricultural production in 1990 to a clearly dominant role with 60% of production in 2002 (Figure 2.9). The slight recovery in the performance of corporate farms observed after 1998 was negligible compared with the magnitude of their fall since 1990. The corporate farms have clearly lost the competition against family farms.

Fig. 2.9. Agricultural production in corporate and individual farms: 1990-2002

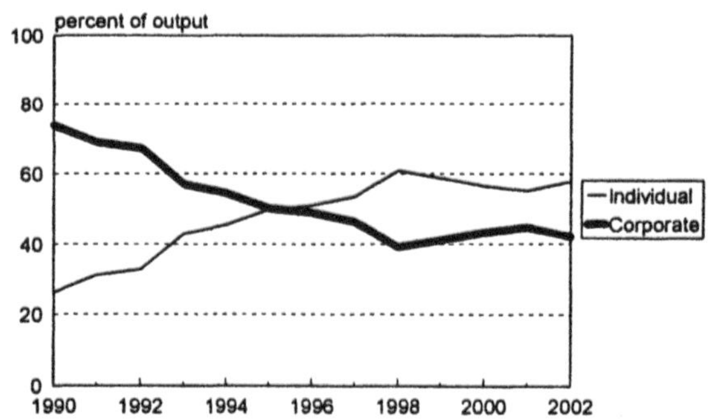

The changing production shares of corporate and individual farms were accompanied by the emergence of clear specialization patterns in the two sectors (Figure 2.10). Corporate farms maintain their lead in grain and industrial crops, while family farms specialize in potatoes, vegetables, and fruits. The individual sector has also emerged as a clear leader in livestock production (except eggs). The specialization of Russian crop farms is diametrically opposite to that observed in the United States, where grain and industrial crops are mainly produced by family farms, whereas potatoes, vegetables and fruits are the domain of large corporations.

The observed specialization of the individual sector is heavily biased toward household plots, which are the majority in this sector by all measures. Peasant farms contribute but a small share of agricultural output (about 4% of the total 60% produced in the individual sector), but their specialization is much closer to that of corporate farms than household plots: peasant farms produce a significant share of grain, sugar beet, and sunflower, while their production of potatoes, vegetables, meat, and milk is negligible. This aspect of the specialization pattern is clearly visible in Figure 2.10. Full data on production shares of corporate and individual farms during 1990-2002 are given in Table 2.19.

Fig. 2.10. Production shares of corporate and individual farms: 2002

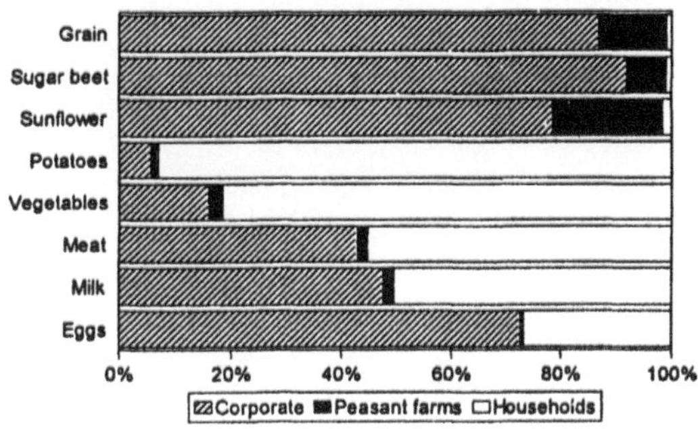

Russian agriculture is driven by two opposing forces. The personal interests of peasant-owners lead to the strengthening of family farms as the main source of food and income for the rural family. Rural entrepreneurs and hired farm employees seek to protect their narrowly specialized jobs (agronomists, veterinarians, mechanics, dairy-parlor operators, engineers, etc.) and are thus motivated to preserve the large corporate structures. The agrarian reform of the 1990s has led to a sharp increase in the role of family farms and a decline in the role of the traditionally dominant corporate farms. The former duality of structure has been softened and Russian agriculture today is characterized by coexistence of corporate and family farms.

The emerging dominance of the individual farms is probably an outcome of their much greater flexibility and faster adaptation to changing market conditions. Family farms were forced to maintain and increase production to satisfy the subsistence needs of the rural population. They were helped in this task by the influx of new labor, as workers dismissed by corporate farms turned to household plot farming as the main source of income. The changing production patterns inevitably lead to further deterioration and ultimate bankruptcy of the corporate farms. This downward process can be arrested only by radical restructuring of the traditional corporate farms into profit-oriented private businesses driven entirely by market forces. The former kolkhozes and sovkhozes are gradually transforming into large private farms, but these are still owned by a very large number of small shareholders (former workers). The incentive structure in these corporate farms is thus different from that in conventional corporations operating in a market environment.

Table 2.19. Share of corporate and individual farms in crop and livestock production (percent)

	1990	1995	2000	2001	2002
Grain					
Corporate	99.7	94.4	90.7	88.2	86.9
Households	0.3	0.9	0.9	0.8	0.9
Peasant farms	0	4.7	8.4	11.0	12.2
Sugar beet					
Corporate	100	95.9	94.4	93.6	91.9
Households	0	0.6	0.7	0.7	1.0
Peasant farms	0	3.5	4.9	5.7	7.1
Sunflower					
Corporate	98.6	86.2	84.4	81.9	78.5
Households	1.4	1.4	1.4	1.9	1.6
Peasant farms	0	12.4	14.2	16.2	19.9
Potatoes					
Corporate	33.9	9.2	6.5	6.3	5.7
Households	66.1	89.9	92.4	92.5	93
Peasant farms	0	0.9	1.1	1.2	1.3
Vegetables					
Corporate	69.9	25.3	19.9	17.8	15.9
Households	30.1	73.4	77.9	79.9	81.5
Peasant farms	0	1.3	2.2	2.3	2.6
Meat					
Corporate	75.2	49.4	40.3	41.0	42.9
Households	24.8	48.6	57.9	57.1	55.2
Peasant farms	0	1.5	1.8	1.9	1.9
Milk					
Corporate	76.2	57.1	47.3	47.2	47.6
Households	23.8	41.4	50.9	50.9	50.3
Peasant farms	0	1.5	1.8	1.9	2.0
Eggs					
Corporate	78.4	69.4	70.9	71.4	72.7
Households	21.6	30.2	28.7	28.1	26.7
Peasant farms	0	0.4	0.4	0.5	0.6
Wool					
Corporate	75.5	53.0	37.5	35.2	31.9
Households	24.5	42.5	57.1	58.8	60.3
Peasant farms	0	4.5	5.4	6.0	7.8

Source: Statistical Yearbook 2000 (p. 402), 2002 (p. 406).

The Impact of Corporate Farms on the Development of Individual Farming

Close interaction or even symbiosis between collective farms and household plots is a long-established feature of Russian agriculture. For instance, potatoes are mainly grown on household plots, but it is the local corporate farm that provides the machinery for plowing and other non-manual work in private potato patches in the village. Corporate farms play an even greater role in private livestock production. The households typically get their young animals (calves, piglets, chicks) from the local corporate farm, either as a convenient outright purchase or, more frequently, as wages in kind. They also receive help with hay cutting and hay delivery. The corporate farm supplies households in "its" village with grain and grain waste to be used as feed for private cattle and poultry. This is normally done free of charge or at preferential prices. The local corporate farm also assists village households with the marketing of livestock products, and especially with transportation and sale of milk. These diverse support links are one of the reasons why rural families make every effort to keep a job at the local corporate farm, even if otherwise their household plot operation is market-oriented and commercialized.

Many Russian scholars and politicians are of the opinion that there is a direct link between the level of development of corporate farms and household plots: the more developed the corporate farm, the higher the level of development of the household plots in the same area.[15] It is argued that family farms rely on inputs and services from corporate farms and cannot exist without them. Yet to be able to provide the household plots with all these services, the corporate farm itself must be sufficiently profitable and productive.

There is also a different view of this relationship. Family farms achieve a higher level of development in a given area because people have lost their jobs in the corporate farm and are forced to concentrate on farming the household plot or create an independent peasant farm. The resources of the deteriorating corporate farms are gradually reabsorbed by the family farms in the area.

The third point of view holds that peasant plots had existed for many centuries before the appearance of collective farms and that they still exist in areas where no corporate farms are to be found anymore. Thus, there is no direct relationship between the two types of farming. To ensure that

[15] For a review of these opinions see, e.g., V. V. Kazarezov and A. N. Rasskazov, *Household Plots: Past and Present* [in Russian], Rosinformagrotekh: Moscow (2002), and G. I. Shmelev, *Agricultural Production by Russia's Rural Population*, Acadmia: Moscow (2002).

large corporate farms operate efficiently, their relationship with household plots must be placed on a strictly commercial basis. Without adequate reciprocal arrangements, the household plots will ruin the corporate farms and pilfer all its resources.

Table 2.20. Regional grouping by percent of unprofitable corporate farms

	All Russia	Percent of unprofitable corporate farms in region			
		up to 40%	40%-60%	60%-80%	more than 80%
Number of regions	77	12	27	31	7
Value of agricultural production, mln rubles					
All farms	774.5	254.9	302.7	192.6	24.4
Corporate farms	333.4	131.3	127.7	68.1	6.4
Peasant farms	23.5	10.4	8.3	4.3	0.5
Household plots	417.6	113.3	166.6	120.2	17.5
Share of farm types in agricultural production, %					
All farms	100.0	100.0	100.0	100.0	100.0
Corporate farms	43.1	51.5	42.2	35.3	26.1
Peasant farms	3.0	4.1	2.7	2.2	2.1
Household plots	53.9	44.4	55.1	62.5	71.8
Rural families, '000	13,642	4,048	5,157	3,852	585
Agricultural production per rural family, '000 rubles					
All farms	56.8	63.0	58.7	50.0	41.7
Corporate farms	24.4	32.4	24.8	17.7	10.9
Peasant farms	1.8	2.7	1.6	1.1	0.9
Household plots	30.6	27.9	32.3	31.2	29.9
Agricultural land, mln ha					
All farms	194.3	70.1	66.1	47.1	10.9
Corporate farms	155.5	56.8	52.4	37.7	8.7
Peasant farms	14.3	5.8	5.6	2.5	0.4
Household plots	24.4	7.5	8.2	6.9	1.8
Share of farm types in agricultural land, %					
All farms	100.0	100.0	100.0	100.0	100.0
Corporate farms	80.0	81.0	79.2	79.9	79.6
Peasant farms	7.4	8.3	8.4	5.4	3.9
Household plots	12.6	10.7	12.4	14.7	16.5
Employed in agriculture, '000	17,890	5,126	7,102	4,934	727
Commercial agriculture, %	44.2	50.2	44.4	39.6	31.0
Subsistence agriculture, %	55.8	49.8	55.6	60.4	69.0

Source: all Russia from Statistical Yearbook 2002; regional data from *Regiony Rossii. Sotsial'no-ekonomicheskie pokazateli 2002*, Moscow: Goskomstat; employed in subsistence agriculture from *Ekonomicheskaya aktivnost' naseleniya Rossii*, Moscow: Goskomstat (2002), p. 106.

To test the hypothesis concerning the relationship between the efficiency of corporate farms and the development of family farms, we have grouped Russia's 77 regions by percent of unprofitable corporate farms.

This grouping is presented in Table 2.20. The total output produced in household plots does not support the hypothesis that households perform better where corporate farms are more profitable: although the household sector output is the lowest in the last group (regions with least profitable corporate farms), household plots in groups 2 and 3 produce greater output than in group 1 (regions with most profitable corporate farms). The table also shows that the value of output per rural family produced in household plots is fairly stable across the profitability groups (30,000-31,000 rubles). Its variability is substantially less than the variability of output per family in corporate and peasant farms.

In the regions where corporate farms are less efficient ("more unprofitable"), their share in agricultural production decreases, while the share of family farms corresponding increases. In the regions that have the smallest percentage of unprofitable corporate farms (less than 40% of farms), corporate farms account for 52% of agricultural production; in the regions with the highest percentage of unprofitable corporate farms (more than 80% of farms), corporate farms account only for 26% of agricultural production (Figure 2.11). However, the increasing importance of the individual sector can be broken down into two distinct components: household plots and peasant farms. In regions where corporate farms are weak, peasant farms are also less developed. It is only the household plots that increase their share of agricultural production as corporate farms shrink. This is also reflected in the share of employed in subsistence-oriented household plots, which rises from 56% of agricultural employment in the regions with best corporate farms to 69% in the regions with least profitable corporate farms.

Fig. 2.11. Role of household plots in regions with varying rate of unprofitable corporate farms

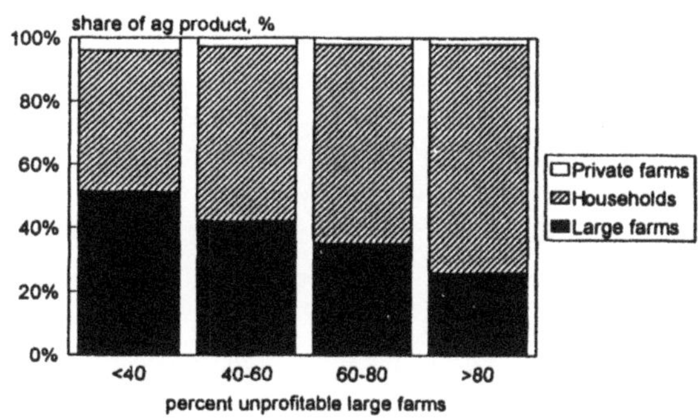

How do the changing roles of corporate and family farms affect the overall agricultural performance of a region? To answer this question, we regressed the regional agricultural product in 2001 on a number of basic

production variables plus a set of organizational variables reflecting the activity of corporate and family farms in each region. The production variables included agricultural land, agricultural employment (in farms of all types), value of fixed assets (in corporate farms only), and subsidies received by corporate farms (all these variables were logged). In addition, the model included a variable reflecting the agricultural conditions in the region ("bioclimatic potential").[16] Organizational variables related to farm type included percentage of unprofitable corporate farms (the basis of the grouping in Table 2.20), the share of agricultural output produced in family farms, the share of agricultural land used by family farms, and the share of privatized agricultural land. The degree of market liberalization was expressed as the share of corporate farm sales through private marketers. The regression results are presented in Table 2.21.

Table 2.21. Organizational determinants of regional production*
Dependent variable: regional gross agricultural product

	Mean value across regions	Coefficient**	*t*-value
Constant	9.0	1.288	3.061
Agricultural land (all farms)	7.2	0.046	1.271
Number of employed in agriculture (all farms)	8.8	**0.452**	6.890
Value of fixed assets (corporate farms)	5.1	**0.439**	5.752
Subsidies received (corporate farms)	4.9	0.077	1.564
Bioclimatic potential	98.8	**0.005**	4.178
Share of unprofitable corporate farms	50.3	**−0.007**	−3.402
Share of privatized agricultural land	57.3	0.001	1.098
Share of agricultural land used by individual farms	20.3	**0.009**	2.691
Share of gross agricultural product from individual farms	61.0	**0.007**	2.405
Market liberalization	66.3	0.000	0.204

*Cobb–Douglas regional production function using data for 77 regions (excluding Chechnya); bioclimatic potential and organizational variables not logged; $R^2 = 0.983$.
**Bold coefficients are significant at $p < 0.05$.

As expected from our grouping analysis, the share of unprofitable corporate farms has a statistically significant negative effect on regional agricultural production. On the other hand, both the share of agricultural land

[16] The bio-climatic potential is a standard measure of biological productivity of soil used since the Soviet times. Other conditions being equal, the bio-climatic potential (on a scale of 1 to 100) is determined by the cumulative availability of heat (the sum of effective temperatures above 10 degrees centigrade), the available moisture (total precipitation during the vegetative period), and the continentality of the regional climate (frost hazard, length of agricultural seasons).

used by individual farms and the share of agricultural production originating in the individual sector have a statistically significant positive effect on regional agriculture. Thus, keeping the physical production variables (land, labor, fixed assets) constant, regions with a higher share of land in individual farms will produce more. Similarly, a region with a larger individual sector will produce more than a comparable region with a smaller individual sector. This means that by transferring resources from corporate to individual farms, we increase regional agricultural production. It is interesting to note that the land privatization variable did not have a statistically significant effect. It is land use, and not legal land ownership, that affects production.

Productivity of Corporate and Individual Farms

Corporate and individual farms face different sets of organizational and management problems. The main differences in the operating characteristics of farms of different types are listed in Table 2.22. Given these operational differences, it is relevant to explore the existence of differences in outcomes as reflected, for instance, in productivity and efficiency levels of corporate and individual farms (see also Chapter 14).

Table 2.23 calculates two partial productivity measures—the productivity of land (output per hectare) and the productivity of labor (output per worker)—for corporate and individual farms. We see that household plots use their land much more efficiently than corporate farms and peasant farms. The order-of-magnitude advantage of household plots may be partially biased upward because the reported land inventory of rural households does not include the commons used for animal feed production and grazing. Yet there is no doubt that household plots use their land more efficiently: this is clear from the outcomes of crop production, where the land inventory is reasonably complete and still the productivity achieved by household plots is an order of magnitude higher than in other farms. The higher land productivity of household plots can be attributed to substantial differences in the product mix: household plots specialize in intensive crops (potatoes, vegetables, fruits), while corporate farms (and also peasant farms) specialize in grain and industrial crops.

The situation is different when we measure the partial productivity of labor. Here, corporate and peasant farms produce significantly more output per worker than household plots. This is a reflection of a well-known phenomenon in development economics: small household farms attract labor to compensate for lack of machinery and equipment, while large corporate farms partially substitute machinery for labor. The use of machinery is not reflected in Table 2.23, but both corporate farms and peasant farms have reasonable access to machinery, while household plots rely on occasional mechanized services from the local corporate farm (see Table 8.5 in Chapter 8).

Table 2.22. Operating characteristics of corporate and individual farms

	Corporate farms	Individual farms
Ownership	Shareholders	Individual, family
Organization	Complex, hierarchical	Simple, nodal
Managerial decisions	General assembly, management team, managing director, division managers	Owner-manager
Accounting systems	Professional accounting, annual financial reports	Simplified accounting and reporting; sometimes none at all
Marketing channels	Sales mainly to processors; limited sales in markets	Sales mainly in rural and district markets; access to city markets difficult due to organized crime
Accessibility of resources:		
Land	Surplus of land	Shortage of land; desire to increase holdings
Labor	Hired workers; reduction of labor force in recent years	Family workers with some hired help; influx of new family labor
Inputs and machinery	Limited by availability of funds	Limited supply options
Credit	Subsidized credit available only to profitable farms	Financial system not geared to small producers

While the productivity ranking of corporate farms and household plots is the same by all measures, the situation with peasant farms is not as clear. In Table 2.23 peasant farms fall between corporate farms and household plots by partial productivity of both land and labor. The opponents of individual farming in Russia often cite the lower partial productivity of peasant farms as evidence of their failure. But these productivity comparisons (95,000 rubles per worker compared to 102,500 rubles per worker in corporate farms; 2,200 rubles per hectare compared with 2,700 rubles per hectare in corporate farms) are based on total output, lumping crops and livestock into one basket. They fail to recognize the distinct advantage that peasant farms have over corporate farms in crop production. Indeed, the crop productivity of peasant farms is higher than the crop productivity of corporate farms both per hectare and per worker (Table 2.23). Relative advantage considerations seem to suggest that peasant farms should strengthen their specialization in crop production and reduce livestock production.

In an attempt to explain the lower productivity of peasant farms we should note that in the corporate sector livestock production is usually concentrated in large industrial "factories" that achieve significant economies of scale. The productivity of peasant farms should be more properly compared with that in relatively small corporate farms that emphasize crop production. Thus, corporate farms with sales of up to 4.1 million rubles in 2002 (47% of all corporate farms) produced roughly the same output as all peasant farms. Yet they employed 58 million hectares compared with only 17 million in peasant farms. The land productivity of peasant farms was thus three times higher than the land productivity of corporate farms in the corresponding group.

Table 2.23. Partial productivity measures for corporate and individual farms, 2001

	All farms	Corporate farms	Peasant Farms	Household plots
Agricultural land, mln ha	194.6	150.4	17.0	27.2
Number of employed, mln	17.7	4.0	0.4	13.3
Gross agricultural output (GAO), bln rubles	1029	410	38	581
Crop production, bln rubles	542	197	28	317
Livestock production, bln rubles	487	213	10	264
Output per hectare, '000 rubles				
GAO	5.3	2.7	2.2	20.9
Crops	2.8	1.3	1.7	11.7
Livestock	2.5	1.4	0.5	9.2
Output per worker, '000 rubles				
GAO	58.1	102.5	95	43.7
Crops	30.6	49.3	70.0	23.8
Livestock	27.5	53.2	25.0	19.9

Source: Production data from Statistical Yearbook 2002, p. 399. Land and employed from previous sections in this chapter.

The diametrically opposite results obtained for the partial productivity of land and labor—higher productivity of land and lower productivity of labor in household plots compared with corporate farms—highlight the need to switch from partial productivities to total factor productivity, i.e., the value of output per unit of a whole basket of inputs. Calculations of total factor productivity include not only land and labor, but also machinery, fertilizer, fuel, and other farm inputs, which are combined into a single basket by the use of appropriate econometrically derived weights. Only total factor productivity (TFP) enables us to decide which farms are more efficient. TFP has not been calculated in the present study, but Chapter 14 discusses the technical efficiency index, which is conceptually equivalent to TFP.

Large-Farm Bias in Government Support Policies

Since the very beginning of reform, the declared policy was to achieve a "multi-organizational" farming structure with a level playing field for farms of all organizational forms. This policy was clearly stated in all agricultural development programs since 1991. However, it has never been really implemented. The actual policy and the great majority of Russian politicians are strongly biased toward supporting large business and restricting small business in agriculture. For instance, the new law on household plots in principle allows regions to limit the maximum plot size. Many regions have taken advantage of this provision and have actually decided to restrict the household plots to not more than 1-1.5 hectares (larger units have to register as peasant farms with all the attendant tax implications). Subsidies from the federal budget are only available to large producers. Household plots are not even mentioned in the budget law and therefore cannot enjoy any of the state subsidies. Peasant farms had access to earmarked funds in the early 1990s, but subsequently this support has evaporated.

Formally, peasant farms are eligible for partial reimbursement of interest expense within the subsidized interest rate credit program. In practice, however, peasant farmers cannot take advantage of this subsidy, because commercial banks refuse to lend to small producers and there are no interest rate subsidies for loans from credit cooperatives—practically the only financial institution that peasant farmers patronize. Corporate farms are thus the main recipient of state subsidies, but even within this sector there is discrimination in favor of the larger farms. Thus, 1.4% of the corporate farms received 23% of all subsidies, averaging more than 10 million rubles ($350,000) per farm. Just three of the largest beneficiaries received more than 200 million rubles ($6.5 million) each. At the other extreme, 15% of corporate farms received no subsidies at all and 20% received 45,000 rubles ($1,500) per farm on average.

The large-farm bias of Russian agricultural policy is clearly illustrated by a comparison of the distribution of subsidy payments to farms of various size groups in the United States and Russia. In the United States, 26,500 largest farms contributing 42% of sales receive 8% of government payments in all farm programs,[17] whereas in Russia, the equivalent group of 24,300 corporate farms contributing about 40% of gross agricultural product receive almost 100% of government subsidies. The small and medium farms in the United States (contributing 50% of sales) receive 87% of sub-

[17] USDA Census of Agriculture 1997 (Chapter 1, Table 52).

sidies, whereas their Russian equivalents—peasant farms and household plots—receive virtually no subsidies.

Table 2.24. Structure of federal and regional support to agriculture (percent)

	Federal budget	Regional budget	Share of regional budget in total support, %
Total support budget, million rubles	5,444	13,080	
percent	100.0	100.0	71
Breeding livestock development	15.7	5.0	43
Development of crop varieties	4.3	3.3	65
Soil conservation/soil fertility programs	11.9	17.4	78
Livestock production	0.7	32.4	99
Crop production	0.6	1.5	85
Feed subsidies	0.2	3.7	97
Disaster relief	33.9	4.9	26
Other operating subsidies	9.6	15.8	80
Interest expense subsidies	10.3	2.2	34
Capital investment subsidies	12.6	9.9	65
Rural social services	0.1	4.0	99

Source: Consolidated annual report of corporate farms.

Subsidies are received from two main sources: the federal budget and the regional budgets. In 2001, regional governments played a dominant role in subsidizing agriculture, as less than 30% of total subsidies came from the federal budget (Table 2.24). Subsidies from the federal budget are mainly targeted to disaster relief and to essential public services in agriculture, such as development of breeding livestock, development of crop varieties, and soil conservation programs. Regional budgets, on the other hand, are intended for production support and also for programs that maintain soil fertility (supplementing federal support for soil conservation programs). Subsidized credit and support for capital investments account for major chunks in both budgets, but it is the regional budgets that play the leading role in investment support.

The subsidy data available for corporate farms in their annual reports point to a highly skewed distribution of state subsidies. Thus, 15% of corporate farms (3,800 units) did not receive any subsidies at all. At the other extreme, another 15% of corporate farms captures 66% of the total subsidy budget (Table 2.25). For these privileged farms, state subsidies reached 17% of sales revenue (compared with 5% average over all corporate farms).

Excluding the farms that do not receive any subsidies, there seems to be a negative relationship between the level of sales and the amount of subsidies per farm: the subsidies increase as the sales revenue decreases (Table 2.26, columns 2 and 3). Among the subsidized farms, smaller sales revenue generally leads to a switch from profits to losses (columns 5 and 6). In some farms (with medium sales) the subsidies are large enough to transform underlying losses into reported profits. In other, still smaller and even more unprofitable farms, the subsidies only serve to reduce the reported losses. In contrast, the unsubsidized farms—also medium sized by volume of sales—are reasonably profitable, reporting sales margins of 2.5% without any subsidies. It seems from this pattern that the government pours subsidies mainly into unprofitable farms (as was the policy during the Soviet period). But even this generous support to weak farms is insufficient to turn farm losses into profits. Government subsidies only serve to prolong the agony of basically nonviable farms.

Table 2.25. Distribution of corporate farms by level of subsidies received

Grouping by average amount of subsidies per farm, mln rubles	Percent of farms	Percent of subsidies received	Share of subsidies in sales revenue, %
No subsidies received	15.2	0	0
0-0.1	19.8	14.0	0.8
0.1-0.5	35.2	15.2	3.2
0.5-1	14.7	17.1	5.6
1-5	13.6	43.7	7.4
More than 5	1.4	22.5	9.5
Total	100	100	5.1
	24,995 farms	15,156 mln rubles	

Source: Consolidated annual report of corporate farms.

Table 2.26. The effect of subsidies on farm profits

Grouping by subsidy in percent of sales	Percent of farms	Sales per farm, mln rubles	Subsidies per farm, mln rubles	Profit margin on sales, %	
				with subsidies	without subsidies
	(1)	(2)	(3)	(5)	(6)
0	15	7.4	0	2.5	2.5
0–5	38	18.6	0.335	16.2	14.0
5–10	22	10.2	0.724	8.7	1.0
10–20	15	7.4	1.014	1.5	−13.4
20–50	8	5.2	1.482	−14.3	−48.7
> 50	3	2.2	1.782	−47.9	−92.0
Total	24,995 farms	11.8	0.602	10.2	4.5

Source: Consolidated annual report of corporate farms.

Regional Typology of Farm Structure

At the beginning of the 1990s all regions had the same sharply dual farm structure, with predominance of large corporate farms and a small share of land in household plots. During the 1990s all the regions followed the same transition path, with decreasing share of corporate farms in agricultural product and land and increasing share of the individual sector (peasant farms and household plots). However, the rates of change varied across the regions. As a result, after more than a decade of transition we observe different farm structures in different clusters of regions.

We distinguish between three types of regional farm structure: corporatized, mixed, and individualized. Regions where corporate farms contribute more than 50% of agricultural production are characterized as having a corporatized farm structure; regions where individual farms produce more than 70% of output (and corporate farms produce 30% or less) are characterized by an individualized farm structure; all other regions are characterized by a mixed farm structure, with corporate farms contributing from 30% to 50% of agricultural production.

Table 2.27 is a grouping of Russia's regions by these three types of farm structure. Corporate farms dominate only in 9 of the 77 regions. Here more than 60% of agricultural production is produced in large corporate farms. Individualized agriculture is observed in 26 regions. Here the share of corporate farms in agricultural production is as low as 23%, while individual farms produce 77%. In Ingushetia the individual sector accounts for 95% of gross agricultural output; in seven other regions (Dagestan, Buryatia, Yakutia, Astrakhan Oblast, North Osetia, Tyva, and Birobidjan) the share of the individual sector is more than 80%. All other regions are characterized by a mixed farm structure with clear dominance of the individual sector (which produces more than 60% of agricultural output).

Table 2.27. Regional grouping by farm structure*

	All Russia	Corporatized	Mixed	Individualized
Number of regions	77	9	42	26
%	100.0	11.7	54.5	33.8
Gross agricultural output, %	100.0	19.5	60.1	20.4
Share in GAO, %				
corporate farms	39.8	60.4	39.2	22.7
household plots	56.5	36.6	57.7	71.3
peasant farms	3.7	3.0	3.1	6.0
GAO in 2002 in % of 1998**	351	393	343	332
Share in agricultural land, %				
corporate farms	80.1	86.3	79.7	78.3
household plots	12.5	6.6	12.8	14.6
peasant farms	7.4	7.1	7.6	7.1

*Excluding Chechnya, for which no statistical data are available.
**In current prices.

In regions with individualized farm structure, corporate farms not only produce a small share of agricultural output; they are also of relatively small size. Thus, in many of the individualized regions, more than 80% of corporate farms generate sales of less than 4 million rubles. In corporatized regions, on the other hand, the proportion of such small corporate farms is 10%-20%.

The main factors that influence this sharp differentiation by regional farm structure appear to be the following (see the first two columns in Table 2.28):

- Natural conditions: individualized agriculture occurs in regions where the conditions for agriculture are less favorable (poor soil, harsh climate). These include the Far East, much of Siberia, and agriculturally less attractive parts of the European Russia, such as the non-chernozem belt (Bryansk, Novgorod, Kostroma, Smolensk, Pskov), the far north (Arkhangel'sk), and the arid south (Astrakhan).
- Availability of arable land: arable land is less abundant in regions with individualized farm structure. On the other hand, non-arable agricultural land, i.e., relatively unproductive land, is more abundant in regions with individualized farm structure than in corporatized regions. This feature again shows that corporate farms are stronger where conditions for agriculture are favorable, while individualized agriculture evolves in less favorable regions.
- Efficiency of corporate farms: individualized agriculture is observed primarily in regions where corporate farms are relatively inefficient and have not adapted to market conditions.
- Ethnographic features: individualized agriculture predominates in regions governed by ethnic minorities (for a more detailed discussion see Nefedova's book[18]).
- Regional policies: corporatized agriculture is characteristic of regions with a high level of subsidies per agricultural worker. Strong regional support of corporate farms is apparently the main reason for the corporatization of agriculture in Tatarstan (an ethnic-minority region) and in Murmansk Oblast or Chukotka (agriculturally poor northern regions). On the other hand, the formation of the individualized farm structure in Saratov and Samara oblasts (where the individual sector produces about 72% of gross agricultural product) is most probably the result of regional policies that explicitly support family farms, as it cannot be explained by natural, land-related, or ethnographic conditions.

---

[18] T. Nefedova, *Rural Russia at Crossroads: Geographical Essays* [in Russian], Moscow: Novoe izdatel'stvo (2003).

Table 2.28. Determinants of regional farm structure

	Individualized	Corporatized	Regression coefficients# (p-value)
Natural conditions: bioclimatic potential	94	102	0.11 (0.06)
Availability of arable land: ha per rural resident	0.6	1.1	4.90 (0.06)
Efficiency: percent of unprofitable corporate farms	55*	42*	−0.27 (0.03)
Regional policies: subsidies per agricultural worker, rubles	700	2,400	0.01 (0.00)
Ethnographic features: national republics/oblasts (of out 22)	11	2	−18.08 (0.00)

Source: *Regiony Rossii 2002*, Goskomstat, Moscow; bioclimatic potential from D. I. Shashko, *Natural Agricultural Regions by Land Quality in USSR* [in Russian], Moscow: Kolos (1975).
*The difference between regional types is statistically significant at 10% level.
#Dependent variable: the share of corporate farms in regional agricultural product (high for corporatized regions, low for individualized regions). Positive regression coefficients signify increasing corporatization.

The basic comparative numbers in the first two columns in Table 2.28 provide only qualitative support for these conjectures, as the differences between the two regional types by these variables are generally not statistically significant in a univariate analysis. Yet a linear model that regresses the share of corporate farms in agricultural product (high in corporatized regions, low in individualized regions) on the four numerical variables and a dummy variable for national republics produces coefficients that are all statistically significant at 10% and has a satisfactory explanatory power (R^2 = 0.77). The signs of all the coefficients support the above conjectures. The negative regression coefficient of the ethnographic factor confirms that corporatized agriculture is much less prevalent in ethnic-minority regions (national republics and oblasts) than in mainstream "Slavic" regions (keeping other factors constant). Regression analysis thus shows that natural conditions, availability and quality of land, efficiency of corporate farms, regional policies, and the ethnographic profile of the region all have a significant effect on farming structure.

Conclusion

Russian agriculture today is characterized by a wide spectrum of organizational forms. Western-type family farms coexist with large private corporate farms, superlarge agrofirms and agroholdings, state and collective farms inherited unchanged from the Soviet era, and patriarchal household plots that have continued to farm over centuries of upheavals. The share of large corporate farms in agricultural resources and production has been steadily declining since 1990, while the small business has become the main producer, the main employer, and the main source of income for the rural population. At the same, the shrinking corporate farm sector is becoming increasingly differentiated by financial health and performance, with the share of the largest "best performers" rapidly growing at the expense of the weak unprofitable farms.

The regional farming structure is becoming sharply polarized. Large corporate farms still account for more than 50% of gross agricultural output in only 9 of Russia's 80 regions. In the other regions agricultural production is dominated by family farms, and in 26 regions their share exceeds 70% of gross agricultural output. The main factors responsible for this polarization of farm structure include availability and fertility of agricultural land, profitability of corporate farms, regional agricultural policies, and the ethnic minority composition of the regions.

Large corporate farms affect the development of individual farms in their vicinity. The resources of corporate farms are one of the main factors affecting the performance of household plots, which rely on the local corporate farm for assistance with services and machinery. Family farms achieve much higher productivity of land than corporate farms, but use labor much less productively.

Russian agricultural policy continues to channel support to the largest farms, ignoring the increasing role of small agricultural business. As a result, government support programs and subsidies have no virtually no impact on the efficiency of agriculture. The government should urgently revise its policy priorities in view of the changing structure of agricultural production.

Chapter 3

Development of Peasant Farms: The Case of Tambov Oblast in Central Russia

Sergei Sazonov and Damira Sazonova

Peasant farmers are a special legally defined category of agricultural producers in Russia. Together with household plots, they form the individual sector of Russian agriculture. These are individual or family farmers who differ from household plot operators at least in two main respects: (a) they are without formal employment affiliation with the local corporate farm (or the local municipal authorities); (b) they receive relatively large land allotments from the state for the explicit purpose of engaging in commercial farming. Peasant farms (called *krest'yanskie (fermerskie) khozyaistva* in Russian, a term coined in the late 1980s during Gorbachev's regime, before the dissolution of the Soviet Union) are often referred to in the literature as "private farms" or "independent farms," to emphasize their separation from large corporate farms.

The first peasant farms in Russia were created as an experiment in two oblasts (Pskov and Orel) in 1987, i.e., still in the Soviet era. The 1988 Law on Cooperatives gave a boost to private business activity and allowed the creation of family farms in the guise of agricultural cooperatives. The full legal framework for the establishment of peasant farms across Russia was set up in the 1990 Federal Law on Peasant Farms and today their number is around 270,000. General features of peasant farms are discussed in Chapter 2. In this chapter we focus on the special case of peasant farms in Tambov Oblast—a province in the fertile central-chernozem region, 400 km southeast of Moscow, between the rivers Don and Volga. The data are primarily derived from a survey of Tambov peasant farms that has been conducted by the authors on an annual basis since 1993 and provides a unique database on the development of peasant farming in this province and in Russia in general.

Agricultural and Land Policy in Tambov Oblast

The development of peasant farms in Tambov Oblast in the early 1990s proceeded without any help or interference from the regional government. As everywhere else in Russia, the Tambov newspapers voiced concerns that the creation of peasant farms would destroy the large farm enterprises from within by depriving them of qualified labor and by encouraging massive theft of seeds, fertilizer, fuel, spare parts, and so on. Yet the regional government neither showed special favoritism nor discriminated against peasant farms. They were simply not at the top of its agenda, and through benign neglect the emerging peasant farms in principle enjoyed the same privileges as the established agricultural producers: they had theoretically equal access to commodity credit, leasing programs, natural disaster insurance, or state procurement quotas. In reality, however, the implementation of agricultural policies was strongly biased in favor of large farm enterprises. Peasant farmers often did not have the information or the know-how to apply to the various regional or federal programs, and as a result all credits and subsidies ended up with the large farm enterprises. The negligible share of peasant farms in Tambov agriculture was usually given as formal justification for this state of things.

Peasant farms in Tambov Oblast experienced a spurt of growth in the early 1990s, peaking at 4,700 farms in 1994. After that, the number of peasant farms steadily declined, reaching 3,200 in 2002. Yet the agricultural land cultivated by peasant farms is steadily increasing, and the average farm size accordingly grew from 37 hectares in 1992 to 89 hectares in 2002 (the Russian average is 70 hectares). The share of peasant farms in agricultural production is also rising continuously, but their relative productivity of land use is low: on more than 10% of agricultural land in the oblast peasant farms produce a negligible 5% of agricultural output. Relatively low productivity of land in peasant farms is a phenomenon observed all over Russia (see Chapters 2, 7).

The number of peasant farms did not stay at the maximum level because various institutional constraints had created an environment not particularly supportive of such new initiatives. Problems with access to credit for private farmers provide one notable example. The first boost to peasant farming probably came when it was decided in 1991 that all financial support to peasant farmers from state sources would be channeled through regional farmers' associations, and not through government organizations. This policy gave the Tambov farmers' association full control over fairly substantial funds during 1991-1993, and the supply of subsidized credit exceeded demand: in the end of 1992, beginning of 1993 the regional association actually had to beg farmers to borrow at preferential rates. Availability of easy credit naturally facilitated and encouraged the creation of new peasant farms during these years. In 1994, however, large-scale financial support from the state budget was discontinued, the oblast government with its indifference to peasant farming did not come up with matching funds, and the resulting shortage of credit could not but have a detrimental effect on peasant farms (see the discussion below on distress asset sales and the switch from profits to losses).

Peasant farms are created on land formerly used by collective and state farms. Back in the early 1990s, the regional government created a redistribution reserve by earmarking 10% of agricultural land within the boundaries of each rural council. This reserve land intended for the establishment of new peasant farms was generally less fertile than the land remaining in collective farms (some estimate the quality gap at as much as 20%) and was relatively far from the villages. According to oblast procedures established in 1992-1993, each farm member was entitled to 20 hectares of land from the redistribution reserve, up to a maximum of 60 hectares for the entire peasant farm (including 40 hectares of arable land). In 1995, when it had become clear that peasant farmers were not rushing to claim agricultural land, the oblast government raised the limit from 60 hectares to 200 hectares per peasant farm (without imposing special restrictions on arable land). Peasant farmers get land free in private ownership up to the district land quota, and additional land has to be leased from the state under a formal contract that specifies a certain annual lease payment.

Buying and selling of agricultural land was prohibited by a 1997 oblast law. This law was amended in January 2004 to allow buying and selling of land in Tambov Oblast and thus avoid contradiction with the new federal laws passed in 2002-2003 (see Chapters 5, 6). Three specific restrictions that remain in the oblast law (echoing the corresponding provisions of the federal law) have a direct impact on the individual sector. First, state and municipal land will not be privatized within 10 years from the passage of the law, i.e., there is a moratorium on land privatization until January 2014. During the moratorium, land can only be leased from the state, not bought. Second, a single person cannot own (jointly with his relatives) more than 10% of agricultural land in a particular district. Third, household plots cannot exceed 1 hectare; excess land must be alienated to the state, or to avoid alienation the family must register as a private farm, accepting all the tax implications of this action.

It is still too early to say how the amended oblast law will affect land market transactions in Tambov Oblast. We can only cite some data on land transactions announced in 2003 on the basis of the federal law, before the adoption of the oblast law. Thus, 47 individuals offered their land shares for sale. The offer covered 333.7 hectares of arable land and 42.3 hectares of pastures and meadows at an average advertised price of 1,250 rubles per hectare. Figure 3.1 is an example of the legally required notice of a sale offer. Another 947 share owners announced their intention to request conversion of land shares into 10,026 hectares of land (11 hectares per share on average), generally for the purpose of leasing it to other producers.

To put these transactions in perspective, we should note that in Tambov Oblast 258,500 rural residents own land shares covering 1,911,600 hectares of agricultural land. A land market is beginning to emerge in the new legal framework, but there is still a long, long way to go.

Fig, 3.1. Legal notice of the intention of five land-share owners to sell their shares in a corporate farm KIM. The notice specifies the size of each land share (7.1 ha arable land, 0.5 ha hay meadows, 0.4 ha pastures) and the asking price (10,000 rubles per share). In accordance with the July 2002 law on land transactions, the notice states that other KIM share owners have a preemptive right to purchase these land shares at the stated price within one month. Published in oblast-level newspaper *Tambovskaya zhizn'*, July 1, 2003.

УВЕДОМЛЕНИЕ дольщиков СХПК «КИМ»

Руководствуясь Федеральным законом от 24 июля 2002 г. № 101-ФЗ «Об обороте земель сельскохозяйственного назначения», мы:

1. Копылов Александр Владимирович, паспорт 6802 № 688321, выдан Никифоровским РОВД 03.10.2000 г.;

2. Дегтярев Виктор Петрович, паспорт 68 02 № 559017, выдан Никифоровским РОВД 03.07.2002 г.;

3. Бобков Сергей Владимирович, паспорт 68 02 № 615137, выдан Никифоровским РОВД 15.08.2002 г.;

4. Шаталова Нина Анатольевна, паспорт 68 03 № 857412, выдан Никифоровским ОВД 24.03.2003 г.;

5. Сидорова Мария Емельянова, паспорт 68 01 № 308896, выдан Никифоровским РОВД17.09.2001 г.;

УВЕДОМЛЯЕМ дольщиков сельскохозяйственного кооператива «КИМ» о своем намерении продать принадлежащие нам земельные доли сельскохозяйственного назначения (каждая доля: 7,1 га - пашня, 0,5 га - сенокос, 0,4 га - пастбище), находящиеся в общей долевой собственности СХПК «КИМ» (кадастровый № 68:11:00000007:0023: единое землепользование Тамбовской обл., Никифоровский р-н, земли СХПК «КИМ»; категория: земли сельскохозяйственного назначения; площадь: 98101000 кв. га) по цене 10000 (десять тысяч) рублей за одну долю. Согласно действующему законодательству дольщики имеют преимущественное право приобретения отчуждаемых земельных долей по цене не менее указанной, которое они могут реализовать в течении одного месяца со дня публикации настоящего сообщения.

* * *

The sections that follow analyze the results of an ongoing survey of a panel of 56 peasant farms in Tambov Oblast that was begun by the authors in 1992. The data generally cover the period 1992-2002, with some details available only for the later years 2000-2002.

Land in Peasant Farms

The average peasant farm in the Tambov survey had 86.1 hectares of arable land in 2001-2002. The land is held in three legal forms of tenure. First, there is the privately owned land, received free in proportion to the number of adult family members on the basis of their land shares. Second, there is land leased from the state (specifically, from the redistribution reserve). And third, there is land leased from private individuals, i.e., owners of land shares from among employees and pensioners of former collective farms or employees of rural social services. Privately owned land is less than one-third of the holdings in the average peasant farm, and leasing from private individuals exceeds leasing from the state (Table 3.1).

Table 3.1. Legal status of land in a typical peasant farm (2001-2002)

	Arable land, ha	Percent	Annual payment, ruble/ha	
			2001	2002
Privately owned*	24.2	28	24	36
Leased from the state*	24.0	28	24	250
Leased from land shares	37.9	44		315
Total	86.1	100		

*Payment for privately owned land is the land tax.
** Initially lease payments to the state were set equal to the land tax. Recently they went up to ten times the land tax.
Source: Tambov survey.

Privately owned land and land leased from the state were typically acquired at the beginning, when the peasant farm was established. The private land component is roughly equivalent to two average land shares—one owned by the farmer and the other by the spouse. The balance of arable land, making up the initial total to the allowed maximum of 40 hectares, was leased from the state. This component slightly increased over time (after the limit had been relaxed in 1995) as viable farmers began taking over bankrupt farms. The curious legal practice in the mid-1990s precluded failed farmers from returning land to the state: to be allowed to go bankrupt, they had to find another farmer who was willing to assume responsibility for their lease contract with the state.

The growth that became possible after 1995 was mainly realized by leasing land shares from rural individuals. The potential for growth was always there: 47 of the 101 farmers surveyed in 2000-2001 indicated that they wished to augment their holdings and identified short supply of land shares as the main obstacle to growth (the other obstacle was shortage of farm inputs). The leasing of land shares by farmers is organized in the following way. First it should be stressed that most land shares are entrusted to the local corporate farm—the former kolkhoz or sovkhoz. This is not really the result of any strategy or intentional manipulation on the part of a self-interest seeking manager: this is probably an true expression of the will of the rural residents, who need someone to take care of their land and at least pay the land tax to the authorities. On the other hand, most corporate farms are unprofitable and insolvent. They do not really need the land shares, because anyhow they do not have the resources to cultivate it productively. The managers therefore view the leasing of land shares as a favor that they are doing to the local population. If such a corporate farm would cease paying the land tax and making the minimal in-kind lease payments, or even worse actually go bankrupt, this would indeed be a major economic tragedy for the local share owners. It is in this setting that we should view the desperate and often ridiculous attempts of corporate farm managers to attract outside investors by giving away everything that they have (land, forests, animals, fixed assets) in return for vague promises to inject funds into production. They are motivated by the most elementary survival instincts, not by greed or corruption. Given the economic hardships of the corporate farms, the managers actually welcome any opportunity to transfer the land shares from their farm to another producer. The peasant farmer is thus not a shrewd competitor: he is a welcome partner.

After securing the agreement of the share owners, the farmer has to coordinate with the manager the specific location of the land plot that he would receive against the leased land shares. This may turn into a serious sticking point, as the decision is obviously rife with conflict. The farmer seeks to acquire the best possible plot, while the manager naturally wants to keep such a plot for his farm to avoid further deterioration. Share owners and farmers often accuse corporate-farm managers of ill-will, but our observations and information received directly from farmers indicate that the attribution of evil or corrupt motives to managers is greatly exaggerated.

So, farmers usually have no problems leasing land against shares from an unprofitable corporate farm. The situation is different in the case of successful, profitable corporate farms. They fully utilize their land and the managers—highly qualified and strongly motivated individuals—have no motivation to surrender any part of it to peasant farmers. It is virtually impossible for a peasant farmer to lease land from such an enterprise. Conversely, there are cases when successful corporate farms actually lease land from local peasant farmers, although no legal provisions exist for such transactions and the lease is necessarily informal.

The lease payment naturally depends on the size of the land share and it normally consists of three components: payment of the land tax for the landowner; in-kind payment in the form of 1-2 tons of grain for animal feed or other commodities; and help with spring plowing and tilling of the lessor's household plot (which is not part of the leasing transaction!). The land tax is a very small amount (around 30 rubles per hectare annually). The payment in kind on average is 1,600 kg of feed grain per hectare of arable land (valued at 215 rubles, or about $7) plus small quantities of other commodities (a 20-liter container of sunflower oil or a bag of flour, valued at 380 rubles for an average 7-hectare land share or an additional 54 rubles per hectare). The third component—help with plowing and tilling of the household plot—is simply a sine qua non for the lessors. For the farmer, this is a fairly arduous condition, because spring is a very busy time on his own farm, but landowners generally do not agree to give up the field-work component in exchange for a higher payment in kind. This insistence on help with the household plot is quite understandable from the lessor's point of view: a household plot is the main source of food and income for many rural residents, especially retirees, even a younger person would not be able to plow a 0.5-1.0 hectare plot manually, and the local corporate farm, because of its economic conditions, is generally unable to continue providing this traditional service even for payment. As a rough estimate of the value of the third component we take only the cost of diesel fuel used in plowing and tilling a 0.5 hectare household plot, and then trucking the products from the household plot to the nearest market (a distance of some 10 km). The fuel consumption is estimated at around 22 liters, or 131 rubles for one leased land shares (19 rubles per hectare). The three components of a typical share-lease contract thus add up to 315 rubles per hectare, or about $11 (average for 2001-2002). The cost of leasing land shares from rural individuals is double the cost of leasing land from the state.

Despite the cost differential, farmers prefer to lease land shares instead of land from the redistribution reserve. First, they can negotiate a better location and higher quality for the land plot that they are going to get from the corporate-farm manager against the leased land shares (the land quality in the redistribution reserve is usually 20% below average and the plots are far from the village). Second, payment in kind, even if higher than payment in cash, is more attractive to farmers, who always suffer from shortage of cash, but usually have enough farm products on hand.

Unfortunately for the landowners, farmers often do not respect the agreed lease terms: they delay payments or arbitrarily reduce the quantity and quality of grain delivered as payment in kind. This is a clear violation, but the landowners are not really in a position to shift their contract to another lessee. All the farmers in the area usually offer identical terms to the landowners, and the corporate farms tend to pay even less than the reduced payment from the farmers. If the landowner were to annul the lease contract because of these violations, he would have to pay the land tax on his own, which is not a very attractive prospect in the cash-poor rural society. Finally, on the supply side, a lot of farmland is available for leasing. Of the

258,500 rural people who received land shares in the early 1990s, only 75,900 remain in the employ of the corporate farms today and are obliged to let the large farms use their land. The remaining 183,000 landowners and their heirs are looking for lessees—and peasant farmers are the preferred option in this market.

Consistently with the general trend in Tambov Oblast, the average farm size in the Tambov survey increased by 130%, from 37.3 hectares in 1993 to 86.1 hectares in 2002 (Figure 3.2). Not all farmland is cultivated, however: one-third of the arable land in peasant farms was idle in 2002. A sharp dip in sown area occurred in 1994-1995, when elimination of credit programs forced many farmers to reduce their operations. Since 1996, the peasant farms have carried about 25 hectares of idle land, and its share in total arable land was steadily decreasing (from 48% in 1995-1996 to 33% in 2002) as both the farm size and the sown area increased at about the same rate.

Fig. 3.2. Average farm size: arable and sown land

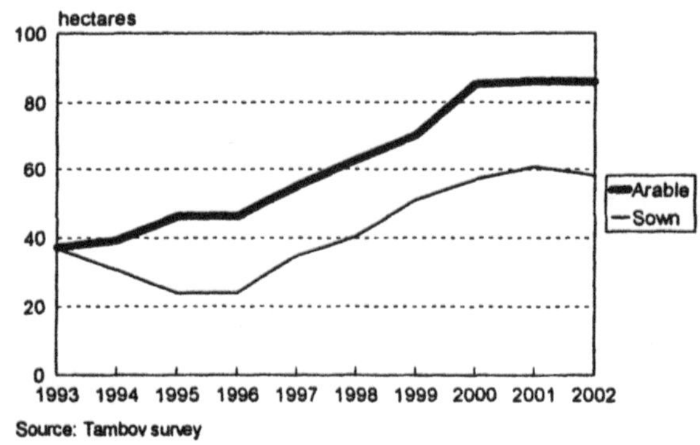

The observed increase in the average farm size is the result of two opposing trends in the distribution of Tambov farms between 1994 and 2002 (Table 3.2). In 1994, 11% of farms did not exceed 15 hectares, and there were no farms larger than 150 hectares. By 2002, the number of smallest farms had risen to 18%, and farms larger than 150 hectares accounted for 14% of all farms. The share of arable land in these two extreme farm size categories changed in opposite directions: the land controlled by the smallest farms dropped from 3.0% to 1.6% of total arable land, while the land in farms larger than 150 hectares rose from zero to 44% of the total. This essentially means that the smallest farms became smaller between 1994 and 2002, while the largest farms became more numerous and grew ever larger over time.

These trends seem to suggest a growing polarization of small and large peasant farms. The dynamic group of large peasant farms (those larger than 100 hectares, say) can be characterized as entrepreneurial or commercial farms, in which the traditional family labor is supplemented with hired workers as needed. The group of shrinking small farms can be characterized as family-oriented subsistence farms that rely entirely on family labor. The growth of entrepreneurial farms is therefore limited by the managerial capacity of the head of the farm and the availability of capital. The growth of family-oriented farms, on the other hand, is primarily limited by the available pool of family labor. The main objective of an entrepreneurial farm is probably profit maximization, whereas the family-oriented subsistence farm seeks to optimize the allocation of time and effort for the family members.

Table 3.2. Distribution of farms by size 1994-2002

Farm size, hectares	Percent of farms		Percent of arable land	
	1994	2002	1994	2002
Up to 15	10.7	17.9	3.0	1.6
16-30	21.4	10.7	13.2	2.6
31-60	57.2	28.6	60.8	16.8
61-100	7.1	17.9	13.8	17.0
101-150	3.6	10.6	9.2	18.0
151-200		3.6		6.3
201-250		3.6		8.5
More than 250		7.1		29.2
Total	100.0	100.0	100.0	100.0

Source: Tambov survey.

Some evidence of this polarization is obtained by comparing the size of farms with and without hired labor. Thus, in 2000-2001, farms that employed hired labor were larger than farms that relied entirely on family labor. Farms reporting hired workers had more land and produced more sales in both these years (the only ones for which data on hired labor are available). They naturally employed more labor, and the entire difference was attributable to hired workers, as the number of family workers was virtually the same in farms of both categories. The profiles of farms with and without hired labor are shown in Table 3.3.

Table 3.3. Profiles of farms with and without hired labor (averages for 2000-2001)

	Farms with family labor only	Farms employing hired labor
Percent in the sample	66	34
Arable land, ha	75	104
Sales revenue, '000 rubles	31	97
Family workers	2.1	1.9
Hired workers (part time)	—	4.8

Source: Tambov survey.

Farm Production

Tambov farmers specialize in crop production. Livestock production in the individual sector is concentrated in household plots, and peasant farms as such have virtually no animals (on average 0.5 cows and 0.4 pigs per farm in the survey). As a result, peasant farms produce a mere 0.5% of meat and 1.1% of milk in the oblast. The main reason is probably low profitability of livestock production. Thus, at the end of 2003, a pig carcass fetched between 2,202 rubles and 3,276 rubles, while the production costs ranged from 1,700 rubles to 2,900 rubles (600-1,400 rubles the purchase cost of a piglet, 1,100-1,500 rubles fattening costs).

Table 3.4. Cropping structure of peasant farms 1993-2004 (percent of arable land)

	Arable land, ha	Sown area	Barley	Wheat	Sunflower	Buck-wheat	Other	Fallow
1993	37.3	99	87	3	3	2	4	1
1994	39.3	78	39	18	11	6	5	22
1995	46.3	52	16	17	15	2	2	49
1996	46.3	52	16	22	6	4	3	48
1997	55.1	63	22	31	2	5	3	37
1998	62.7	65	13	47	3	2	0	35
1999	70.4	73	15	30	24	3	2	28
2000	85.3	67	12	26	14	14	2	33
2001	86.2	70	18	25	9	17	2	30
2002	86.1	68	18	29	15	3	3	32
Average	61.5	69	26	25	10	6	2	31

Source: Tambov survey.

Tambov farmers are thus primarily crop producers. The main crops are wheat, barley, and sunflower, which jointly accounted for 60% of arable land in farms between 1993 and 2002 (Table 3.4). A new addition is buckwheat, which began to capture substantial areas in 2000 with the launching of a modern processing plant in the oblast.[1] The share of idle land fluctuated over time, stabilizing at around one-third of arable land in recent years. This is much more than the "scientifically recommended" fallow area, which is around 10% of arable land. Peasant farms also ignore legumes and feed crops, all of which are important for proper crop rotation. Similar deviations are observed in corporate farms, but they are much more extreme in peasant farms (Table 3.5).

The reasons for the deviation of the actual cropping pattern from the recommended norms are easily understood. Peasant farmers do not grow feed crops because they do not have livestock. They do not sow legumes

[1] The sharp drop in areas under buckwheat in 2002 is a random fluctuation attributable to unfavorable weather conditions that had a highly damaging effect on pollination.

(lupine, vetch), because they do not need seeds for feed crops. Everybody emphasizes sunflower because of the stable demand for sunflower seeds from vegetable oil processors (including new additions in the private sector). There is a lot of idle land, because farmers do not have enough funds to cultivate their entire allotment. However, despite the deviation from scientific recommendations, the actual cropping pattern in peasant farms is very close to the classical Russian three-field system: 30% fallow, 30% winter grain, 40% spring crops. This traditional pattern is the result of physical constraints imposed by the natural farming calendar in Russia, and it provides strong evidence of highly extensive use of land in peasant farms. It is probably for this reason that, in the absence of productivity improving resources, peasant farmers always seek to acquire more land.

Table 3.5 Cropping structure: recommended and actual in 2002 (percent)

	Recommended	Corporate farms	Peasant farms
Grain	50	47	50
Industrial crops	10	15	16
Feed crops	25	14	0
Legumes and other	5	1	0
Other	0	3	2
Fallow	10	20	32
Total	100	100	100

Source. Tambov survey.

Yields per hectare are very difficult to determine for peasant farms. There are no scales on the farm, and quantities are roughly estimated by volume (combine bunkers, truck loads, storage bins). In the absence of professional information and accounting systems, farmers generally do not bother to measure exactly their cropped areas and many simply ignore the questionnaires sent out twice a year by official statistical organs. Tax authorities primarily focus on sales, and sales information is much more accurate than production information because it is usually collected from the grain elevator, the processors, or the wholesaler. Local farmers' associations often try to fill in the gaps by volunteering information of their own. Unfortunately, this information is unverifiable and generally biased. There is a strong suspicion that association officials make every possible effort to show that the yields achieved by peasant farmers are not lower (or at least not much lower) than the yields in corporate farms.

Because of these difficulties, we naturally expect to observe substantial differences between the official yields and the yields that we calculate based on our (hopefully more accurate) surveys of Tambov farmers. Table 3.6 shows that for all main crops (except buckwheat) the average yields reported by peasant farmers are substantially lower than the yields in corporate farms. Peasant farmers are often accused of systematically underreporting their yields for tax reasons. To monitor this possibility, we conduct annual spot checks of a small group of farmers, comparing the yields calculated from survey data with the yields reported by the same farmers to official statistical organs. Our spot checks show that the survey yields of grain and sunflower deviated from official statistics by anywhere from +59% to −74%. Moreover, the actual survey yields were higher than the official yields in 38% of the cases and lower than the official yields in 62% of the cases.

Table 3.6. Yields of main crops in peasant farms and corporate farms 2001-2002

	Peasant farms				Corporate farms
	Sown, ha	Harvested, ha	Harvest, kg	Yield, kg/ha	Yield, kg/ha
Wheat	23.0	22.2	299,900	13,500	28,600
Barley	15.4	15.2	191,800	12,600	17,900
Sunflower	10.3	9.0	38,800	4,300	8,300
Buckwheat	8.7	1.7	16,600	9,800	5,200

Source: Tambov survey for peasant farms; official oblast statistics for corporate farms.

The Tambov surveys therefore find no evidence to support the accusation of underreporting by peasant farmers. On the other hand, there are suspicions (voiced by district agricultural officials) that corporate farms report exaggerated yields, primarily because this is good for the manager's image, may attract creditors or outside investors, and boosts the prestige of the district governor. So far, however, this is strictly anecdotal hearsay. Yet there is sufficient uncertainty regarding the reliability of yield data to make any comparison (like that in Table 3.6) highly suspect.

Factors of Production in Peasant Farms

Labor and Human Capital

The farmer's family is the main source of labor for peasant farms. Most farms employ between 1 and 3 family members (Figure 3.3), and the average number of family workers is 2.25 per farm (data for 2001-2002). The main labor contribution comes from the head of the farm, who is typically a male 35 to 50 years old (mean age 44). He is responsible for organizing all the production activities, maintaining relations with suppliers and buyers, interacting with tax authorities, etc. He personally operates the farm machinery in the field. Other family members play a supporting role and are typically assigned to unskilled tasks.

Nearly 16% of farmers are without any previous experience in agriculture; the rest had worked in agriculture for 11 years on average before they became a peasant farmer. Table 3.7 presents the educational attainment and professional qualifications of the farmers. In view of the previous description of the farmer's role on the farm, it is important to note the high proportion of those with mechanical skills.

Fig. 3.3. Distribution of working family members per farm

Source: Tambov survey

Table 3.7. Educational attainment and professional qualifications of peasant farmers

	Percent of respondents
Educational attainment	
Higher education	15
Vocational and technical secondary school	22
General secondary education	58
Uncompleted secondary education	5
Total	100
Professional qualifications (multiple answers allowed)	
Tractor operator	58
Driver	50
Agronomist	7
Engineer	5
Mechanic	10
Metal worker	29

Source: Tambov survey.

Farmers do not have any formal managerial or economic training, which in the long run is an obstacle to their ability to run an efficient farming operation. Yet these highly qualified and well-educated people surprisingly do not recognize the need for further study and improvement. Only 15% expressed a desire for further training, although farmers often complain about difficulties with accounting and reporting, lack of clarity in tax regulations, and complications with the legal aspects of running a farm as a business. The prevailing opinion is that the environment has to change to fit the farmer, and the farmer does not have to learn the "rules of the game."

The attempts to involve other family members in skilled on-farm jobs have generally proved unsuccessful. Farmers' wives who had attended special accounting courses proved unable to cope with the job, especially because of the rapidly changing regulations and procedures that make any formal training obsolete within a matter of weeks. Even wives who had previously worked as professional bookkeepers in the collective farm cannot handle the farm accounts because of their specificity and frequent changes in requirements. Peasant farmers thus have two options: ignore systematic bookkeeping and risk the wrath of the taxman, or hire a professional accountant to deal with all records and tax returns.

As a general observation, we note that farmers prefer their family members to work off the farm whenever possible. The extra 1.25 persons assisting the farmer do that because they have no alternative. Family members with outside jobs cling to them even if they go unpaid for months and months (as many employees of corporate farms do). Farmers try to find off-farm jobs for their children and send them to school to train for non-agricultural occupations. An exception to this prevailing attitude is the case of farmers' wives who are professional agronomists. This appears to be the only instance when a wife with a marketable skill stays with her husband full time on the farm.

Family labor is supplemented by hired workers, but the practice of hiring labor decreased dramatically between 2000 and 2001 due to deteriorating financial situation. Nearly 50% of the surveyed farms used some hired labor in 2000, whereas in 2001 their number dropped to 20%. While the number of family workers remained at around 2, the number of average hired workers dropped from 7 in 2000 to 3 in 2001. Moreover, the duration of employment of the average hired worker dropped from 2.4 months in 2000 to a mere 1.1 months in 2001 (Table 3.8). Since the average monthly wage did not change significantly between 2000 and 2001, this change of hiring policy produced substantial cost reductions.

We clearly distinguish three types of hired workers in the survey. The largest group includes unskilled seasonal labor hired as needed to perform specific activities in the field. These are workers hired to weed sugar beet or to dig potatoes. The average number of such unskilled workers per farm is quite large (18 unskilled seasonal workers per farm in 2000, 10 unskilled seasonal workers per farm in 2001), but they are only employed for short periods of time, for about one month in a year.

The other category of unskilled labor includes occasional workers hired to help the farmer with a variety of jobs on the farm, such as loading and unloading, sowing and harvesting, taking care of farm machinery, etc. This contingent is on average 1-2 people working cumulatively for 1-2 months in a year, in short spurts ranging from a few hours to a couple of days at a time. Because of their versatility, these occasional workers are hired by the largest number of farms in the survey, and the relations between the farmer and one or two village workers may develop into a semi-permanent arrangement. The wages of the occasional workers are paid in services that the farmer provides (plowing the household plots, cutting hay, delivering firewood and coal, etc.) or in kind (feed grain or grain waste for household animals). The payment in kind is usually made after the harvest is brought in, so that the worker in effect extends a certain credit to the farmer by voluntarily allowing wage arrears. In some instances, the farmer "pays a retainer" by satisfying the village worker's "alcohol dependency" with year-round supplies of home-made spirits, and the worker "repays his debt" by helping on the farm when needed.

Table 3.8. Peasant farms with and without hired labor 2000-2001

2000	Percent of respondents	Family members	Hired workers	Months worked	Wages
Farms without hired workers	52	2.0	—	12	—
Farms with hired workers	48	1.7	6.7	1.5	872
Permanent hired workers	8	1.5	2.3	7.3	1,048
Seasonal field workers	13	2.0	17.9	1.1	816
Occasional hired help	28	1.6	2.6	1.6	1,008
Wages in corporate farms*					1,049
2001	**Percent of respondents**	**Family members**	**Hired workers**	**Months worked**	**Wages**
Farms without hired workers	80	2.2	—	12	—
Farms with hired workers	20	2.0	2.8	1.0	822
Permanent hired workers	—	—	—	—	—
Seasonal field workers	4	2.0	10.0	1.0	710
Occasional hired help	16	2.0	1.0	1.1	1,100
Wages in corporate farms*					1,265

*Including social deductions.
Source: Tambov survey.

Finally, there is the contingent of "permanent" skilled workers who are hired to operate farm machinery, i.e., tractors and combines owned by the farmer. These workers are employed during the entire agricultural season on the same farm (7 months) and their wages are comparable to the average wage paid by corporate farms (including social deductions). The survey shows, however, that the use of permanent hired labor virtually ceased in 2001. Wages of permanent skilled workers constituted the largest component of farm labor costs, and had to be eliminated when farmers began to experience financial difficulties due to the sharp drop of grain prices in 2001.[2]

About 12% of farmers in the survey report that they have a need for skilled help, but no such help are available for hire. The attitudes of the rural population toward hired work have undergone a deep change since 1990. After seven decades of socialism, working for a private individual used to be regarded as a severe degradation, especially for skilled workers of collective farms. In an interview in the spring of 1992, a highly qualified farm-machinery operator told the author, "My breast is covered with medals, and you want me to be a day laborer for a private farmer?" In the early 1990s, the very notion of day laborer was a horrible bogey for the Russian peasant, who grew on stories on injustice and deprivation in czarist times. But during the 1990s the corporate farms—the traditional mainstay of rural life—began to fail in the new environment, many employees had to go for months without being paid, and many simply lost their jobs. Between 1990 and 2000, the number of employed in corporate farms in Tambov Oblast dropped by more than 50% and 83,500 people lost their permanent place of work. Under these conditions, skilled workers are willing to do any kind of work, showing great flexibility regarding the amount of wages, the payment schedule, and even formalization of contractual arrangements. There is a potential for a considerable increase of hired labor in both unskilled and skilled jobs in peasant farms, but 3,000 peasant farms cannot solve the employment problems of Tambov Oblast.

Farm Machinery and Other Assets

Farm assets reported by peasant farms in their balance sheet are 90% machinery and 10% buildings and other production infrastructure (Figure 3.4). This is in sharp contrast to the structure of assets in the pre-reform period, when farm buildings and infrastructure in collective farms exceeded farm machinery by 4 to 1. The dearth of farm buildings is forcefully underscored by the detailed inventory of the surveyed farms. Only 9% of respondents report covered grain-storage sheds, 5% report hard-surfaced platforms for grain preprocessing, and 2%-6% of respondents have any technical facilities, such as repair and maintenance shops or machine storage sheds. The physical availability of farm machinery (regardless of depreciation status) is shown in Table 3.9.

---

[2] The price of wheat in 2001 was 2,058 rubles per ton compared with expectations of 2,700-3,000 rubles per ton based on the rising trend for the previous three years, when wheat prices had steadily increased from 660 rubles per ton in 1998 to 2,200 rubles per ton in 2000.

Is this sufficient? Our engineering estimates indicate that the average peasant farm requires a machinery fleet consisting of one chain tractor, one wheel tractor, one grain combine, and one truck. The actual availability of machinery is thus 30%-40% below the technical norm for tractors and harvesters and about 15% below norm for trucks. Farmers participating in the survey claim that their true machinery needs are somewhat greater than the estimated technical norms for harvesters and trucks (1.15 harvesters and 1.10 trucks per farm) and somewhat less for tractors (0.89 units compared with a technical norm of 1 tractor). The deviations are small, however, and the technical norm may be accepted as a correct benchmark of farmer machinery needs.

Fig. 3.4. Structure of farm assets (percent of balance-sheet value)

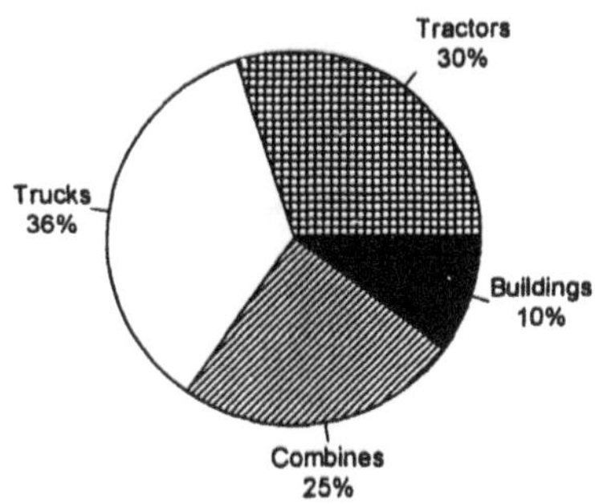

Source: Tambov survey

Table 3.9. Physical availability and sufficiency of farm machinery in peasant farms 2002

	Chain tractors	Wheel tractors	Grain combines	Trucks
Number of units per farm	0.70	0.61	0.61	0.86
Of which fully depreciated	0.63	0.07	0.22	0.00
Technically estimated machinery requirements, units per farm	1	1	1	1
Machinery requirements estimated from the survey, units per farm	0.89		1.15	1.10
Sufficiency ratio, % of technical norm	66		61	86
Sufficiency ratio, % of farmers' expressed requirements	74		53	78

Source: Tambov survey.

Most of the machinery is 8-10 years old, as investments in farm assets were mainly made between 1992 and 1995, when subsidized credits were still available to peasant farms. These credits petered out after 1995, and the relative volume of investments (as measured by the ratio of investment to farm sales) plummeted (the bars in Figure 3.5). In recent years, the small remaining investments have been financed almost totally with own funds. The abolition of subsidized credit forced many farms to start selling their machinery with the purpose of augmenting the annual cash flows. This phenomenon reached its peak in 1994-1996, when asset sales contributed 28% of total farm sources (50% of agricultural sales revenue) and half the farms were liquidating their assets below depreciated book value. Subsequently, the situation stabilized and distress sales of machinery dropped to 7% of farm revenue in 1997-1998 and less than 1% in 1999-2002. As a result of these financial pressures, the real value of the stock of farm assets (adjusted for inflation) steadily declined after 1995 (the curve in Figure 3.5).

Fig. 3.5. Investment and real stock of assets 1992-2002

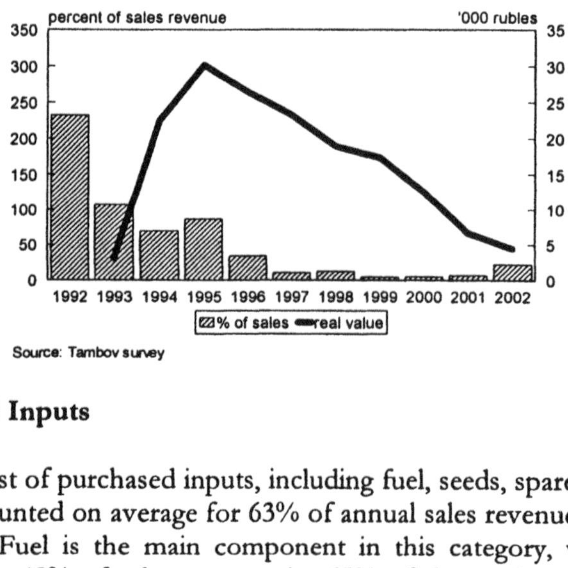

Source: Tambov survey

Purchased Inputs

The cost of purchased inputs, including fuel, seeds, spare parts, and fertilizer, accounted on average for 63% of annual sales revenue between 1992 and 2002. Fuel is the main component in this category, with fuel costs representing 42% of sales revenue (or 65% of the total cost of purchased inputs). Figure 3.6 shows the structure of input purchase costs averaged for 1992-2002. In this figure, the component shares are in percentage of the total cost of purchased inputs.

While the decision to buy seeds or fertilizer is largely discretionary, fuel is a daily necessity and it is purchased by virtually all the farmers in the survey. The level of fuel purchases naturally fluctuates over time, but on average peasant farms purchased 81 kg/ha between 1992 and 2002. We estimate the optimum fuel consumption in peasant farms at 100 kg/ha given their cropping pattern and the specific production technology. Thus, farmers do not squander fuel. They actually buy less than what is optimally needed and are forced to skip certain stages in the technological chain, leaving 30% of their land idle.

Fig. 3.6. Structure of input purchase costs (average for 1992-2002)

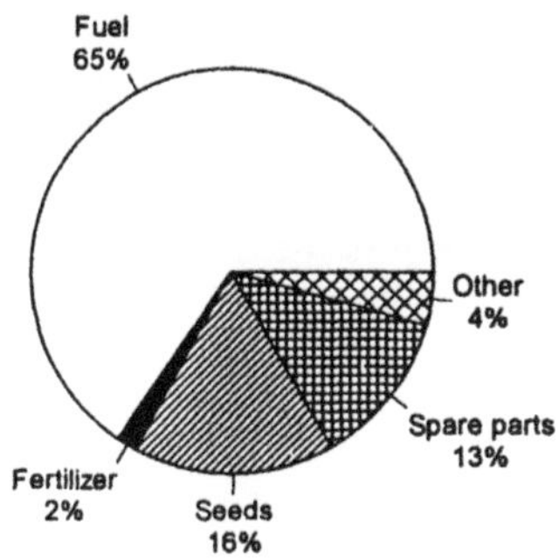

Source: Tambov survey

The financing mix that farmers use to pay for their fuel changed repeatedly between 1992 and 2002 (Table 3.10). In 1992-1993 most fuel purchases were financed with subsidized credit. Then subsidized credit was abolished and in 1994 no credit of any kind was available for fuel purchases: farmers had to pay everything out of pocket. Since 1995 farmers were generally able to get fuel in return for grain as part of commodity credits programs, so that between 1995 and 2002 fuel purchases were financed on average with 45% credit and 55% own funds.

Table 3.10. Fuel purchases and financing sources 1992-2002

	Fuel purchases, kg/ha	Subsidized credit	Commodity credit	Commercial credit	Disaster insurance	Own funds	Idle land, % of arable land
1992	141	100	0	0	0	0	
1993	142	73	0	0	0	27	1
1994	22	0	0	0	0	100	22
1995	57	0	0	48	0	52	49
1996	56	0	46	31	0	23	48
1997	94	0	27	54	0	19	37
1998	70	0	17	0	0	83	35
1999	139	0	7	0	0	93	28
2000	54	0	18	0	0	82	33
2001	60	0	38	8	0	54	30
2002	57	0	20	38	9	33	32
Average*	81	—	22	22	1	55	36

*1992-2002 for fuel purchases; 1995-2002 for financing sources. Source: Tambov survey.

Farmers purchased the largest quantities of fuel per hectare when subsidized credit was available. The purchase quantities decreased when farmers had to pay with own funds. Figure 3.7 illustrates the strong negative correlation between fuel purchases and the percentage of own financing (the coefficient of correlation is −0.71 and it is significantly different from zero). The regression line in Figure 3.7 has been estimated without the outlier year 1999. In that year, Tambov farmers were obliged to purchase excessive quantities of unneeded fuel following the failure of a joint business venture that they had established with a commercial supplier in the expectation of reducing their fuel costs.

Fig. 3.7. Fuel purchases versus financing mix

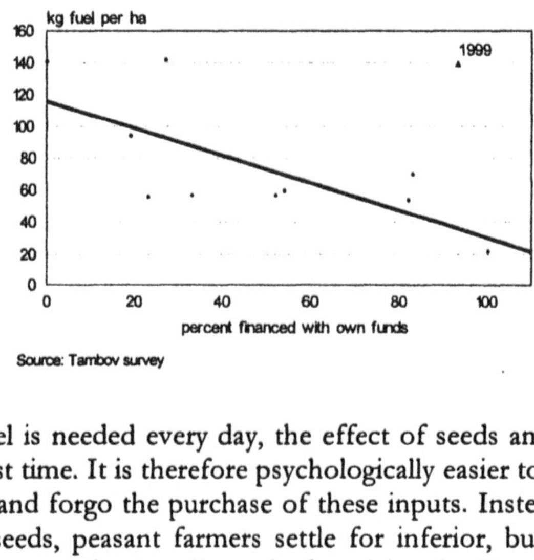

Source: Tambov survey

While fuel is needed every day, the effect of seeds and fertilizer is felt only at harvest time. It is therefore psychologically easier to give in to financial pressure and forgo the purchase of these inputs. Instead of purchasing quality elite seeds, peasant farmers settle for inferior, but still serviceable seeds from own production. Instead of applying fertilizer, peasant farmers rely on the natural fertility of Tambov chernozems and save the cost. This behavioral pattern is clearly demonstrated in Table 3.11, which shows the percentage of respondents who purchase seeds and fertilizer for their farms.

Fertilizer purchases are particularly infrequent and in 4 of the 11 years surveyed peasant farmers did not buy any fertilizer at all. The relatively high frequency of fertilizer purchases in 1993 is directly attributable to the subsidized credit schemes that operated in that year, while the sudden burst in fertilizer use in 2001 is the outcome of a commodity credit program launched that year (through the Tambov Food Corporation) to save a large local manufacturer from bankruptcy. There is no shortage of fertilizer supply in the oblast, only shortage of money to buy it. Indeed, farms that do purchase fertilizers are relatively large and generate relatively high sales revenues (Table 3.12; all differences are statistically significant at $p < 0.10$, except profit).

Table 3.11. Frequency of seed and fertilizer purchases among private farmers

	Percent of farmers purchasing seeds	Percent of farmers purchasing fertilizer
1992	61	6
1993	53	39
1994	15	0
1995	20	0
1996	14	7
1997	21	0
1998	11	0
1999	18	3
2000	18	7
2001	14	21
2002	18	7
Average	24	8

Source: Tambov survey.

Table 3.12. Fertilizer profiles: 2001-2002 averages

	Farms that do not apply fertilizer (86%)	Farms that apply fertilizer (14%)
Arable land, ha	84	146
Revenue, '000 rubles	39	84
Land productivity, rubles/ha	470	650
Profit, '000 rubles	-31#	-37#
Yield, kg/ha	1,360	1,640

Difference between the two groups not statistically significant.
Source: Tambov survey.

On top of everything, Tambov farmers probably do not experience an acute need for fertilizers. First, the chernozem soils in Tambov Oblast are phenomenally fertile. Their humus content is estimated at 264 ton/ha in the arable layer and annual depletion is negligible. Second, calculations show that a farmer can break even on fertilizer costs if 1 kg of nutrient material increases the yield by 7-9 kg per hectare. Such gains have never been attained in Tambov even on experimental fields. The third, and probably most important reason is that Tambov farmers keep one-third of their soil fallow and such extensive land use practices compensate to a considerable extent for fertilizer shortage. Yet the inadequate use of fertilizer has taken its toll on the accumulated content of nutrients in the soil, which decreased from 950,100 tons in 1976-1990 (before the transition) to 238,200 tons in 1991-1995 and continued to decline after that due to insufficient fertilizer consumption. This is a general problem of Tambov agriculture and it is not restricted to peasant farms. Ultimately, solutions for fertilizer use may need to be found in the framework of soil conservation programs.

Despite these reservations, the analysis of the available data for 2001 shows that farms reporting fertilizer application achieved higher grain yields than farms not using fertilizer (1.7 ton per hectare with fertilizer and only 1.3 ton per hectare without fertilizer; the difference is statistically significant at $p < 0.01$; see also Table 3.12, which shows yields averaged for 2001-2002). For the relatively small subsample of farms reporting fertilizer use (16 observations in 2001-2002), the yields increase with the amount of fertilizer used: an increase of 1,000 rubles on fertilizer expenditure per hectare raises grain yields by 300 kg per hectare. However, the market value of this additional grain is about 400 rubles, much less than the cost of fertilizer.

Production Function Analysis

To assess the efficiency of peasant farms, we used the Tambov survey data to estimate a Cobb–Douglas production function with sales revenue (expressed in thousand rubles) as the output. The factors of production (the inputs) included arable land, farm machinery, labor, and purchased inputs (fuel, spare parts, seeds, fertilizers, and outside services). Table 3.13 gives the mean values of the production variables and the estimated coefficients. All the coefficients are positive (and statistically significant), so that an increase in the use of a particular input (keeping all other inputs fixed) leads to an increase in output. In line with the accepted interpretation of the Cobb–Douglas production function, each coefficient is the percentage increase in output associated with a 1% increase of the corresponding input. The coefficient of arable land, for example, is 0.201, which implies that a 1% increase in arable land (measured in hectares) increases the output (sales revenue in thousand rubles) by 0.2%.

Given the estimated Cobb-Douglas production function

$$Y=aX_1^{b1}X_2^{b2}\ldots X_k^{bk},$$

the value of marginal product of factor i, i.e., the incremental output (in rubles) associated with a unit increase in the use of input i, is calculated as

$$VMP_i=\Delta Y/\Delta X_i=b_iY/X_i,$$

where Y/X_i is the mean output (in rubles) per unit of input i and b_i is the estimated coefficient of input i (the elasticity of input i). The value of marginal product (VMP) for each of the inputs included in the estimated production function for peasant farms is shown in the last column in Table 3.13.

Table 3.13. Production function estimation for peasant farms 2001-2002

	Sample mean (*n*=112)*	Estimated coefficient **	VMP at factor mean, '000 rubles
Output: Farm sales, '000 rubles	42.8		
Inputs:			
Arable land, hectares	86.2	0.201	0.110
Farm machinery, pieces	2.8	0.074	1.264
Fuel, '000 rubles	30.5	0.315	0.486
Spare parts, '000 rubles	8.9	0.134	0.712
Seeds, fertilizer, services, '000 rubles	11.1	0.153	0.652
Labor, number of workers	2.25	0.223	4.673
R^2		0.898	

*A pooled sample for 2001 and 2002 was used (56 observations in each year) because the coefficient of the year dummy variable in a preliminary regression was not statistically significant for $p < 0.20$.

**Coefficients obtained by OLS estimation of a Cobb–Douglas production function. All coefficients statistically significant at $p < 0.01$.

Production is efficient (i.e., maximizes the profit) when the value of marginal product of input i is equal to the marginal cost of input i (or, in practice, to the market price of input i). As long as the value of marginal product is greater than the market price of input i, by increasing the use of this input the farm will increase its profit. In this range we say that input i is underused. When the value of marginal product is less than the market price of input i, it is not efficient to increase the use of this input, as the cost of each extra unit of input is greater than the incremental value of output that it produces (see the univariate example of fertilizer use in the previous section). In this range we say that input i is overused.

Table 3.14 compares the value of marginal product for each input with some estimate of the market price of that input and classifies the inputs as underused, overused, or efficiently used according as $VMP_i > P_i$, $VMP_i < P_i$, or $VMP_i = P_i$. The easiest comparison to make is for the purchased inputs, where each input is measured in money units and its value of marginal product is simply the additional sales revenue produced by one extra ruble of input costs. For all purchased inputs (fuel, spare parts, seeds, fertilizer, services), the value of marginal product is less than 1, which means that, from profit maximization considerations, it simply does not pay to increase the output by increasing the use of these inputs. This conclusion for the combined bundle of all purchased inputs obtained in a multivariate setting with all other factors of production is consistent with the result obtained in the previous section for fertilizer on its own. The conventional interpretation of this result is that the purchased inputs are overused. Conversely, however, it can be argued that the farms use these inputs with optimal technological efficiency, but the inputs are overpriced because of market imperfections (i.e., monopolistic fuel suppliers, fertilizer manufacturers who prefer to sell their products for export, etc.). If the prices are not determined in a competitive market, it is impossible to decide who is inefficient—the farmers or the markets.

Table 3.14. Efficiency of input use and optimal input quantities

Factors of production (inputs)	VMP at mean input use, rubles	Price, rubles	Efficiency conclusion	Mean input use in the sample	Optimal input use
Arable land, hectares	110	315	Overused	86.2	23
Farm machinery, pieces	1,264	1,139	Nearly efficient	2.8	3.1
Fuel, rubles	486	1,000	Overused/ overpriced	30.5	10.7
Spare parts, rubles	712	1,000	Overused/ overpriced	8.9	6.0
Seeds, fertilizer, services, rubles	652	1,000	Overused/ overpriced	11.1	6.7
Labor, number of workers	4,673	4,616*	Used efficiently	2.25	2.29
		14,300**	Overused		0.53

*Average value of income received from the farm per family worker (calculated as the mean per worker income over all farms in the sample). It consists of 4,082 rubles cash withdrawals plus 534 rubles worth of feed grain and barley received for own consumption from the farm (see Table 3.15 below). **Average salary paid by corporate farms in Tambov Oblast.

The other inputs are measured in physical units (hectares, pieces of machinery, number of workers) and their value of marginal product needs to be compared to the price of a physical unit of input. Since buying and selling of land is not widespread in Tambov Oblast (or anywhere else in Russia for that matter—see Chapter 5), additional land is generally acquired by leasing, and lease payments are an appropriate proxy for marginal cost of land. Lease payments average around 315 rubles per hectare in the sample, whereas the value of marginal product of land is 110 rubles per hectare. The output from an additional hectare of land does not cover its cost, which means that in terms of efficiency land is overused. In the past farmers could get land for free from the state, and the land tax was the only relevant cost of land. Free allocation of land has been discontinued, however, and the land tax is no longer a relevant measure of the marginal cost of land.

The average cost of one piece of farm machinery in the survey is 11,390 rubles, but machinery purchase is a long-time investment that cannot be treated as annual expenditure. Annual depreciation is the conventional mechanism that converts capital investment stock into expenditure flows. Assuming straight-line depreciation of machinery over 10 years we estimate the annual depreciation charge at 1,139 rubles per "average" machine. This is very slightly below the value of marginal product (1,264 rubles per piece of machinery), which implies that machinery is just slightly underused or is used almost efficiently.

The picture for labor is less clear. The marginal product of labor in Tambov peasant farms is about 5,000 rubles per worker per year. What is the cost of one farm worker? The wages in corporate farms average 14,000 rubles per year, and judging by this alternative labor is grossly overused on peasant farms. It can be argued, however, that this is not a valid alternative, because corporate farms are not hiring new workers: they are trying to shed labor and in any case the wages are often recorded in the books without actually being paid in cash. Moreover, on a psychological level, farmers may not regard wage employment on a corporate farm as an alternative to the opportunity of enjoying independence and exercising individual initiative as a self-employed operator.

Instead of looking at market wages, we can look at what family members actually "earn" on the farm as measured by cash withdrawals from farm revenues and by the value of farm products consumed in the household. The corresponding calculation is presented in Table 3.15, which shows that these earnings averaged 4,616 rubles per farm member during 2001-2002, which is practically equal to the value of marginal product of labor (4,673 rubles per worker). If on-farm "earnings" are a valid measure of the cost of family labor, which really has nowhere else to go, then labor is used optimally in peasant farms. The gap between on-farm "earnings" and the average agricultural wage is truly astonishing. Yet it is consistent with the general impression of material hardships among farmer families and the lack of real alternatives. The authors hear the same refrain from the farmers every autumn: "Enough is enough, I cannot go on like this, we will stop farming." But there is nowhere to go, and therefore come spring everything starts anew.

Table 3.15. "Payment" to family labor (in rubles per worker per year)

	2001	2002	2001-2002 average
Cash withdrawals	4,956	3,208	4,082
Farm products for family consumption:			
Wheat	313 (352 kg)	176 (441 kg)	245 (397 kg)
Barley	314 (483 kg)	264 (753 kg)	289 (618 kg)
Total in-kind	627	440	534
"Payment" per family worker per year	5,583	3,648	4,616
Average annual wages in agriculture*	12,036	16,646	14,340
Ratio of farm "wages" to average wages in agriculture, %	46	22	32

*Based on Tambov Oblast statistics for the corresponding years.

We can look at the efficiency of input use from a different (though equivalent) angle of quantities, not prices. Choosing one specific input in the Cobb–Douglas production function as variable and keeping all other inputs fixed (at their mean values), we obtain a reduced single-variable form that describes the output as a function of the chosen input. This function actually describes the value of marginal product over the entire range of values of the chosen input. VMP in Table 3.13 is simply the value of this function at the point corresponding to mean input use. The descending curve in Figure 3.8 is the marginal product function of land obtained by keeping all the variables except land fixed at their mean values. The mean value of land in the sample is 86.2 and the function value at this point is 0.110, which is equal to the VMP of land in Table 3.13. If we now draw a horizontal line at the level corresponding to the price of land (0.315), it intersects the function at a point where the land endowment is 23.0 ha. This is the optimal use of land given its market price. Since the actual use of land is 86.2 ha, we clearly see that land is overused.

Fig. 3.8. Marginal product of land and optimal land use

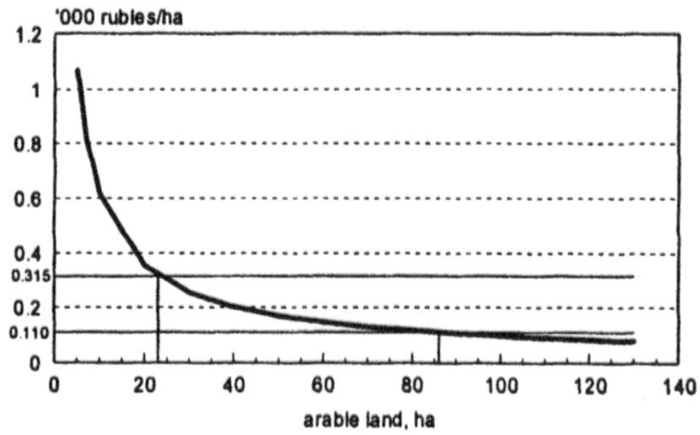

Source: Tambov survey

Similar calculations were carried out for all the inputs entering the production function. The optimal quantities obtained in this way for the given market prices are shown in the last column in Table 3.14. Their comparison with the mean quantity of inputs used in the sample confirms the efficiency conclusions based on VMP and price data.

In conclusion, land and all purchased inputs are overused, i.e., these inputs are in excess supply and do not constrain farm production. Machinery, on the other hand, is underused or used at the efficiency point, which means that farms are probably constrained in their access to machinery. The conclusions regarding labor are ambiguous: there is no information on the relevant price of labor and therefore no clear decision can be reached about the efficiency of its use.

Effect of Farm Size

The estimated coefficients of the Cobb-Douglas production function in Table 3.13 sum to 1.1, which a priori provides an indication of increasing returns to scale. However, the sum of the estimated coefficients is not significantly different from 1 and to assess the effect of scale on productivity we examined the behavior of value added per hectare as a function of farm size. Value added is the farm profit calculated in the section on *Financial Results*. It allows for all production costs (including the cost of using farm assets, as represented by annual depreciation charges), but does not include payments to labor. In case of constant returns to scale, value added per hectare would be independent of farm size. In case of increasing returns to scale, we expect the value added per hectare to increase for larger farms.

Fig. 3.9. Partial productivity of land

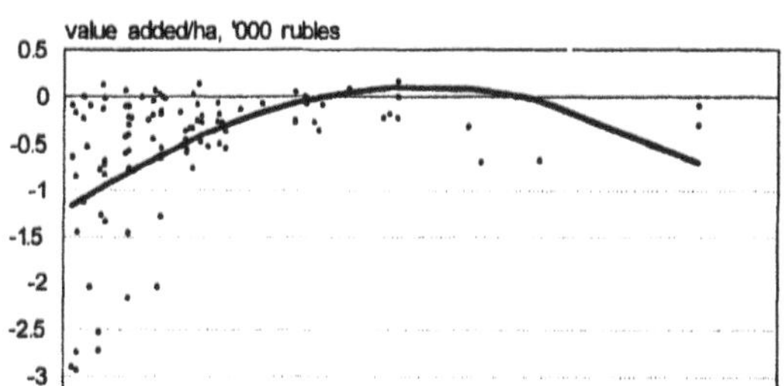

Source: Tambov survey

Value added per hectare increases with size for smaller farms (up to 200 hectare of arable land) and then stabilizes and begins to decrease for larger farms (with more than 200 hectares). The overall dependence on farm size is nonlinear, and fitting a quadratic function to pooled survey data for 1998-2002 we obtain the curve shown in Figure 3.9 (the coefficients are statistically significant at $p < 0.01$, $R^2 = 0.2$). Returns to scale thus do not increase over the entire range of farm sizes, and their behavior is different for small and large farms. The value added per hectare for the group of smaller farms (where returns to scale are increasing) is about −700 rubles, while for the larger farms (where returns to scale are constant) it is about −200 rubles (the difference is statistically significant at $p < 0.01$). The value added per hectare is negative because all peasant farms report losses in recent years, but the larger farms have smaller losses per hectare.

Commercialization

Tambov farms are clearly commercial operations. The family consumes less than 5% of the output, and the rest is sold, used as intermediate inputs in farm production (seeds), or goes to lease payments in kind (Table 3.16, Figure 3.10).

Table 3.16. Shares of output sold and consumed on the farm (2001-2002 averages, percent)

	Output, kg clean weight	Sold	Used on farm	Lease payments in kind	Seeds	Family consumption
Wheat	27,050	63	37	4	29	4
Barley	17,400	38	62	28	25	8
All grain	44,450	53	47	13	28	6
Buckwheat	1,530	84	16	0	16	0
Sunflower	3,440	87	13	3	10	0
All crops	49,420	57	43	12	26	5

Source: Tambov survey.

Fig. 3.10. Allocation of output to sales and on-farm use

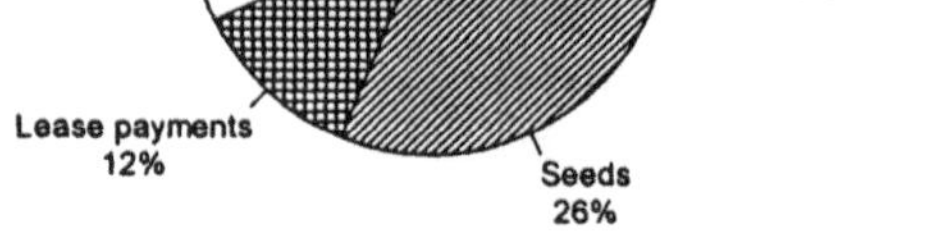

Source: Tambov survey

The low consumption rate is understandable given the farmers' product mix: families do not eat wheat, barley, buckwheat, or sunflower seeds. They eat potatoes, vegetables, meat, and milk—and these products are not part of the production bundle in peasant farms. Farmer families probably satisfy their need for these products from the small household plot, like the rest of the rural population in Russia. It is notable that only grain is used in own consumption. This is so because farmer families need grain to feed the livestock they keep as part of their household plot, which is maintained as a separate operation, mainly for reasons of taxation and registration (there is practically no livestock on the peasant farm as such). Lease payments are also mainly made in barley, which is used to feed lessors' household animals. In principle, of course, farmers could mill some of their grain into flour for family consumption, but this is not a common practice. From price (and quality) considerations farmers today prefer to sell their farm products and purchase food for their family in the market—as in a normal market economy. This constitutes a significant change compared to the early 1990s, when shortages and high food prices justified small-scale processing of own products for family consumption.

Table 3.17. Allocation of farm output to various uses

	Output, kg clean weight	Family consumption	Available for sales and on-farm uses	Sold	Lease payments in kind	Seeds
"Small" farms (<50 ha)	16,260	13	87	56	2	29
"Medium" farms (51-150 ha)	52,721	5	95	54	15	26
"Large" farms (>150 ha)	127,000	3	97	51	18	28

Source: Tambov survey.

A consistent finding in transition agriculture is that the level of commercialization of individual farmers increases with farm size.[3] A similar finding is observed for Tambov peasant farmers. The consumption of grain by the farmer's family is fairly constant in absolute quantities (about 2,500 kg per household), and as a result the percentage of output consumed by the family sharply declines with the increase of farm size. A higher percentage of output is thus left for other uses, which include sales and lease payments in kind. As we see from Table 3.17, the percentage of output allocated to lease payments increases sharply for larger farms, which achieve their size primarily by leasing land (see Table 3.1). If lease payments had been made in cash, not in kind, these quantities of grain would have been sold in the marketplace and the share of output sold would have shown a substantial increase with farm size, from 58% for the "small" farms with up

[3] See, e.g., Z. Lerman, C. Csaki, and G. Feder, *Agriculture in Transition: Land Policies and Evolving Farm Structures in Post-Soviet Countries*, Lanham, MD: Lexington Books (2004).

to 50 hectares of arable land to 69% for the "large" farms with more than 150 hectares.

Contrary to a popular myth, farmers do not conceal their sales. There is no tax incentive for concealment because farming income is exempt from taxes (at least until January 2006) and farm operations anyhow report substantial losses (see next section). On the other hand, there have been cases of farmers approaching retirement age who artificially exaggerated their sales revenue to show profit and thus become eligible for making deductions to the pension fund (the need for these manipulations was eliminated in 2001, when the law allowed contributions to the pension fund even in years with reported losses; prior to 2001, social deductions for family members were calculated in percent of farm profit, because expenditures did not include any payment to family labor).

Table 3.18. Distribution of sales by marketing channels (averages over four main commodities 2001-2002)

Sales channel	Share of total value, %	Share of total quantity, %	Percent by quantity			
			Wheat	Barley	Sunflower	Buck-wheat
Food Corporation	20	13	18	7	0	0
Private intermediaries	38	40	27	42	92	79
Wholesalers	24	25	30	26	8	7
Enterprises	6	6	6	12	0	0
Processors	7	10	16	0	0	14
Direct sales to consumer	4	6	4	13	0	0
All channels	100	100	100	100	100	100
Quantity sold, kg		55,910	34,120	13,250	5,960	2,580

Source: Tambov survey.

The specific marketing channels used by farmers change with commodity and with time. Private traders and wholesalers (primarily grain elevators) are the main marketing outlet for peasant farms in Tambov. They take up more than 60% of farmer's output (both by volume and by value), including virtually all buckwheat and sunflower (Table 3.18). Another major marketing channel is the Tambov Oblast Food Corporation, which in 2001-2002 purchased 13% of farm products by volume and 20% by value. The Food Corporation, a state company established in 1995 to assist agricultural producers and ensure supply of cheap bread products to the population (Box 3.1), purchases only grain (wheat and barley), and even that in relatively small quantities—just sufficient to repay the commodity credits allocated to peasant farmers.

Other uptakers include processors (dairies, flour mills), agricultural enterprises (fish farms, poultry complexes), and of course private individuals in the village or the area markets. In aggregate, these three channels absorb about one-third of the output by volume (but less than 20% by value). Although there is sufficient buckwheat and sunflower processing capacity in Tambov Oblast, direct deliveries of these commodities to processors are very small because their remote locations significantly increase the farmers' transportation costs.

Box 3.1. How a Food Corporation Operates

Food corporations exist in Russia on all levels: the federal Food Corporation is reinforced by regional food corporations virtually in every oblast. The Tambov Oblast Food Corporation is owned and controlled by the oblast government. Its objective of assisting agricultural producers is met in two ways: by providing farm inputs (mainly fuel and fertilizer) in the form of commodity credits to be eventually repaid with grain and by maintaining relatively high purchase prices for grain and barley. The conflicting objectives of high grain prices to producers and low bread prices to the population are theoretically achieved by purchasing grain (at relatively high prices) when the market price is low in the expectation that the market prices would rise before the grain is sold to flour mills and bakeries (in addition, the markup to flour mills and bakeries is kept at an absolute minimum of about 3%). By maintaining the role of a major player in the oblast grain market, the Food Corporation should be able to juggle the two conflicting motives in an optimal manner.

The loans to the corporation are guaranteed by the oblast government, which also determines the list of inputs to be allocated to producers in the form of commodity credits, approves the specific producers entitled to these programs, and authorizes the corporation's entry into various investment projects. As a result of government intervention, the Food Corporation is burdened today with a large overdue debt from agricultural producers who are unable to repay the commodity credits. The corporation's economic situation has been further aggravated because the grain market prices went against the expectations. As a result of its financial difficulties, the corporation was forced to reduce its purchases in 2002. This also affected the sales of peasant farmers to the Food Corporation, which went down from 24% of their output in 2001 to 4% in 2002.

There are significant differences in prices paid by various marketers (Figure 3.11). The Food Corporation pays the highest prices as part of its charter to support agricultural producers in the oblast. Indeed, in Table 3.18, the share of sales by value to the Food Corporation is much higher than the share of sales by volume. Private traders and agricultural enterprises pay the lowest prices. Wholesalers, processors, and the market represent the average price across all channels and commodities.

Farmers do not have the option of storing their grain after harvest until the market prices rise to more attractive levels. There is no storage capacity and no grain cleaning equipment in peasant farms. The only alternative is to store the grain in oblast elevators, but their charges for 6-8 months of storage wipe out or even exceed potential price gains. The only attempt to establish a farmers' cooperative grain elevator failed in the mid-1990s.

Fig. 3.11. Relative prices paid by marketers

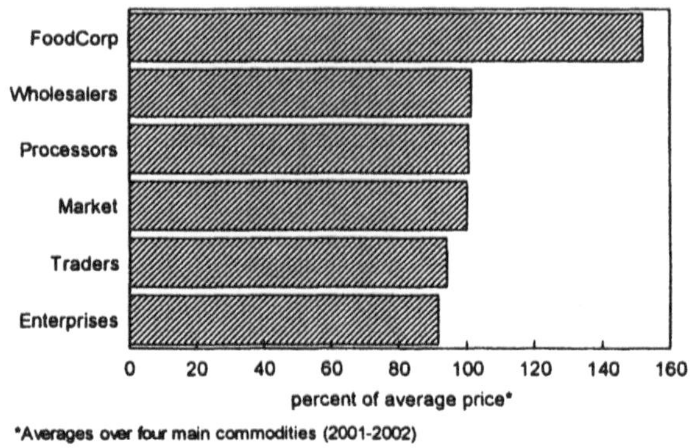

The channels also vary by their quality requirements. The Food Corporation, wholesalers, and processors subject the deliveries to stringent laboratory tests and require the products to meet all state standards. Sales to consumers in the open market are the exact opposite, with absolutely no formal quality standards to be observed. In terms of price–quality tradeoffs, it therefore seems that the sales to consumers bring the highest effective (though not nominal) prices. Despite the stringent quality control, the Food Corporation is the farmers' favorite outlet because of the high prices. Yet the capacity of this channel is restricted and farmers cannot increase their deliveries at these attractive prices beyond the limits set by repayment of trade credits.

Financial results

Peasant farms derive their business income from three main sources: sale of farm products, non-agricultural business activity, and occasional sale of assets. However, non-agricultural business diversification is minimal. The bulk of the revenue is generated by sales of farm products (86% of total business revenue on average between 1992 and 2002), while non-agricultural business contributed a mere 5% of revenue. A surprisingly hefty 9% came from sale of assets, which occurred mainly in the financially stressed years 1994-1995 after the abolition of subsidized credit (in these two years asset sales reached 35% of revenue).

Costs are monitored in the survey according to the same categories, and we can therefore calculate the gross profit for each of these three lines of activity. Agricultural production has been unprofitable every year since 1995: costs were roughly double the revenues and the average loss rate reached 90% of sales between 1995 and 2002. Thus, in 2002, sales revenues of 35,000 rubles incurred production costs of 73,000 rubles producing a negative gross profit of −38,000, or −108% of sales. Non-agricultural business activities, on the other hand, were profitable with margins reaching 36%. Asset sales were also generally profitable, although in 1996, at the peak of distress sales after the elimination of subsidized credits, assets were being sold at half their book value, generating substantial losses. Nevertheless, the overall impact of non-agricultural activities and asset sales on profits was negligible because of the marginal role of these sources in total farm activity. The average loss rate between 1995 and 2002 was −76% of total business revenue.

Fig. 3.12. Costs and revenues: 2002 prices

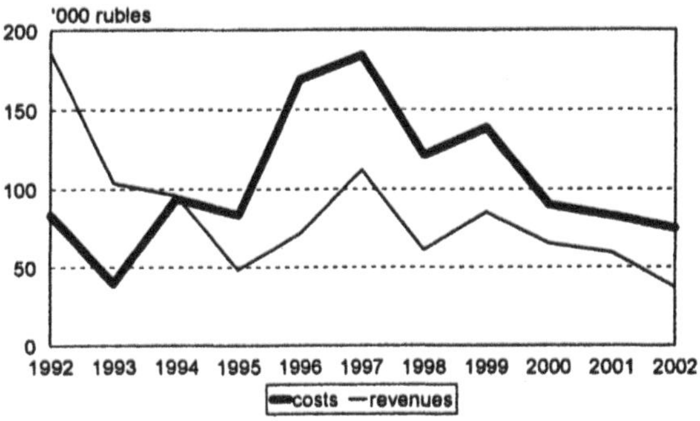

Source: Tambov survey

The changes in performance over time are shown in Figure 3.12, where costs and revenues have been adjusted for inflation and are expressed in constant 2002 prices. Real revenues continuously decreased between 1992 and 2002. Real production costs increased steeply up to 1997 and then began to decline. These trends are a reflection of the deteriorating terms of trade in agriculture: while grain prices decreased (in real terms), the cost of manufactured inputs (fertilizers, fuel, machinery) increased rapidly because of price liberalization policies. As a result of these differential trends, profits reported by peasant farms during the first two years changed to losses after 1994. The inflation-adjusted losses peaked in 1996 and then began to decline, but peasant farms have not returned to profitable operation so far. Since 1995, the percentage of profitable farms fluctuated around 10%; the remaining 90% of peasant farms were unprofitable.

The structure of production costs is shown in Table 3.19. Cash-flow based expenditures on purchased inputs and services (variable costs) constitute 55% of production costs. Payment for labor is not included, as peasant farms primarily rely on unpaid family labor. The remaining 45% is depreciation, which is a non-cash expense. Depreciation has a huge impact on the reported profit of peasant farms. Between 1992 and 2002 production costs averaged 155% of sales revenue, leading to an average loss rate of 55% of sales. Without depreciation, cash-flow based production costs were 75% of sales, leading to a positive profit margin of 25% of sales. However, even without depreciation, peasant farms did not cover their variable production costs in three of the eleven years.

Table 3.19. Structure of production costs and effect of depreciation on profit margins

	Average for 1992-2002	Average for 1992-1994 (startup years)	Average for 1995-2002*
Purchased inputs	44	44	44
Fuel	29	31	28
Spare parts	6	1	8
Seeds	6	10	5
Fertilizer and other inputs	3	2	3
Services	4	5	4
Other costs**	7	20	2
Depreciation	45	31	50
Total production costs	100	100	100
Production costs as % of sales	155	58	191
Variable costs	75	33	90
Depreciation	80	25	101
Reported profit margin	−55	+42	−91
Profit margin excluding depreciation	+25	+67	+10
Percent of profitable farms	27	70	11

*This period excludes the three "good" startup years 1992-1994, when the peasant farms on the whole were profitable.
**Includes financing expenses, which were very high in 1992-1994 due to galloping inflation.
Source: Tambov survey.

Profitability (measured by profit margin—the ratio of profit to sales) generally increases with farm size, but up to a point: the largest farms (those with more than 200 hectares of arable land) are less profitable than the smaller farms with 60-200 hectares. Since 90% of Tambov peasant farms report losses (after depreciation), it is difficult to speak of "increasing profitability:" instead it makes more sense to speak of "decreasing losses." In this context, the largest farms with more than 200 hectares report higher loss rates than the smaller farms with 60-200 hectares (Figure 3.13). The difference between the large farms (>200 ha, profit margin −107%) and the better performing medium-sized farms (61-200 ha, profit margin -45% in a coarser aggregation) is statistically significant by both parametric (ANOVA) and nonparametric (Wilcoxon) tests.

Fig. 3.13. Profit margin by farm size (average 1998-2002)

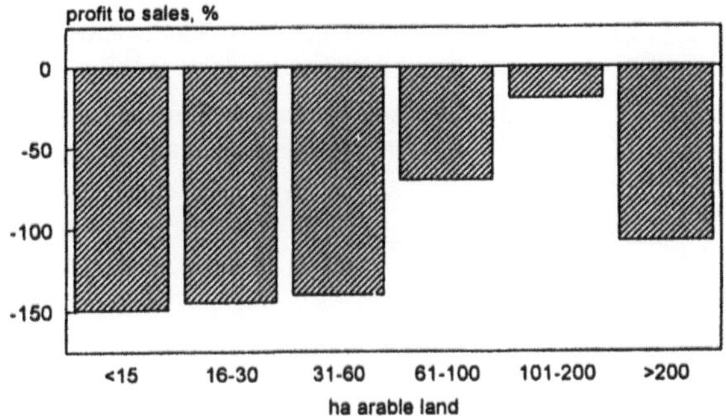

Source: Tambov survey

Source and Uses

Analysis of sources and uses of funds during the entire period 1992-2002 shows that operating revenues (farm and non-farm sales) exceeded operating expenses, with enough left over to cover investment needs (Table 3.20). Since family members do not receive formal wages from the farm, the remuneration for their work on the farm is reflected in cash withdrawals by the family, which average 16% of all uses over the entire period. These withdrawals are partly offset by own funds that are additionally injected by the farmer into farm cash flows (about 10% of all sources); on the other hand, they are augmented by the value of own consumption of farm products (wheat and barley), which add about 20% to cash withdrawals.

Table 3.20 Sources and uses of funds: average for 1992-2002

	Percent		Percent
Sources	100.0	Uses	100.0
Operating revenues	75.1	Operating expenses	57.8
Cash balances	7.5	Net investment	14.8
Borrowing	7.6	Family withdrawals	16.2
Own funds	9.8	Surplus	11.1

Tambov farmers routinely borrow from banks to supplement the cash needs of their farms. However, they typically follow a debt rollover strategy as new loans are largely offset by annual repayment of maturing loans. As a result, we do not observe any alarming debt accumulation in peasant farms. Net borrowing fluctuated around zero during the entire period (with the notable exception of 2002, which was a particular bad year), and the total volume of debt remained fairly constant over time. Since sales revenues were steadily growing, the level of indebtedness measured by the ratio of debt to sales declined sharply from an average of 1.1 in 1992-1995 to 0.1 in 1996-2002. A debt-to-sales ratio of 0.1 implies that less than a month and a half of sales is sufficient to repay the total debt of the average peasant farm in Tambov Oblast. These positive developments are shown in Figure 3.14.

Fig. 3.14. Indebtedness indicators: net borrowing and debt-to-sales ratio

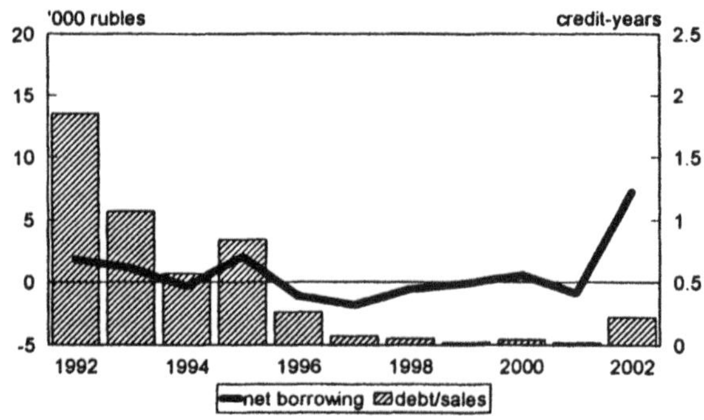

Source: Tambov survey

Conclusion

The development of peasant farming in Russia is far short of the original hopes and expectations. This is usually blamed on the government, which is accused of providing inadequate legal and administrative support on both federal and regional level, as well as reneging on its promises of subsidized credit for working capital and investments. The Tambov case certainly supports these accusations, but it also highlights the importance of two negative factors that come "from the bottom," not "from the top." The first factor is the unfavorable experience of peasant farmers that went bankrupt. In Tambov Oblast nearly 50% of the farmers created between 1991 and 2002 ceased to operated, went into liquidation. The second factor is the extremely poor profit performance of peasant farms, which survive practically only by "eating" their depreciation reserves. The percolating evidence of failure at the farm level creates an unhealthy resonance in the rural society and prevents many prospective farmers from even trying.

Yet those who farm will not consider the option of abandoning their farm despite the hardships. There are simply no alternatives: they cannot go back to the local corporate farm, which is trying to shed as much labor as possible, there are no non-farm job opportunities in rural areas, and they have no skills for urban employment. In the end of the day, they have no intention of giving up their newly found status of an independent, if impoverished, operator.

Initially, the huge multitudes of household plots in Russia were regarded as a pool for potential expansion of private farming. What could be more natural than to predict that the successful household plots will acquire more land, expand their commercial sales, and gradually evolve into the ideal family farm. This vision has not materialized. Instead, we are often witnessing conversion of peasant farms into household plots. The blame for this is squarely on government policies: household plots are not taxed, while peasant farms are subject to a whole range of taxes; household plots can operate virtually without any restrictive registration requirements, while peasant farms need to register with several levels of administration; finally, regional governments may impose size limits beyond which household plots no longer enjoy tax exemption and thus severely curtail the expansion of even the most successful household plots. On balance, it seems that government has seriously impeded the development of private farming through both action and inaction.

While the number of registered peasant farms remained fairly constant at around 3,200 during the last few years, we expect that the new legal requirements for re-registration of agricultural producers will lead to a substantial decrease in the immediate future. According to existing estimates and prevailing opinions, a significant number of registered peasant farms exist only "on paper:" they do not farm and do not produce for a variety of reasons. Tax authorities put the number of non-functioning farmers at 40%. In our view, this is an overestimate, because it includes all peasant farmers who do not submit their tax returns systematically and in a timely manner. A more realistic estimate is 10%-20% non-functioning farmers, who will be struck off the register when the new legal formalities are completed after 2004. The number of peasant farms in Tambov Oblast will then stabilize below 3,000.

Chapter 4

Agroholdings: Russia's New Agricultural Operators

Dmitri Rylko, Irina Khramova, Vasilii Uzun, and Robert Jolly

In the second half of the 1990s, observers of Russian agriculture began to note the emergence of a new organizational form, which took root alongside the three conventional farm types in post-Soviet agriculture: former collectives, peasant farms, and subsistence-oriented household plots. This new organizational form became known as "agroholdings" in the Russian research literature and in the media. In most cases, agroholdings were created by non-agricultural enterprises making investments in agricultural production. Agroholdings were in fact a subcategory of corporate farms, but they showed strikingly different characteristics—huge size and non-agricultural core activities being two of them—which justified the introduction of a separate classification category. The interest in agroholdings intensified as the persistent decline since 1991 changed to growth in 1998-1999. The end-of-century agricultural recovery was certainly attributable to a whole spectrum of factors, including the 1998 devaluation of the Russian ruble that eliminated the price advantage of food imports and encouraged domestic production. Yet researchers were attracted by the possible link between the agricultural recovery and the emergence of agroholdings.

Agroholdings represent a dramatic strengthening of the subsector of corporate farms in Russian agriculture and a deviation from the paradigm of individual or family farming—the expected outcome of transformation of Russian agriculture.[1] The sheer size and scope of operation of agrohold-

[1] This expectation of Western scholars, economists, and politicians was articulated, e.g., in *Food and Agricultural Policy Reforms in the Former USSR: An Agenda for the Transition*, Studies of Economies in Transformation No. 1, Washington, DC: World Bank (1992).

ings clearly justifies in-depth study of these structures. In this chapter, we describe the modes of formation and the operating principles of agroholdings and provide a brief characterization of the specific organizational structures. We treat this new phenomenon in Russian agriculture as a manifestation of institutional innovation that is shaping the sector's frontiers. Our goal is to try and explain how and why agroholdings are developing in Russian agriculture and what these changes mean for the future in terms of productivity, competitiveness, and social welfare.

The chapter starts with brief definitions and discusses some of the economic forces that might lie behind the formation of agroholdings. We then provide an empirical description of the phenomenon drawing on a unique database of more than 100 agroholdings, two recent surveys, and a number of case studies. The empirical section discusses the scope and size of agroholdings, their origin and growth since 2001, their reasons for entering agriculture, their entry paths and functions with regard to the agricultural subsidiaries, and finally their management practices. The empirical discussion ends with a brief evaluation of the economic performance of agroholdings compared to the rest of agriculture. Are agroholdings more profitable than conventional farms? Are they less indebted? Do they create value, or are they simply the result of rent seeking by powerful interest groups? Do they achieve higher productivity? Unfortunately, the available data provide only tentative answers to these pivotal questions. The third part pulls together the various dimensions and features of agroholding operation based on several case studies in different regions and assesses their potential role in transforming in Russian agriculture. The chapter ends with a brief list of further research priorities.

Drawing the Boundaries of Agroholdings

Agroholdings vary considerably in their origin, structure, and economic objectives. We start with a description of several agroholding archetypes—to better define the term by demonstrating variability and to provide motivation for the analysis that follows.

Agroholding Profiles[2]

MTS Olimpia: "custom harvester." In the early 1990s, a group of young Russian entrepreneurs started a firm serving as a debt collector on behalf of energy

[2] The descriptions that follow are based on case studies conducted by IKAR—Institute for Agricultural Market Research in Moscow.

firms in southern Russia. In 1998, following Russia's financial crisis, they identified some apparent profit opportunities in grain production. However, they also observed that expansion of grain production was constrained by the supply of agricultural machinery, especially combines. Using their connections with financial institutions, they purchased several modern combines. They created a limited liability company named Agricultural Machinery-Technological Station (MTS) Olimpia to provide custom farming services to corporate farms (former collectives) in Krasnodar and Rostov regions of southern Russia. After a few years of operation, MTS Olimpia was custom farming 3,500-4,500 hectares of small grains and row crops. MTS Olimpia charged relatively high custom rates by western standards and earned reasonable rates of return. However, other custom operators have now entered the market reducing custom margins. In early 2003, MTS Olimpia began to seek out opportunities for leasing land and entering agricultural production.

EFCO: "food processor." During the 1990s, EFCO emerged as one of the leading private oilseed crushing and processing firms in Russia. In the spring of 2000, the company was asked by the governor of Belgorod Oblast to participate in the farm restructuring process. EFCO was primarily interested in securing a reliable source of sunflower seeds for its vegetable-oil operations. The company initially acquired the assets of more than twenty technically bankrupt former collectives and entered into land lease agreements with hundreds of individual land share owners.[3] As the lease of the land shares shifted from the failing former collectives to EFCO, rural people in Belgorod Oblast began receiving stable in-kind lease payments for the first time since the early 1990s. EFCO also expanded into two neighboring regions, ultimately leasing more than 100,000 hectares for sunflower and other crops. EFCO established an agricultural management subsidiary to handle farm operations. All key investment and production decisions, however, had to be approved at headquarters. Since its inception, the company has made significant investments in crop-production machinery and equipment. They have drastically reduced unprofitable livestock operations associated with the former collectives. Each season, they would enter into large-scale purchase agreements with input suppliers and obtain necessary quantities of fertilizers and pesticides at competitive rates. A field and farm security system has been launched to control theft. The number of farm employees has been gradually reduced. Most of the former farm managers have been replaced. However, EFCO also provides health care, social service programs, and technical training for workers. Recently, EFCO has organized cooperatives in each former collective farm's area to manage social

[3] See Chapters 2 and 5 for details of this land leasing mechanism.

services. The land rental payments plus fixed company contributions have been used to create a "social benefits" package. The local cooperative decides how to spend this money. The first results of this innovative social scheme are promising.

Agrovoronezh: "agricultural venture entrepreneur." In 2003, a group of private investors formed an agricultural production company in Voronezh Oblast in the fertile central chernozem region. The investor group consisted of several local and Moscow-based individuals with ties to the oil and gas and construction business. During the next several months, the company negotiated with the local administration to acquire the non-land assets of a bankrupt former collective farm in one of the districts. At the same time, the firm organized the purchase of land shares from the former collective farm's employees. By the autumn of 2004, some 10,000 hectares of land had been purchased and ownership rights had been registered in the local land office. Another 10,000 hectares are currently being converted from land lease to land ownership. To facilitate this process, Agrovoronezh has agreed to continue land lease payments for another four years. Ultimately the firm intends to acquire up to 50,000 hectares. At the same time, the firm has dramatically restructured the former collective's assets, organization, and management. The inefficient livestock herd was cutback sharply. The number of workers was reduced from more than 1,000 to 375. The number of agricultural managers and technical employees was reduced by over one-half, while the company hired better trained agricultural managers and offered salaries and bonuses that were well above local levels. One of the company owners actually relocated from the city to the farm. With the good harvest results expected in 2004, the company owners remain optimistic about further expansion. One of the opportunities under discussion is to pledge company-owned land as collateral for loans from a local bank to finance additional investments.

Bank Avangard: "supply chain management." Bank Avangard is a large Russian financial group heavily involved in the domestic malting business. The firm requires 300,000 metric tons of malting barley annually. In order to assure a reliable supply, Avangard-Agro, the bank's agricultural entity, contracted with two corporate farms in Orel and Voronezh oblasts to grow malting barley on 10,000 hectares. The company provides seeds, pesticides, and fertilizers to these farms. It also guarantees to buy the malting barley at the current market price plus a fixed premium. If the farms fail to deliver malting quality barley due to unfavorable weather, the company still guarantees to purchase their feed barley at the current market price, but without premium. The bank has entered into an agreement with the Russian subsidiary of Syngenta, a multinational agrochemical company, to provide chemicals and agronomic technical assistance. In 2004, Avangard-Agro leased five additional farms.

These four cases illustrate some of the general patterns observed in the emergence of agroholdings in Russia.

1. The entrepreneurs are responding to perceived profit opportunities in production agriculture—primarily in the former collective sector.

2. The agroholdings are founded by outsiders with access to credit, processing services, or end-user markets.

3. The agroholdings focus, to a large extent, on overcoming input market imperfections, supplying machinery services, credit, pesticides and fertilizers, managerial services, skilled labor among others.

4. The agroholdings make significant investments and technological improvements in production agriculture.

5. The agroholding founders work closely with administrative officials in developing and managing their businesses.

6. The social consequences in rural areas appear to be mixed—layoffs are common, yet all firms try to guarantee competitive land lease payments to rural people and some firms have attempted to provide extended packages of social services previously offered by the collectives.

Terminology and Definitions

We broadly define agroholdings as commercial farms established, owned, or controlled by entities whose core business is outside primary agriculture. One key feature of agroholdings is the participation of the non-agricultural controlling entity in farm production decisions and in farm risk sharing. Another key feature is a clear commercial orientation of the farming activities as opposed to farming activity oriented toward internal consumption. In our view, an agricultural producer has to pass both tests to be classified as an agroholding. A third common feature is a multi-farm structure. An agroholding typically consolidates several large farms within a common ownership or control structure. A single-farm entity however, could qualify as agroholding if the two former conditions—ownership by a non-agricultural entity and commercial agricultural activity—are met. Figure 4.1 is a schematic illustration of our view of an agroholding.

The dividing line between agroholdings and "traditional" corporate farms (i.e., former kolkhozes and sovkhozes) is somewhat arbitrary. Many "traditional" corporate farms follow the same aggressive agricultural business pattern as agroholdings: they acquire neighboring farms, lease land from other landowners, develop custom farming services, or diversify into non-agricultural activities. Yet they are not owned by non-agricultural business entities, a nd thus fail the first test of being an agroholding.

Fig. 4.1. Defining the concept of agroholding

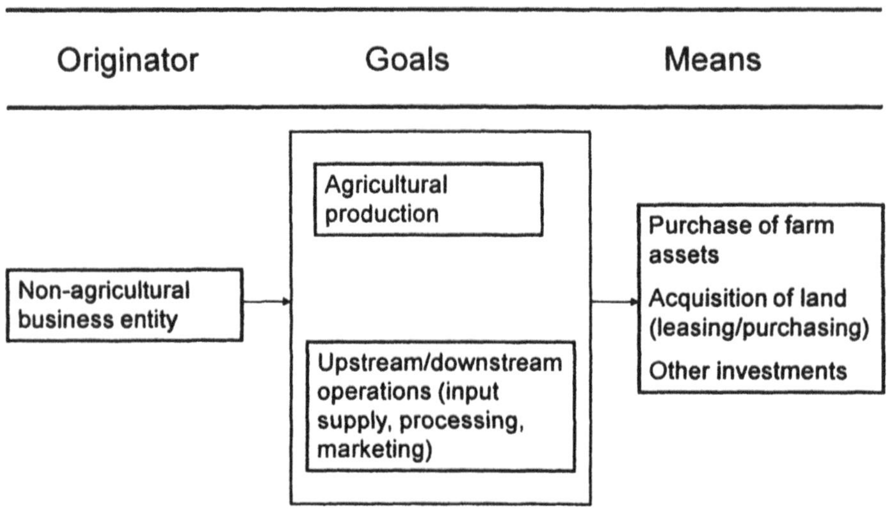

By origin, the agricultural operations of agroholdings may be broken down into farms owned by input suppliers, food manufacturers and traders, banks, manufacturing and commercial conglomerates outside the food industry. The main modes by which agroholdings enter agricultural production include custom farming operations, joint production agreements with existing producers, leasing of farm land from other farmers, taking over an existing farm or several farms, and various combinations of these approaches. Another common operating mode by which Russian agribusinesses enter agricultural production is through the provision of commodity credit, i.e., advance supply of farm inputs in return for delivery of farm products at harvest time. Although commercial pre-harvest financing does not imply engagement in direct farming activities by creditors or lenders, in Russia it is typically associated with a high level of production risk and thus entails a high level of creditor involvement in production decisions. For instance, a fuel supplier "lending" fuel against repayment with grain after the harvest will naturally attempt to safeguard the commodity loan by closely monitoring all stages of farm production. For these reasons, pre-harvest commodity creditors in Russia often fall in the agroholding category.

Reasons for the Emergence of Agroholdings

Two economic phenomena characterize the emergence of agroholdings—entry and growth. Outside firms without ties to production agriculture enter the sector, and these entering firms generally enlarge the overall size of the acquired farming operations by extending their vertical or horizontal boundaries. In this section we briefly sketch out some of the underlying economic forces that can help explain the emergence of agroholdings. We will focus on four motives for expansion—horizontal integration, vertical integration, diversification, and rent seeking.

Horizontal expansion generally refers to the growth of firms within the same line of business. The motives for horizontal expansion are primarily to achieve economies of scale or scope, which are typically an industrial phenomenon and are not very common in production agriculture.[4] As our case studies illustrate, agroholdings, once established, do grow horizontally. However, by definition, the agroholding "mother firm" is not involved in production agriculture. Consequently entry does not represent an expansion of the mother firm's core businesses. There are examples of agroholdings, such as Agrovoronezh, that grow solely within the production agriculture sector. In this case the motive for entry appears to be acquisition of access to underpriced assets—land and labor—to improve management and to resolve financial and human capital constraints. This entry and growth path resembles that of the nineteenth century bonanza farms in the United States (Box 4.1). In a broader sense, this entry and growth strategy is similar to conglomerate mergers in which unrelated firms are acquired with the expectation that firm performance can be significantly enhanced by a new management team. In the United States, conglomerate mergers have a checkered history.

Vertical expansion—whether through ownership or business relationships—is undertaken for a number of reasons. Two key motivations are (a) to improve the coordination of a linked production process through administrative means rather than through arm's length market transactions; and (b) to protect critical production assets or information from exploitation by business partners. The case studies illustrate that vertical expansion motives appear to be important in the creation of some agroholdings. Food processors and marketers trying to secure their sources of raw materials, fertilizer manufacturers, and suppliers of farm machinery or other inputs looking for

---

[4] See, e.g., H. Binswanger, K. Deninger and G. Feder, "Power, distortion, revolt and reform in agricultural land relations," in: J. Behrman and T. N. Srinivasan, eds., *Handbook of Development Economics*, Vol. III, Amsterdam: North Holland (1995).

ways to consolidate and increase their markets benefit from vertical integration with primary agricultural producers, especially in imperfect markets.

Diversification of risks also offers an explanation for the emergence of agroholdings. Managers may seek to combine firms with uncorrelated or negative returns in an effort to reduce the volatility of returns. With well functioning capital markets, shareholders can diversify their portfolios much more effectively than managers can diversify their firms. However, there may be a diversification benefit to conglomerate-type mergers when capital markets are absent or inefficient.

Box 4.1. "Bonanza" Farms in the United States

Large-scale farms that rely on hired labor exist worldwide and have existed for centuries. Agroholdings can be viewed as another variant on this time-tested organizational form, which, depending on language and geography, is variously known as hacienda, Junker estate, or wage plantation among others. Perhaps one of the closest historical analogs to agroholdings are the bonanza farms of the Upper Great Plains in the United States.[5] Here are a some of the key issues describing their rise and fall:

- The bonanza farms developed in the 1870s and were effectively gone fifty years later.
- A typical bonanza farm cultivated 8,000-25,000 hectares—150 times as large as a "family farm" of that era. One farm was reputed to have over 40,000 hectares.
- The bonanza farms produced one crop—wheat.
- The key economic drivers behind the creation of the bonanza farms were cheap land—a result of the financial collapse of the Northern Pacific railroad—and plentiful labor.
- The bonanza farms were fostered by public officials and railroad and banking magnates as a (temporary) economic development alternative to small holder settlement.
- The bonanza farms were financed by outside investors. Farm size and productivity were critical elements in attracting capital.
- Labor was hired. Seasonal labor was recruited from the lumber camps in Minnesota and Wisconsin.
- Rising land values and falling commodity prices were key contributors to the decline of bonanza farms.
- Rising land values encouraged investors to sell to family farmers.
- Family farms were inherently more productive than the bonanza farms.

The bonanza farms accelerated the development of a family owned and operated agriculture in the United States. Economic and political forces led to their creation and to their ultimate demise. Russia's agroholdings may follow the same path.

A diversification strategy that seems appropriate for agroholdings is to manage risk associated with volatile or thin commodity markets. For example, a farm producing feed grains may integrate forward into livestock production. The returns to both enterprises would be negatively correlated if

[5] H. Drache, *The Day of the Bonanza; A History of Bonanza Farming in the Red River Valley of the North*, North Dakota Institute for Regional Studies (1964).

feed grains are a major revenue source for the farm and a major cost for the livestock operation. An example of potentially beneficial diversification is provided by the case of grain producers and poultry farms in Russia.[6] The earnings of these two farm sectors showed zero (or even slightly negative) correlation during 1994-1998. A grain farmer investing in poultry production could expect to stabilize profits. Zero correlations were also observed in this period between the earnings in oil and gas industry and the earnings of pig enterprises, possibly justifying the diversification of these non-agricultural companies into livestock production.[7] The relative importance of the diversification motive is an empirical question. Our case studies give only limited evidence that agroholdings emerge to allow the non-agricultural parent firm to diversify by entering into production agriculture.

The final motive for agroholding emergence is rent seeking. Rent seeking behavior can occur in conjunction with any of the horizontal, vertical, or diversification motives. Rent seeking would occur when powerful elites are able to shift the terms of trade in their favor by exploiting relationships with public officials in a way that makes other groups less able to compete for resources or denies them access to markets.[8] Again it is an empirical question whether or not the rise in agroholdings can be related, for example, to the inadequate development of family farms in Russia or the increasing frequency of bankruptcies among former collective farms.

Table 4.1 summarizes the discussion of economic motives for the formation of agroholdings. We conclude that firms pursuing a horizontal expansion strategy would tend to operate entirely within production agriculture. Vertical integrators would emphasize transactions between group members and active management of their portfolio. Diversifiers would be more likely to be passive toward the management of firms in their portfolios in order to maximize the benefits from uncorrelated returns. Moreover, vertical integrators would tend to invest in technology and production facilities within the group, setting their sights on long-term benefits. Diversifiers, on the other hand, would be much more flexible regarding their investment horizon and would tend to allocate funds to acquisition of new additions to the portfolio instead of investments in existing portfolio components. Their main emphasis would be on financial control and redistribu-

---

[6] I. Khramova, "Agroholdings' operation in Russia: a case study," in: Ye. Gaidar, ed., *Economics of Transition: Collected Studies 1999-2002*, Moscow: IET, Izd. Delo (2003), pp. 675-714.

[7] Correlations calculated on the basis of industry profitability data from Russia's Statistical Yearbook 2003.

[8] For more details on rent seeking see K. Deininger and H. Binswanger, "Rent seeking and the development of large-scale agriculture in Kenya, South Africa, and Zimbabwe," *Economic Development and Cultural Change* 43(3): 493-523 (1995).

tion or appropriation of the profits generated by each component in their portfolio. Rent seekers would exploit political contacts and access to resources and information.

Table 4.1. Mechanisms, motives, and testable characteristics of agroholding creation

Strategy	Motive	Organizational behavior
Horizontal expansion	Economies of scale and scope	Grow within single industry
Vertical expansion	Improved coordination of production processes; protection of relationship-specific assets; reduction of transaction costs and market barriers	Mergers, acquisitions, contracts, alliances; value chain management (emphasis on intrafirm transactions); long-term investments
Diversification	Reduction of financial and production risks	Acquire an equity position in non-core businesses; emphasis on central appropriation and redistribution of provits; short-term investments
Rent seeking	Elites seek to capture excess returns, limit competition	Focus on lobbying and other influence activities

Empirical Description of the Agroholding Sector

The empirical section is the analytical core of the chapter. It uses unique data collected by the authors to describe the scope and the size of agroholdings, their origin and growth, the reasons for entering agriculture, the entry paths and functions of the parent companies with regard to the agricultural subsidiaries, and finally the management practices in agroholdings. The empirical discussion ends with a brief evaluation of the economic performance of agroholdings.

The Data

Agroholdings are a poorly defined category that is not covered by official Russian statistics. Less than five years ago many researchers doubted their very existence. Today nobody doubts the existence of agroholdings, but due to inadequate data there are huge differences in estimating the scope of the phenomenon.

The only publicly available source of quantitative information on agroholdings, i.e., non-agricultural companies with subsidiaries ("daughter companies") engaged in agricultural production, is the rating of the largest agricultural producers in Russia published by the Agrarian Institute (VIAPI) in Moscow in cooperation with MinAg. The VIAPI rating contains data on 33 agroholdings for a single year (2001), obtained using a mail survey and the firms' consolidated financial reports.[9] The Institute for Agricultural Market Research (IKAR) in Moscow has augmented the VIAPI database with information on additional agroholdings and supplemented it with qualitative data about the origin of each company, ways of entering agriculture, management structure, and other topics. In this way, the list of agroholdings gradually has grown from VIAPI's original 33 to 148, with data covering the years 2001-2003 collected through personal interviews with company managers, several survey questionnaires, and company-level case studies. Attempts to extend the database are continuing.

Unfortunately, the 148 agroholdings in the IKAR database are a highly biased sample determined by considerations of accessibility and personal contacts. The database includes information from some 25 oblasts. It is known, however, that agroholdings exist in significant numbers in provinces that so far are not represented in the database (Novosibirsk, Kaluga, Ryazan', Novgorod, Nizhnii Novgorod). For some of the provinces in the database the information is quite sporadic (Moscow, Kursk, Bryansk, Smolenks, Leningrad, Kurgan, Tambov, Vladimir). Even in provinces that are reasonably well covered, the actual number of agroholdings is probably substantially greater than in the database (e.g., Rostov, Belgorod, Voronezh, Stavropol', Volgograd, Krasnodar). The quantitative data in the following sections are therefore not statistically representative, and they are given only as a partial illustration for a subsegment of Russian agroholdings.

Much of the analysis that follows derives from IKAR's unique database of 148 agroholdings (we refer to it as the IKAR Agroholdings Database 2004). Another statistically nonrepresentative database comprising 33 agroholdings has been provided by VIAPI—the Nikonov Institute of Agrarian Problems and Informatics in Moscow (the VIAPI Agroholdings Database).[10] Additional analyses use data from two original field surveys: the 2002 Rostov survey—an in-depth survey of 14 agroholdings in Rostov Oblast carried out by the Analytical Centre for Agri-Food Economics (AFE) in Moscow in cooperation with the Rostov Institute of Economics and Standards, and the 2003 BASIS survey, which included 9 farms (among 142 respondents) that could be identified as belonging to an agroholding (in each of these 9 farms, an outside corporation had a controlling interest, i.e.,

---

[9] *Reitingi naibolee krupnykh i effektivnykh proizvoditelej sel'skokhozyaistvennoi produktsii v Rossii za 1999-2001 gg.*, Moscow: MinAg (2003), pp. 27-28.

[10] *Reitingi naibolee krupnykh i effektivnykh proizvoditelej sel'skokhozyaistvennoi produktsii v Rossii za 1999-2001 gg.*, Moscow: MinAg (2003), pp. 27-28.

owned more than 50% of equity). In the 2002 Rostov survey the respondents were agroholding officials; the 2003 BASIS survey interviewed farm managers.

Scope and Size

As of August 2004 IKAR had identified 148 agroholdings operating in 25 Russian provinces. Of these 148 agroholdings, 118 provided basic size information for 2002-2003, which is summarized in Table 4.2. The 118 agroholdings control a total of 6.1 million hectares of agricultural land, which is more than 5% of all agricultural land in the corresponding regions (109 million ha). The average agroholding has 52,000 hectares—a huge area even by former Soviet standards. It is an agglomeration of 8 large corporate farms, each typically organized as a legal entity within a holding ownership structure. These farms (each with 7,000 hectares—roughly the size of a traditional kolkhoz) constitute the agricultural business component of the holding company ("the agricultural project"), justifying its classification as an agroholding.

The size distribution of the agroholdings in the IKAR database is highly skewed upward. While 50% of agroholdings are not larger than 30,000 hectares, 3% of agroholdings control from 200,000 hectares to more than 500,000 hectares each (Figure 4.2). This small number of very large agroholdings is responsible for the relatively large average size of 52,000 hectares.

Fig. 4.2. Size distribution of agroholdings 2002-2003

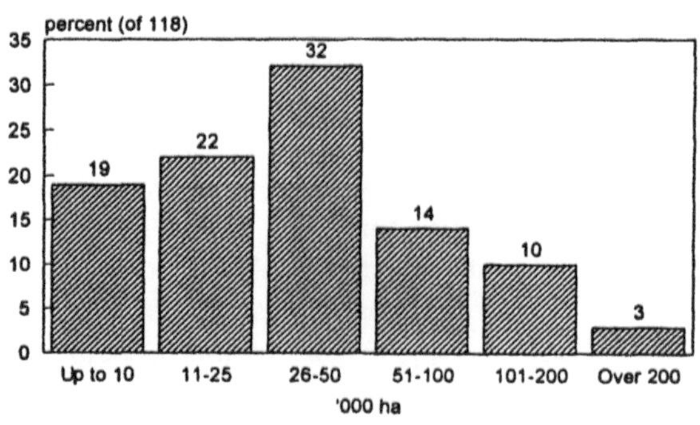

Source: IKAR Agroholdings Database

The agroholdings in the IKAR database are mainly concentrated in the Central and Southern Federal Regions of Russia (Table 4.2). While the Southern Region is traditional farming country with fertile chernozems, the

Central Region has poorer soils. Exceptional agricultural productivity does not appear to be a decisive factor for the formation of agroholdings. A remarkable phenomenon is the concentration of at least 10 agroholdings in the relatively remote Altai Oblast in the Siberia Region. Altai is an important grain producer in the Russian part of the virgin land belt developed in the 1950s-1960s, and grain elevators appear to be popular candidates for acquisition by agroholdings. Finally, 27 agroholdings are classified as interregionals, with their operations spanning several provinces. These interregional agroholdings are significantly larger than the agroholdings in other four regions. Their large size is manifested in total agricultural land and in number of farms per company. Moreover, they are characterized by a much larger volume of sales (per agroholding and per farm) and, most significantly, by a much larger output per hectare—a partial measure of productivity.

Table 4.2. Agroholdings in 2002-2003

	Total	Central	South	Volga	Siberia	Interregional
Number of agroholdings	148	55	37	11	18	27
Land per agroholding, '000 ha	52.2	55.8	33.2	31.9	37.4	94.9
Number of farms per agroholding	8	9	5	7	4	15
Land per farm, '000 ha	7.0	6.6	7.2	5.7	9.4	6.3
Sales per agroholding, mln. Rubles	540.9	365.6	321.2	—	478.6	1,033.6
Sales per farm, mln. rubles	60.0	38.1	78.2	—	81.6	65.3
Sales per ha, '000 rubles	12.0	8.0	9.4	—	5.8	19.9

Note: Reasonably complete data on land and number of farms are available for 118 out of 148 agroholdings. Sales data are available for 47 agroholdings only.
Source: IKAR Agroholdings Database 2004.

Annual sales revenue also can be used as a measure of size, but unfortunately sales data are reported in the database only for 47 out of 148 observations. For the agroholdings that report sales, 28% fall in the small category with annual sales of up to 100 million rubles, 40% fall in the medium category with sales between 100 million rubles and 500 million rubles, 17% fall in the large category with sales between 500 million rubles and 1 billion rubles, and fully 15% have sales over 1 billion rubles. Because of this strongly skewed distribution of farm sales, the mean is 540 million rubles, while the median is only 200 million rubles. There is naturally a strong positive relationship between annual sales and agricultural land in agroholdings (correlation coefficient 0.6). Figure 4.3 illustrates this relationship using grouped data for the four sales categories (each bar shows the average size, in thousand hectares, for the corresponding sales category).

Fig. 4.3. Agricultural land by farm sales categories
(46 agroholdings, 2002-2003)

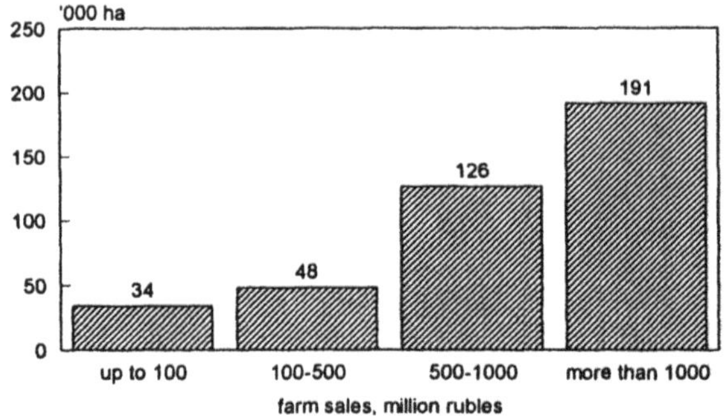

Source: IKAR Agroholdings Database

Table 4.3. Comparative size characteristics of agroholdings and conventional corporate farms in Rostov Oblast (per farm averages, 2001)

	Agroholdings (n=14)	Conventional corporate farms (n=949)
Cultivated land, '000 ha	31,641	5,965
Land under wheat, '000 ha	18,223	1,276
Wheat production, metric tons	98,600	3,800
Yield, kg/ha	2,940	2,990
Fertilizer consumption, kg/ha	135	21
Production costs, ruble/ton	1,245	1,092

Source: 2002 Rostov survey.

To get some sense of how large the agroholdings are, it is useful to compare the basic size characteristics of agroholdings to corresponding regional averages (Table 4.3). In the 2002 Rostov survey covering 14 operators, the average agroholding cultivated 32,000 hectares, compared with 6,000 hectares for the average corporate farm in Rostov Oblast. The average agroholding harvested 99,000 metric tons of wheat, compared with 4,000 metric tons for the average corporate farm. Yet the yields per hectare and production costs per ton were practically the same in agroholdings and traditional corporate farms. This suggests that the Rostov agroholdings did not achieve any significant economies of size or improvements in management or technology. The regional role of Rostov agroholdings is commensurate with their size. In a region with nearly 1,000 farms, just 14 agroholdings control 7% of agricultural land and are responsible for 30% of wheat and 10% of sunflower production. The agroholdings in the IKAR database similarly control about 6% of all agricultural land in the regions where they operate. Recall, however, that each agroholding is a conglomeration of many farms, and the numbers for each member-farm separately (both in Rostov and in the IKAR database) are not much different from the national averages based on all 27,000 corporate farms in Russia (Box 4.2).[11]

Box 4.2. Agroholdings in Rostov Oblast

Rostov Oblast is undoubtedly one of the leaders among regions with a significant presence of agroholdings. The Rostov agroholdings were created entirely by private initiative, with very little involvement of regional authorities. This sets Rostov apart from some other regions, such as Orel and Belgorod, where regional authorities were actively involved in agroholding creation. The Rostov agroholdings specialize in grain and oilseeds, which are the most competitive local crops. Four of the Rostov agroholdings rank among the top 15 Russian grain exporters. Their cumulative share in the Russian grain exports was 20% in 2002-2003. Three out of the four companies have their own grain export terminals. Another Rostov agroholding is the largest Russian sunflower oil processor with a share of 20% in the bottled vegetable oil market.

The IKAR database includes 17 Rostov-based agroholdings, more than in any other single oblast. They incorporate 81 farms and control 320,000 hectares—more than 5% of all agricultural land in the oblast. This does not include the farms controlled by two interregional agroholdings with presence in Rostov. However, the agroholdings are distributed very unevenly across Rostov's 50 districts. The highest concentration is observed in key grain and oilseed producing areas. Our data on two of these districts show that agroholdings control 34% of farmland in Tselinskii Raion and 23% in Salskii Raion. On the other hand, there are raions with zero or a very limited presence of agroholdings. These areas are characterized by a different specialization or with marginal land productivity.

Fig. 4.4. Specialization profile of agroholdings 2002-2003

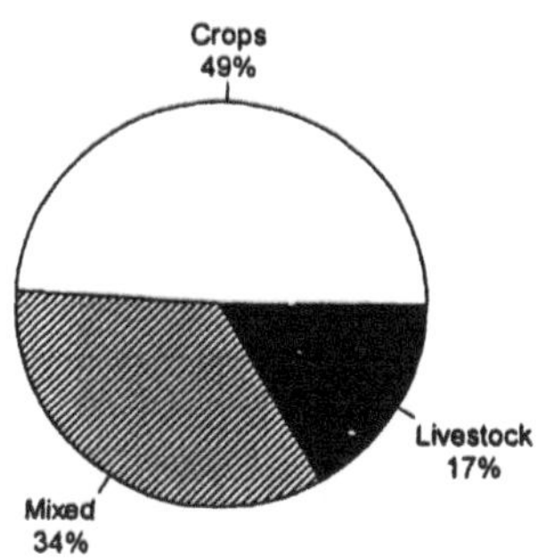

IKAR Agroholdings Database; percent of 118

What do agroholdings produce on their farms? G iven the huge accumulation of land, the focus is obviously on crop production—mainly grain,

[11] In the 2003 BASIS survey, the average size of a farm that belongs to an agroholding (4,300 ha) is not statistically different from the average size of all other farms (4,100 ha).

oilseeds, sugar beet, in some cases rice. Many agroholdings supplement crops with livestock production, continuing the Russian tradition of mixed crop-livestock farming. A relatively small percentage of agroholdings specialize in livestock and poultry (Figure 4.4). However, agroholdings with a livestock specialization are often leaders in their respective industries. Thus, in the poultry industry, five vertically integrated holdings supply 35% of Russia's broiler output through 24 farms.

Origin and Growth

An agroholding is typically created when a non-agricultural company, a term that broadly covers everything outside primary agricultural production, enters production agriculture. The classical mode of entry is through the acquisition of existing farms or establishment of new farms. Entry into service-oriented agricultural activities, such as custom farming, input supply, or product marketing, usually creates an agribusiness, not an agroholding in the sense used in this chapter.

Agroholdings are generally owned by private capital, although there are prominent cases of state-owned and quasi-state agroholdings. For example, Gazprom, the leading state-owned gas company in Russia, has a large number of farms, which originally were intended to act as subsidiary farms producing food products for Gazprom employees. These farms gradually developed into full-fledged commercial farming operations. Gazprom is thus a state-owned agroholding with a cluster of fairly independent farms. The large agroholdings in Bashkiriya and Voronezh Oblast are regional state companies. These farms enjoy relative independence within a hierarchical holding management structure. The agroholdings in Orel Oblast are examples of quasi-state companies, which entered agricultural production with the encouragement and active involvement of regional authorities. Rostov is a much less interventionist oblast, but even here one-third of the companies have some share of state ownership (2002 Rostov survey).

Most of the agroholdings in the database were created by non-agricultural companies with business links to agriculture. Thus, fully 50% of agroholdings in the IKAR database have their roots in agribusiness, including the food industry, food marketing, and farm input supply (Figure 4.5). Nevertheless, some agroholdings were established by non-agricultural industrial conglomerates, financial companies as well as state-owned corporations without any prior agricultural affiliations. Companies with a primary focus in production agriculture are only 6% of the agroholdings in the database.

Fig. 4.5. Distribution of agroholdings by origin

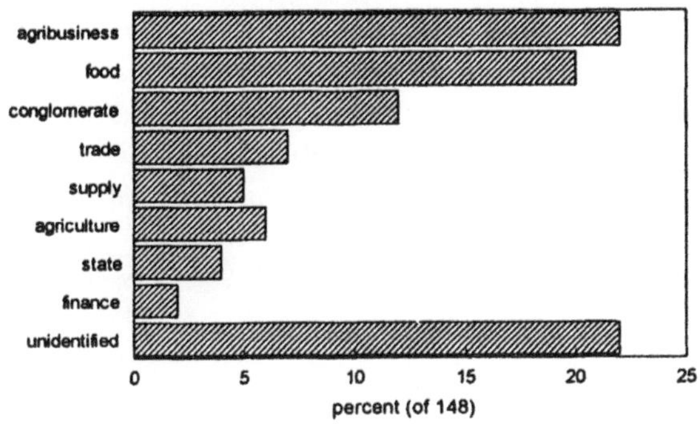

Source: IKAR Agroholdings Database

The average land size of agroholdings is generally independent of origin. Across all origin categories (except state-owned) the average agroholding operates between 30,000 and 60,000 hectares of agricultural land (Table 4.4; the differences are not statistically significant). The average state-owned agroholding, on the other hand, controls 270,000 hectares of agricultural land, and is significantly larger than the agroholdings in other origin categories. The six state-owned agroholdings are distributed fairly uniformly between a minimum of 23,000 hectares and a maximum of 507,400. The median size 235,000 hectares, is practically equal to the average, 270,000 hectares. Moreover, of the four agroholdings with more than 200,000 hectares in the database, three are state-owned companies.

Table 4.4. Agroholdings by origin and size

Origin	Number of agroholdings in database	Number of agroholdings reporting land	Average land size, '000 ha
Diversified agribusiness	32	29	52.4
Food processing	29	23	40.6
Trade (agricultural commodities)	11	10	26.3
Supply (farm inputs)	8	8	39.3
Agriculture (primary)	9	8	28.3
Industrial conglomerates	17	15	45.5
Banking and finance	3	2	90.5
State-owned (fully or partially)	6	6	267.7
Unidentified	33	17	25.1
Total	148	118	52.2

Source: IKAR Agroholdings Database 2004.

Agroholdings began to form in the mid-1990s. A surge in startups occurred in 2000-2001, just after the 1998 financial crisis, when domestic farm products suddenly became highly competitive due to devaluation and began to replace imported food. It seems that increased agricultural profitability in the wake of the 1998 devaluation encouraged the entry of profit-seeking businesses into agriculture and led to the second wave of agroholding creation. However, new entries did not stop after the post-crisis devaluation effect had been exhausted. Agroholding creation, after 2001, has continued at a strong rate (Figure 4.6). This suggests that agroholding formation is due to more factors than currency devaluation and import substitution.

Fig. 4.6. Distribution of agroholdings by date of creation

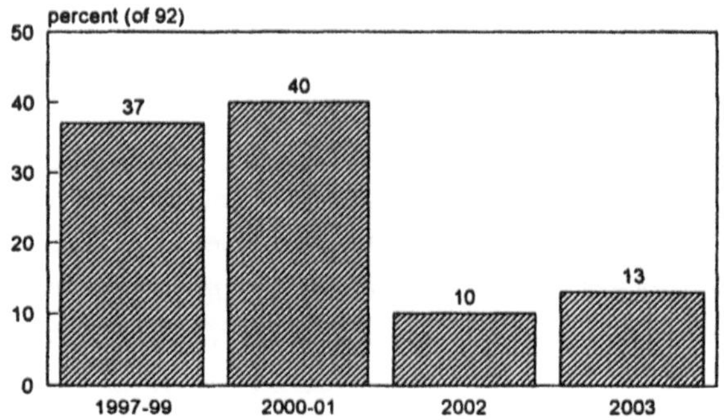

Source: IKAR Agroholdings Database

Agroholdings are a highly dynamic sector. Of the 67 agroholdings that existed in 2001 and for which land data are available, 31 increased their total land holdings during 2002-2003 from 1.3 million hectares to 2.3 million hectares (Table 4.5). A relatively small number of companies (10 out of the original 67) reduced their operations in 2002. One small agroholding, Ogo, which initially controlled only 8,000 hectares but was planning to increase its holdings to 30,000 hectares, pulled out of crop farming in 2002, although it has kept the poultry operations and remains in custom farming. The state gas company, Gazprom, although still in agriculture in 2003, announced plans to divest all of its 76 farms and 500,000 hectares in the near future. The agroholding sector as a whole showed strong growth in 2002-2003, adding 22 new entrants with more than 700,000 hectares of agricultural land.[12]

---

[12] Of the 104 agroholdings reporting land in 2002-2003, many are simply new additions to the database, not newly created entities. The 13 agroholdings listed here as new entrants were in fact created in 2002-2003.

Table 4.5. Land in agroholdings 2001-2003

	Number of agroholdings reporting land	2001 ('000 ha)	2002-2003 ('000 ha)	Percent change
Existed in 2001	67	3,785	4,385	16
Expanded in 2002-2003	31	1,322	2,271	72
No change in 2002-2003	26	1,661	1,661	0
Shrank in 2002-2003	10	802	454	−43
New in 2002-2003	22	—	716	—

Source: IKAR Agroholdings Database 2004.

These aggregated data highlight an apparent tendency toward more land in agroholdings between 2001 and 2003. But did the average agroholding grow over time? Taking matched samples of 55 agroholdings that report their agricultural land for both 2001 and 2003 (out of the 118 agroholdings with land in the database), we note that the average size increased over these two years from 45,200 hectares to 55,200 hectares. For a larger subsample of agroholdings reporting land in 2001 and 2002 (67 out of 118 cases), the average size increased from 56,500 hectares to 65,700 hectares (Table 4.6). However, the differences between these means are not statistically significant for either period. The size variation in the two subsamples is so large that the 95% confidence intervals practically overlap for the two years. Analyzing the growth by different size groups, we note that the smallest agroholdings, those with up to 30,000 hectares in 2001, show statistically significant growth both in the two-year period 2001-2003 and in the earlier one-year period 2001-2002. The medium and large agroholdings, on the other hand, do not show a clear increase in mean size, and the changes are not statistically significant. We can thus conclude with any degree of statistical confidence that only the smallest agroholdings grow by adding more land, whereas the larger agroholdings keep a fairly constant size.

Table 4.6. Growth of agricultural land in agroholdings since 2001, by size group# (mean size, '000 ha)

	Total sample	Small	Medium	Large
Two-year change 2001-2003	n=55	n=32	n=21	n=6
2001	45.2	13.4*	53.4	186.9
2003	55.2	30.3*	56.6	183.3
One-year change 2001-2002	n =67	n =36	n =23	n = 9
2001	56.5	14.5**	53.8	230.9
2002	65.7	25.5**	55.1	252.1

Small: agroholdings with up to 30,000 ha in 2001; medium: 30,000-100,000 ha in 2001; large: more than 100,000 ha in 2001.

*Difference statistically significant at 5%; ** Difference statistically significant at 10%.

Source: IKAR Agroholdings Database 2004.

Reasons for Entering Agriculture

The reasons for entering agriculture were explored in the 2002 Rostov survey. Most companies (12 out of the 14 surveyed) gave multiple reasons for entry. The need to secure sources of raw materials for core activities—such as food processing was the most common reason cited by more than half the respondents. It was cited with the same frequency among pre-1998 and post-1998 entrants into agriculture. We would have expected this reason be the most important for companies that entered agriculture after the 1998 financial crisis. It was then that imports were sharply curtailed by devaluation and companies began to shift to domestic inputs.

Nearly half the companies (6 out of 14) justified their decision to take over existing agricultural producers by the need to secure repayment of outstanding debt. Processors traditionally extended cash advances and in-kind credits to farms in order to guarantee future supplies using a form of production contracts. The repayment of these advances was jeopardized by the increasingly difficult financial situation of the farms (see Chapter 9). Finally, a similar number of companies (5 out of 14) reported that they entered primary agriculture as a way of expanding the market for their products. In the 2003 BASIS survey, most respondents who were farm-level managers, not agroholding officials, also identified securing the raw materials and market expansion as the main reasons for the farm acquisition by an outside corporation (Table 4.7).

Table 4.7. Reasons for entering agriculture: results of two surveys (percent of respondents)

	2002 Rostov survey (14 agroholding officials)	2003 BASIS survey (9 farm managers)
Secure raw-material base	57	56
Expand market for non-farm products	36	33
Ensure collection of outstanding debt	43	NA
Enter agriculture because of its profit potential	36	22
Reduction of financial risk	21	NA
Recommendation of regional authorities	50	NA

As concrete examples of these motives we can cite three instances from the VIAPI Agroholdings Database:

- Agroholding "Mikhailovskii" created by the Cherkizov Meat Processing Plant, includes a feedlot operation and broiler factories, all of which act as sources of raw materials for the meat processing unit.

- Agroholding "Malino" acquired a large farm to secure the supply of vegetables to its wholesale depots and retail markets, vertically integrating downstream along the entire chain from production to marketing.
- Agroholding "Ogo-Agroprom" was created by a large grain trading company, which owned feed mixing plants. The company then acquired several poultry operations with the objective of expanding the market for its concentrated feed. Ogo has since discontinued all crop production, but has kept the poultry units as a buyer for its non-agricultural output.

These examples and the reasons cited in the two surveys suggest that agroholding management essentially thinks in terms of a vertically integrated production chain that starts with the financing of working capital for production and goes through the purchase of raw materials for processing, expansion of processing output, and sales growth. Only 3 out of the 14 respondents in the 2002 Rostov survey cited diversification—reduction of financial risk—as a reason for entering agriculture.

On a conceptually different level, encouragement or recommendation of regional authorities was cited as the reason for entering primary agriculture by half the respondents in the 2002 Rostov survey. This reason was particularly important for the post-1998 entrants. This implies that, despite all the political changes and reforms, regional authorities continue to exert significant influence on the production decisions of companies doing business in their region. The practice is observed even in Rostov Oblast, which is generally regarded as one of the less interventionist regions in Russia today.[13]

Our case studies also demonstrate that the importance of vertical integration in agroholding development depends on the originating firm and the industry. State and quasi-state companies, banking corporations, and conglomerates tend to restrict their activities to production agriculture as an industry. They are not generally a part of vertical supply chain and they appear to be driven mainly by diversification motives. The relatively small regional agroholdings maintain close links between agricultural production and their farm supply, marketing, or processing businesses. Large, diversified interregional agribusiness companies also employ vertical integration strategies, but because of their sheer size the actual integration of their agri-

[13] See, e.g., E. Serova and I. Khramova, "Emerging supply chain management in Russia's agro-food markets," ZEF Discussion Paper No. 14, Series: The Russian Food Economy in Transition, University of Bonn (July 2000), p. 6. In a ranking of 7 oblasts (see Chapter 8, Table 8.9), Rostov's regional policies get a liberalization score of 1.7, in the middle between Vologda (0.4, most interventionist) and Chelyabinsk (2.8, least interventionist).

business assets and farming units may be quite modest. For example, in both the grain and sugar beet industries, where agroholdings are among the industry leaders, the level of captive supply within agroholdings remains quite modest—from 5% to 25% depending on the specific company. Even large integrated farming operations are not always able to satisfy the processing demand of the giant interregional companies. EFCO, discussed earlier in this chapter, possesses a huge multi-farm operation of 100,000 hectares. However, this large agricultural component supplies only 7% of EFCO's raw sunflower-seed requirements.

Table 4.8. Purchases of fertilizer, fuel, and machinery through vertical-integration arrangements by corporate farms in Rostov Oblast

	Supplied by parent company		Supplied by buyer of farm commodities	
	Percent of quantity purchased	Percent of respondents reporting purchases	Percent of quantity purchased	Percent of respondents reporting purchases
Fertilizer (*n*=56 farms)	8.5	7	5.7	9
Gasoline (*n*=68 farms)	16.4	6	23.9	28
Diesel fuel (*n*=69 farms)	7.2	6	29.1	35
	Percent of acquired value	Percent of new machinery transactions	Percent of acquired value	Percent of new machinery transactions
Machinery (*n*=44 transactions by 17 farms)	42.8	27	—	—

Source: 2003 BASIS survey.

The 2003 BASIS survey highlights the role of the "parent company" and the "raw-material buyer"—both potentially parts of a vertically integrated system—as suppliers of farm inputs and machinery to Rostov farms (Table 4.8). Only 4-5 farms report buying inputs (fertilizer and fuel) from the parent company. This is consistent with the relatively small number of farms formally identified as belonging to agroholdings (a total of nine in all three oblasts surveyed). Input purchases from farm commodity buyers are reported much more frequently, especially for fuel. About one-third of Rostov farms report that 20%-30% of their fuel is purchased from a buyer of the farm's commodities. This arrangement is typical of a vertical integrator seeking to secure its raw-material base (although, of course, it may also be triggered by credit constraints). A different pattern is observed for purchases of farm machinery. Here commodity buyers have no role at all. The parent company accounts for 43% of total value of new machinery acquisitions in Rostov Oblast in 2002 and 27% of the number of corresponding transactions (these transactions, however, were carried out by only 6 farms).

Agroholdings are major producers in specific industries. In the poultry industry, five vertically integrated leaders supply 35% of Russia's broiler output through 24 former collectives and newly established farms. In the grain industry, six of the ten leading grain exporters accounting for 35% of Russia's grain export have grain production operations. In the sugar beet industry, all ten leading companies accounting for about 85% of refined sugar production have sugar beet production enterprises. Large scale processors or exporters enter agriculture in order to strengthen their vertical supply chain. Some use agriculture as a diversification or horizontal growth strategy. For example, Prodimex—a leader in Russia's sugar market—has entered swine production. Roskhleboprodukt and Razgulyai, both grain industry leaders, have entered poultry farming, which provided portfolio risk reduction with growth opportunities outside their main asset focus and geographic location.

Entry Patterns and Functions

We distinguish five principal paths for the entry of non-agricultural companies into production agriculture:

- become a financial "sponsor" of an existing farm
- enter into joint production agreements with an existing farm
- provide custom farming services
- lease agricultural land from existing farm, landowners, or the state
- acquire a whole farm (including land and other productive assets)

These entry paths form a hierarchy of increasing involvement in farm operations (Table 4.9). *Sponsors* are cash-rich non-agricultural entities that assume the farm's financial risk. They do not directly control or own the farm assets and do not manage the farm on a daily basis. Their main role is to provide guarantees for the input supplier and the bank, or to lend directly for working capital and investments. By relieving the farm management of financial pressures, a sponsor often assists in the transformation of the "traditional" farm into a modern operation. Sponsored farms are typically selected from among the best corporate or individual farms. Under *joint production agreements*, outside entities participate in farm production decisions, but without assuming ownership of the farm's assets. Under *custom farming*, the outside firm supplies the missing production functions (e.g., plowing, harvesting) to an existing farm and partly bears the risks of crop failure and negative market developments. Under *land leasing*, the outside firm takes over production and assumes the full range of agricultural risks to the extent that they apply to the leased land. *Whole-farm acquisition* makes the firm a full-fledged agricultural producer.

Of the five entry paths described in Table 4.9, land leasing and whole-farm acquisition are the two main entry modes that produce "classical" agroholdings. We call entry by land leasing and whole-farm acquisition an acquisitive mode. It is distinct from the non-acquisitive modes of sponsorship, custom farming, and joint production agreements. In these arrangements, the ownership and control by an outside non-agricultural entity is missing and its overall involvement in agricultural production decisions is less pervasive. Strictly speaking, the non-acquisitive entry modes do not produce agroholdings at all.

Table 4.9. Classification of entry channels and economic functions of agroholdings

Functions	Sponsorship	Custom farming	Joint production agreements	Land leasing	Whole-farm acquisition
Legal ownership and control	No	No	No	Yes*	Yes
Financing	Yes	No	Yes, partly	Yes*	Yes
Input supply	No	Yes, short term, narrow range	Yes, comprehensive solutions	Yes*	Yes, long-term, wide range
Production services	No	Yes, short term, narrow range	Yes, comprehensive solutions	Yes*	Yes, long-term, wide range
Management decisions	Participate in strategic decisions	No	Yes, together with farm managers	Yes*	Yes, full range
Marketing of farm products	Strategic agreements with marketers	Yes (limited to harvest share)	Production contracts, strategic agreements with marketers	Yes*; rely on vertical integration	Yes, through vertical integration

*Limited to leased land.

In reality, however, the dividing line between the different entry modes is not as sharp as the table suggests. In the 2002 Rostov survey, most agroholdings were created when the non-agricultural parent company began to lease farmland (6 out of 14) or to acquire whole farms (4 out of 14). Thus, three-quarters of the respondents surveyed were indeed created according to the classical pattern of entry into agriculture. However, the same number of respondents also indicated that they were providing inputs, commodity credit, or even cash loans to farms. Most of the working-capital transactions (7 out of 10) overlapped whole-farm acquisitions. In these cases, the parent company supplied inputs and credit to its daughter farms, but in 3 out of 10 cases, working-capital financing was carried out independently of asset acquisition or land leasing. Custom farming and joint production were also represented among the responses, but they were secondary in importance to whole-farm acquisition and working-capital financing.

At a first glance, the dividing line between land leasing and whole-farm acquisition is quite clear. And yet the substantive differences are very fuzzy. Non-land assets purchased as part of whole-farm acquisition may be useless and the entire transaction may be motivated by acquisition of land through leasing. Therefore, some agroholdings may have intentionally ignored the option of purchasing non-land assets and focused on acquiring land lease contracts. Although formally not part of the whole-farm acquisition category, their managerial decisions are the same. In other words, agroholdings that acquire the land of an entire farm, even if they did not acquire the non-land assets, should be included under whole-farm acquisition. The land-leasing category as such should be reserved for agroholdings that lease only a part of a farm's land and possibly shift their lease contracts from farm to farm in different years.

Table 4.10. Entry paths into agriculture: Rostov Oblast

	Respondents (out of 14)	Non-overlapping responses
Land leasing	6	6
Asset acquisition	4	4
Acquisitive modes	**10**	**10**
Working-capital financing	10	3
Inputs, in-kind credit	7	
Cash credit	3	
Custom farming, joint production	6	1
Non-acquisitive modes		**4**

Source: 2002 Rostov survey.

The results of this classification by entry modes are summarized in Table 4.10, which shows that the non-acquisitive modes account independently for only 4 cases out of the 14 surveyed. Nevertheless, these 4 non-acquisitive companies are classified as agroholdings, because they all exercise control over the operation of their farming units. As with other agroholdings, these 4 companies enforce centralized decisions regarding product pricing, input purchasing, profit distribution, and investments. They exercise control by appointing the managers of their agricultural subdivisions.

Control is thus accomplished through the provision of financing, rather than asset ownership. This conclusion accordingly elevates "sponsorship," in addition to land leasing and whole-farm acquisition, to the category of entry modes that may lead to the creation of agroholdings.

In addition to land, the agricultural enterprise also needs farm buildings, machinery, and equipment to operate. Most respondents in the 2002 Rostov survey (8 out of 14) indicated that they had acquired the non-land assets for their agricultural activities in the course of bankruptcy proceedings of a failed farm. This is apparently much less costly than the acquisition of a successful farm as a going concern. The second most common form of acquisition reported by 4 respondents out of 14 was the creation of a joint enterprise. This option is also characterized by relatively low transaction costs, since it only requires the signing of an agreement and the transfer of farm shares into the equity capital of the joint venture.

There are indications that some regional authorities in Belgorod and Orel oblasts, for example, facilitate bankruptcy-driven takeovers by encouraging over-indebted farms to transfer their assets to a new debt-free entity. This entity is then acquired as a whole by an outside investor. The debt remains with the original legal entity. Shagaida has identified this trend by noticing the increasing number of old farm enterprises without any land (see Chapter 5, Figure 5.6). These appear to be defunct debt-carrying shells that remain after the transfer of all assets, including land, to a new entity that is positioned for a takeover by an agroholding.

How Agroholdings Manage Their Farming Operations

The agricultural operations of the agroholdings included in the 2002 Rostov survey are organized in independent farming divisions. The number of divisions ranges from 1 to 6 in different agroholdings—substantially less than the average of 11 farming divisions in the IKAR national database. The divisions are registered as separate legal entities (subsidiaries) and are managed locally. No cases of central operating management are reported. This independent legal format is apparently a requirement for maximizing the tax benefits available exclusively to agricultural producers and, no less importantly, ensuring eligibility for direct subsidies, both federal and regional, which are paid only to producers that derive more than 70% of their sales revenue from farming. Indeed, 11 of the 14 agroholdings reported receiving direct subsidies, mostly as subsidized credit and partial compensation of fertilizer costs.

Each agricultural division has its own local manager, who generally, though not always, is responsible for farm-level hiring and firing. The managers themselves, however, are typically appointed by the holding company

(Table 4.11). This, of course, is quite natural in a holding structure, but in Russian agroholdings the centralized control of the agricultural divisions is quite pervasive despite their "legal" independence. The holding company appropriates the profits of the farming divisions and redistributes them to satisfy the working-capital and investment needs of the farms in the group. Occasionally profits may also be used outside primary agriculture for group development programs, headquarters expenses, or for investment in processing. The practice of profit redistribution essentially implies central control of financing and investment.

Table 4.11. Centralization of management and control within agroholdings

Centralized functions	Number of respondents*	Percent (of n=14)
Redistribution of farm profits by holding company	13	93
For farm uses	13	93
For other uses	5	36
Centralized product sales	11	79
Control of selling prices within the group	14	100
Obligatory intrafirm transactions	12	71
Centralized input purchasing	11	79
Inputs supplied by holding company at transfer prices	7	50
Inputs purchased at market prices with intermediation of holding company	4	29
Appointment of farm managers	12	86
Intervention in labor hiring	4	29

*Multiple answers allowed.
Source: 2002 Rostov survey.

Input purchases are generally controlled or coordinated from headquarters (Table 4.11). In most cases, inputs are actually supplied by the holding company at internal transfer prices. Sometimes the holding company facilitates delivery of inputs at market prices, generally charging a handling fee. The sale of farm products is also centralized and selling prices are set by headquarters for all agricultural divisions.

One of the central functions not mentioned in Table 4.11 is facilitating the access of farms to bank credit. By pooling collateral and presenting a low-risk profile to the banks, agroholdings can obtain credit more easily and at lower interest rates. Some of this credit finds its way to the farm level for financing working capital and investments (see Box 4.3).

We have previously hypothesized that some features of the management process in agroholdings can be used to differentiate between vertical integration and diversification motives (see the discussion around Table 4.1). These features include central appropriation of profits and emphasis on intrafirm transactions within the group. Unfortunately from the viewpoint of testing between the different motives, centralization of profits is practiced by virtually all Rostov companies in the survey (Table 4.11). Setting of obligatory quotas for intrafirm transactions and institution of obligatory business dealings within the group are less universal, however. Two of the fourteen surveyed companies centralize profits but, contrary to the remaining twelve, they do not insist on group members dealing with one another. The vertical integration characteristics are thus less pronounced for these two agroholdings and they may be regarded as an example of companies that are driven more strongly than others by diversification motives.

Box 4.3. Agroholding Structure Facilitates Access to Credit

A fuel and gas supplier owns a 75% stake in a 9,000 ha farm, which is organized as a subsidiary. Like all agricultural producers, the farm is exempt from profit tax and some other taxes. The owner collateralizes both on-farm and off-farm assets (including fuel inventories) with the purpose of raising credit for the farm. The collateral is used to secure subsidized-rate working-capital loans for the farm from a commercial bank recently launched within the holding group. The farm receives these loans within a federal support program for agriculture at 7% annual interest rate (compared with 25% Central Bank refinancing rate and 18%-20% inflation forecast). Only two other farms out of 23 corporate farms in the district are eligible to participate in this credit program through banks. In addition, the bank uses the collateral to issue medium-term guarantees (up to 3 years) to agricultural machinery suppliers. The cost of the guarantee is 2% compared with 5%-6% for well-capitalized food processing companies and 8%-10% for successful independent farms.

Source: IKAR case studies (2001 data)

Are Agroholdings Successful?

We noted in Table 4.3 that, for a small sample of Rostov agroholdings, wheat yields per hectare and production costs per ton of wheat were essentially the same in agroholdlings and in independent corporate farms. Yet agroholdings may enjoy synergies or benefits of integration that make their overall performance more than the sum of the agricultural components. Unfortunately, agroholdings surround themselves with a deep shroud of secrecy and confidentiality, so that no data are available for conducting a comprehensive performance analysis. A tentative evaluation can be obtained using a small nonrepresentative sample of 33 agroholdings included in the VIAPI database, which provides some basic performance data combining both physical and financial variables for 2001. Table 4.12 presents the results of a preliminary analysis based on these data.

Not all agroholdings in the VIAPI database are profitable—9 of the 33 agroholdings report losses and one company has zero profit. Some agroholdings have very large losses, reaching 40% of sales and more. At the high end of the distribution, some agroholdings report profits at more than

20% of sales. The average profit margin for agroholdings reporting positive profits is a respectable 11% of sales. However, the unprofitable agroholdings pull the mean profit margin for the entire sample to a mere 2.2% of sales, substantially below the average profit margin for all corporate farms in Russia (8.0%). Nevertheless, the percentage of unprofitable companies in this sample of agroholdings appears to be much less than the percentage of unprofitable corporate farms across Russia (Table 4.12).

Table 4.12. Main indicators of agricultural projects in agroholdings (2001 data, averages per agroholding)

	All sample (*n*=33)	"Solvent" (*n*=20)	"Insolvent" (*n*=8)	All corporate farms** (*n*=29,696)
Financial ratios				
Sales, million rubles	415	552	142	10
Profit margin, %	2.3	6.2	−11.0	8.0
Farms reporting losses, %	27	15	63	46
Months of payables in sales	12	8	23	11
Share of overdue payables, %	28	18	55	47
Partial productivity of land and labor				
Agricultural land, ha	83,900	97,600	48,400	4,760
Workers	4,000	4,800	2,200	143
Sales per ha, '000 rubles	9.1	12.2	3.1	4.5
Sales per worker, 000 rubles	122	150	76	69

* "Insolvent" agroholdings are those with payables of at least 18 months of sales and overdue payables of at least 50%.
Source: VIAPI database for agroholdings; Goskomstata database of annual financial reports for corporate farms.

The average level of payables is 12 months of sales, comparable to all corporate farms across Russia. For nearly three-quarters of the agroholdings in the sample, payables are less than one year of sales, and the median is a relatively acceptable 7 months. Some agroholdings (4 out of 33), however, have accumulated payables of more than 18 months of sales, and two of these seem to have unsustainably high levels of current debt (29 and 79 months of sales, respectively).

The mean share of overdue payables is 28%, far below the average for Russian corporate farms. But again there is a group of agroholdings, 5 out of 33, with more than 50% of payables overdue. Higher payables are positively associated with higher overdue payables. In addition, farms with weak financial performance tend to have a higher level of payables and debt.

Another characteristic that emerges from Table 4.12 points to a clear size differentiation between the two groups of agroholdings. The solvent agroholdings are larger than the insolvent ones—by sales, by land, and by number of workers. This finding is reminiscent of the observations regarding the grouping of all corporate farms by financial health (or solvency) measures. As discussed in Chapter 2, farms with a high financial health score are on average larger than financially weak farms.

Fig. 4.7. Partial productivity of land and labor: agroholdings and all corporate farms (2001)

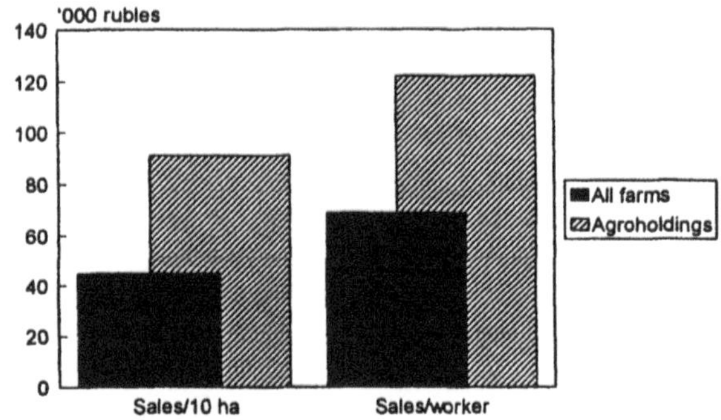

Source: VIAPI Agroholdings Database

Fig. 4.8. Partial productivity of land and labor by size of agroholdings (2001)

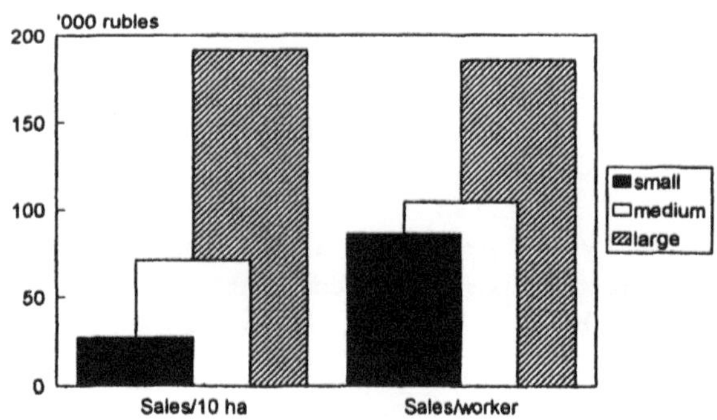

Source: VIAPI database; size groups defined in terms of sales

In addition to standard measures of financial performance, the VIAPI database enables us to calculate the partial productivities of land and labor—the sales per hectare and per worker achieved in agroholdings. Solvent agroholdings achieve productivity levels that are substantially higher that those achieved by insolvent agroholdings. But it is perhaps more relevant to note that agroholding as a group achieve partial productivities that are about double the average for all corporate farms in Russia (Figure 4.7; Table 4.12). Both productivity measures show a tendency to increase with farm size. This is clear from Figure 4.8, which shows the partial productivity of land and labor for agroholdings grouped into three size groups by the volume of sales in 2001 (small with sales up to 100 million rubles, medium with sales between 100 million rubles and 500 million rubles, and large with sales greater than 500 million rubles). Note that no such relationship is observed when size is defined in terms of land or is taken as a continuous variable (sales or land).

Summarizing this limited analysis, we conclude that the financial performance of agroholdings is uneven. The small sample of 33 agroholdings in the VIAPI database contains both good and bad performers, and the same picture probably carries over to the entire population. In terms of production, on the other hand, agroholdings appear to be more efficient, with productivities of labor and land (based on the aggregate value of production) substantially higher than those of the average corporate farm in Russia. There is an apparent inconsistency between this result and the previous finding that wheat yields and wheat production costs are not statistically different between agroholdings and other corporate farms (Table 4.3). The discrepancy should not surprise us, because the two sets of results were obtained for two different samples (both very small) utilizing different measures of output (aggregate value of farm production in rubles and wheat production in tons). There is obviously a very strong need to collect information for a detailed performance analysis of this new sector.

Managerial Dilemmas and New Solutions

Our case studies and interviews with top managers highlight some managerial dilemmas characteristic of agroholdings. The decision-making structure of the mother company is typically rooted in an industrial, trading, or financial management culture. In the Russian context this means an emphasis on economies of scale, standardization, and top-down control. This management approach dramatically conflicts with the biological nature of agriculture, which requires timely, expert decisions that must be made at specific physical sites. The potential benefits of centralization and economies of scale can be offset in rigidly hierarchical organizations by the lack of individual initiative and inability to make timely decisions at the local level. This characteristic of production agriculture is generally used to explain the apparent superiority of family farming compared to corporate farming. Yet when the holding company tries to delegate responsibility, increase motivation and align incentives to allow operational freedom for local agricultural managers, this policy can increase the risk of misuse and pilfering of resources. These agency costs, which appear to be particularly high in agricultural operations, result in rapid turnover of managers at all levels of the vertical chain. Turnover and a lack of accountability make the situation worse and are incompatible with the need for "perpetual stewardship" of land.

Box 4.4. Family Farms within the Agroholding Structure: First Attempts

- Among EFCO's two dozen farming units, one is viewed as the most successful. It is located in Rostov Oblast in a remote, sparsely populated steppe district. The farm specializes in crop production and is run by a small team of workers. This is an example of an "internal" quasi-individual farm.
- Kubanagroexport encourages the creation of family farms within its farming enterprise. Three family farms have been allotted land under sublease agreements. The holding company provides financing with crop production inputs and is responsible for post-harvest marketing. The company would like to extend this scheme to the rest of its agricultural project.
- Aleiskhleboprodukt, a major flour mill in the Altai region, has included a family farm in its holding structure alongside several corporate farms.

Source: IKAR case studies

So far, there are no convincing examples of a satisfactory managerial solution to this problem. One potentially promising approach that has been tried by several agroholdings is the creation of farms run by a small group of individuals or even a single family within the agroholding structure (Box 4.4). These approaches attempt to combine the benefits of centralization and economies of scale with the well-known advantages of family farming. The organizational shift toward "internal" individual farming that we are beginning to detect among Russian agroholdings is reminiscent of the factors that led to the demise of the huge "bonanza farms" in the American Midwest, which broke up and transformed into family farms despite their resounding initial success (Box 4.1). The transformation of the bonanza farms in the United States probably indicates that the new "Russian latifundia" may not be sustainable in their present form.

Agroholdings as an Agent for Economic Development and Transformation

Throughout Russia and indeed most of the countries of the former Soviet Union, the agricultural sector has been slow to respond to economic reforms. The emergence and growth of Russia's agroholdings offers a dramatic contrast to the "hopeless stagnation" that has dogged the agricultural sector. Organizational innovation, new management and new money are now flowing into the sector through the agroholdings. At first glance, agroholdings seem to offer a way to develop and transform Russian agriculture along the entire value chain.

The organizational innovation of the agroholdings may offer a way to break out of the sector's productivity trap. Outside of the agroholdings, the productivity of the farm sector is so low that it is incapable of servicing short or intermediate term debt. Even with adequate collateral, external credit rationing would occur in this situation. Inadequate credit leads to shortages in inputs and technology as well as suboptimal firm size, further exacerbating the already low productivity. Inefficiency in input and product markets along with internal and external factors reinforces the trap. For example, restrictions on the movement of grain imposed by regional authorities can result in significantly lower local market prices and revenue. Poorly defined property rights for land or other assets make leasing arrangements costly and limit firm size. Poorly developed supply channels for key inputs can result in higher producer costs or increased risk due to an unreliable supply. Obsolete food processing firms may transmit higher costs or reduced revenue from lower quality products back to producers.

In Table 4.13, we list the "traditional" or prevailing practices inherited by Russian corporate farms from the socialist economic system. These old practices create and sustain the productivity trap. They are incompatible with the new market-oriented environment. In effect they constitute market imperfections or barriers to normal market behavior of Russian farms. The new ways of doing business, the organizational responses of agroholdings emerged, in part, as a consequence of the imperfections implicit in the traditional practices.

The table demonstrates that the entry of agroholdings into the agricultural sector of companies is accompanied by substantial changes in how the farms are operated (internal change) and how they interact along their vertical and horizontal boundaries. The table suggests the following generalizing of the organizational consequences of agroholdings.

- The farm decision-making process has become centralized. All key strategic decisions are elevated from the farm level to the headquarters of the holding company. This leads to much stricter performance requirements from farm top management. The price of wrong or inadequate decisions dramatically increases.
- Farm top management, farm technicians, and farm labor ultimately become hired employees of the agroholding company. The agroholding must confront the coordination and incentive problems inherent in any large enterprise.
- Farm labor is being reduced. This may not be accompanied by increased unemployment—at least not initially because rural labor is absorbed in household-plot farming.
- Land lease contracts become formalized. "Virtual" leasing of paper land shares is gradually transformed into long-term leasing of physical plots.

Table 4.13. Market imperfections and organizational change by agroholdings in Russian agriculture

	Prevailing practice	Organizational response by agroholdings
Inputs		
Fuel	District fuel stations and intermediaries enter into barter agreements with farms. Regional authorities arrange reciprocal clearance of payables between fuel suppliers and farms; farms get fuel against deliveries to regional stocks	No substantial change: fuel still purchased through district fuel stations, but cash transfers through banks improve trading and guarantee fuel quality
Seeds	Seeds (usually cheap varieties) are purchased through intermediaries.	High-volume purchases at agroholding HQ. Quality seeds are purchased through licensed dealers or directly from international seed companies. Own multiplication centers are established.
Feed	Primitive on-farm production and/or purchase of standardized concentrates from independent feed mills	Feed mills incorporated into vertically integrated company. Farms order formulas specified according to current market situation and livestock needs. Holding company places large orders for high protein feed and premix with international suppliers
Fertilizer	Purchased by farms for cash from local distributors. Distributed by regional administration through barter arrangements at highly disadvantageous terms	Centralized large purchases between agroholding and international supplier. Partial prepayment combined with open credit. Final payment in money, not in kind
Chemicals	Same as for fertilizer, but less intervention from regional administration, as chemicals are viewed secondary input	Same as for as fertilizers. Agroholdings also invite bids from local distribu tors of large chemical firms, achieving substantial cost reductions
Farm machinery	Large farms own machinery and provide leasing or custom services. Machine availability ranges from excess capacity to missing links in the technological chain	Professional custom farming companies operate at local, regional and interregional level. Rates becoming progressively more competitive
Credit	Weak business planning, lack of transparency, and shortage of collateral restricts access to bank credit and leads to high dependence on extra-banking sources (local authorities, intermediaries)	Centralization of financing decisions at HQ. Business planning formalized and standardized. Collateral pooled across the system: non-agricultural collateral used to secure farm credit and farm commodities used to collateralize credit for non-agricultural activities
Labor	Generally excess of low-skilled, undisciplined labor; shortage of skilled workers	Excess farm labor gradually reduced; temporary engineers and other professionals hired as needed

Table 4.13 continued

	Prevailing practice	Organizational response by agroholdings
Land	Large farm leases land shares from owners; lease contracts often not formalized, lease payment sporadic	Long-term leases of physically marked and registered plots. Lease rates proposed by farm management and approved by HQ; regular lease payments. Attempts to collateralize long-term lease contracts
Services		
Management structure	Traditional Soviet-style management structure with two extreme scenarios: either a strong autocratic manager (seldom), or a weak manager dependent on workers and local authorities	Highly hierarchical management structure with up to four levels: Moscow-based HQ—regional or industry management company—regional farming project—farm. Random reorganizations between vertical industry and regional cluster structure. High turnover of top farm managers. High risk of undermining reasonable and timely decision making at farm level. Outsourcing narrow agricultural specialists to solve concrete managerial and technical problems
Tax management	Unpaid tax arrears, inability to obtain VAT refunds, overpayment of taxes. In regions with a single farm tax, farms are getting into a seasonal "tax trap"	Sophisticated tax optimization. Professional teams of in-house lawyers deal with payment of single farm tax
Crop insurance	Practically unavailable to corporate farms	Attempts to negotiate comprehensive crop insurance deals at HQ level. Substantial difficulties due to inadequacy of state support for crop insurance
Sale of farm products	Predominance of spot and "out-of-pocket" cash sales. Reliance on regional government credit programs and commodity credit from input suppliers, which require repayment in kind and limit the share of output available for free commercial sales	Developing "captive market" strategies: farms become part of vertical supply pipeline; uses marketing agreements with reputable food processors and exporters
Legal protection	Farm powerless in front of regulatory bodies and business counterparts	Lawyers at HQ protect the company portfolio, including farm assets, claims, and obligations

Source: Based on IKAR case studies and interviews.

- Centralized purchase of farm inputs through agroholding company improves the bargaining position of the farms. The degree of change in input supply practices ranges from minimal for fuel to a complete transformation for chemicals and farm machinery.
- Farms owned by agroholdings enjoy much better access to bank credit and to financial markets in general than "traditional" corporate farms.
- Agroholdings are also able to mitigate input and capital market imperfections by developing strategic partnerships with commodity merchandisers and processors who agree to purchase a fixed amount of the agricultural commodity at a specified price and in some cases assume part of the operator's nonperformance risk. The guarantee may be offered to the farm directly or to a commercial bank that can then attract domestic and foreign bank loans to finance working capital. Strategic partnerships with trustworthy commodity end users and foreign traders are an important factor in determining the success of a farm or the entire agroholding.
- Arrangements involving insurance or re-insurance companies, international surveyors, and commercial warehouses are being launched to exploit the benefits of commodity collateral. Such arrangements depend, however, on the presence of a reputable end user (food processor) or foreign trader in the region.
- Farms owned by agroholdings have much stronger legal and institutional protection than "traditional" corporate farms. This may be one of the most important achievements under Russian conditions today.

Farms owned by agroholdings have much stronger legal and institutional protection than "traditional" corporate farms. This may be one of the most important achievements under Russian conditions today. In the current business environment in Russia the "traditional" corporate farm—a successor of a Soviet collective or state farm—is not well suited to deal with the new administrative and institutional forces. Many independent corporate farms fall easy victims of the inadequate legal system and extremely limited enforcement of the "rule of law" in the Western sense of this term. Entry into a strong agroholding therefore makes perfect sense for a corporate farm in the new environment. The agroholding structure is thus mutually beneficial both "from the top" and "from the bottom."

The agroholding captures the benefits of vertical integration or diversification for the outside investor at the top of the pyramid and at the same time it provides a legal and financial shield to the independent farm at the bottom. The existence of such mutual benefits may account for the success of this phenomenon, at least for the time being.

Will all independent corporate farms eventually disappear and be absorbed in agroholdings? It is difficult to speculate, and much depends on the development of managerial and organizational structures that ensure a proper balance between the advantages of farm-level independence and the benefits of centralized support. This balance is probably essential if Russian agroholdings are to avoid the fate of American "bonanza" farms.

Further Research

Agroholdings in Russia are an emerging phenomenon. Their structure, economic role, and extent are not yet well understood. The data for our preliminary analysis is assembled from case studies, interviews, and small surveys. Such ad hoc information is helpful but is not adequate for the study and monitoring of new organizational forms.

Russian agricultural and agribusiness statistics must be adapted to reflect the new production patterns and functions that emerged during the 1990s. U.S. agricultural economists, for example face similar problems as they try to conceptualize recent developments in the U.S. farm sector. The USDA has recently proposed changes in the U.S. farm typology used in examining farm structure and economic performance.[1] This redefined farm typology groups U.S. farms into the following categories: small family farms (including as subcategories limited-resource farms, retirement farms, residential/lifestyle farms, farming-occupation farms, lower-sales farms, and higher-sales farms), large family farms, very large family farms, and non-family farms. The U.S. farm typology resembles, to an extent, the farm structure in Russia (and throughout the rest of the CIS). A key feature of the proposed U.S. classification is the emphasis on identifying the actual operator and the operator's status on the farm. In Russia it is also imperative that a legal and statistical definition of agricultural operator be adjusted in order to target agricultural policies including eligibility for tax benefits and government subsidies.

We have reviewed the present tendency for creating huge agricultural operations. Is the creation of latifundia or "bonanza farming" an inevitable or beneficial outcome in Russia? What should be the legal limitations on huge agrarian corporations, regardless of whether they are owned by agricultural interests or non-agricultural companies? If we accept the inevitability of such huge structures, what is their "optimal" internal corporate organization? The issues of manageability and "perpetual stewardship" of land are of key importance. What incentives and motivations must be designed

[1] Hoppe R., Johnson J., Perry J., Banker D., *A New Farm Typology for a Diverse Ag Sector.* Washington, DC: ERS USDA, 2000.

for managers and employees along the vertical decision-making chain? What governance structures are necessary for stockholders and landowners to control the company and its operations? And finally, the U.S. historical example has demonstrated the unsustainability of "bonanza farms." Is it likely that the huge agroholdings will be transformed into individual farm farms?

In addition to these broader questions, a number of specific research issues are apparent. The following list is illustrative.

- How do the new types of farm businesses differ in terms of agricultural technologies, resource allocation, output, and productivity?
- Does the emergence of the new farming model improve the liquidity of farm assets? How do they improve access to capital and input markets?
- Do the new farming models improve contract enforcement mechanisms?
- How do the new agroholdings influence the performance of the agricultural value chain?
- What new risks emerge in Russian agriculture as the result of the entry of agroholdings? What new risk management tools are needed in the new environment?
- Why is the development of new farming models so different even in neighboring regions?
- What are the interrelationships between the agroholdings and the "traditional" farming sector? Will the agroholdings foster change in other agricultural businesses?
- What are the social consequences of agroholdings?
- Can we identify a natural "life cycle" for agroholdings?

Conclusion

In a relatively short time, agroholdings have become a powerful component of Russian agriculture. Their agricultural value at risk is enormous. Domestic farm policy—from official agricultural statistics to agricultural support programs—must be re-examined in the light of this new phenomenon. Specifically, the official definition of an agricultural producer must be adjusted to reflect the new reality. Currently, a highly capitalized custom farming operator is not considered an agricultural producer and is not eligible for farm support programs, while a technically bankrupt former collective farm may still enjoy various producer privileges.

Given the acute decapitalization of existing farms, agroholdings are forced to collateralize their non-agricultural assets to support investment in primary agriculture. This process diverts resources from the firm's core activities and exposes the entire organization to significant risks in the event of poor performance in the agricultural divisions. To mitigate the risks of the new agricultural investors, the government should amend the hopelessly outdated mortgage and collateral legislation. This would include introducing an efficient and enforceable system of grain warehouse receipts, legalizing lien rights, launching collateral registration offices in rural areas, enabling simplified out-of-court collection enforcement procedures. In addition the government should modify the current crop insurance program to make it more applicable for a wide range of agricultural investors. Specific steps might include the development of alternative subsidized insurance packages, a shift from all-inclusive to specific risk coverage packages, transition to investment-based valuation, and better attention to reinsurance issues. Serious attention is required to land ownership rights and land leasing mechanisms. Functioning land markets and streamlined leasing procedures are essential to enable the flow of land from the less efficient traditional farms to the (hopefully) more efficient new producers (see Chapter 5).

What is the prognosis for agroholdings? Will they continue to grow bigger and stronger, or will they suddenly disappear like the bonanza farms in the United States? Already there are signs that some agroholdings are turning to organizational solutions that allow greater operational freedoms to smaller subunits run by a small number of individuals, going as far as the creation of family farms within their holding structure. Transformations of this type move Russia away from a plantation agriculture toward family-owned structure linked through markets and contracts to the entire supply chain, as is common in all market economies.

From a global perspective, the agroholdings have the potential to bring Russia's vast agricultural resources into world agricultural markets. Efficient production methods, improved quality, and value chains will transform Russia into a significant agricultural competitor, both in export markets and in domestic import substitution. We are witnessing the first steps of an organizational change in Russian agriculture, and it remains to be seen if these changes will bring about significant new growth.

Part II

Developing Factor Markets

Chapter 5

Land Reform and Development of Land Markets

Natalya Shagaida and Zvi Lerman

Russia ranks fifth in the world by its agricultural land area (after China, Australia, USA, and Brazil).[1] At 220 million hectares, Russia's agricultural land matches that of the whole of Europe and is about 40% of agricultural land in China or in all 15 countries of the former USSR combined. Russia usually has a reputation of being a land-rich country (*mnogozemel'naya* is the Russian adjective describing this characteristic). Russia's endowment of agricultural land is 5.5 hectares per rural resident, which is actually less than in the United States (6.3 hectares per rural resident), but much more than in China (0.7 hectares per rural resident) or in India (0.3 hectares per rural resident). Availability of agricultural land is thus not a constraint in Russia (see Chapters 10, 11), and yet land ownership is an emotionally charged issue that cuts across all segments of the population.

This chapter examines the dramatic changes in land ownership and land tenure that have occurred in Russia since the dissolution of the Soviet Union at the end of 1991. We start with a discussion of the legal framework for land reform that crystallized in the early 1990s and has continued to evolve dynamically ever since, consider the impacts of land reform on privatization and ownership structure of agricultural land, and analyze the development of land market transactions. The analysis is based on official national sources and the results of the 2003 BASIS survey. We conclude with a review of the existing constraints on land transactions and some policy recommendations that may stimulate the development of land markets

[1] The international perspective presented in the first paragraph is based on data from FAOSTAT—FAO's online statistical database [www.fao.org].

and ultimately help Russian agriculture improve its productivity by simplifying farm size adjustment.

Legal Framework and Land Reform Outcomes

Russia's agricultural land area has remained remarkably stable at 220 million hectares since 1990. On the other hand, agricultural land used by producers (both corporate and individual farms) dropped from 214 million hectares in 1991 to 195 million hectares in 2003—a decrease of 16% during the transition. Most of the "missing" land was transferred to the state land reserve, which is a pool of land available for allocation to producers but not currently in use. Reclassification of land from farm use to rural municipalities also contributed to this decrease. The main change in agricultural land use occurred around 1998-1999, when reserve land and land controlled by municipalities (and other users) abruptly doubled from 13 million hectares to 26 million hectares.

Table 5.1. Structure of agricultural land by users 1991 and 2003 (beginning of year)

	million hectares		percent	
	2003	1991	2003	1991
Total agricultural land	220.9	222.1	100.0	100.0
Used by farms	194.6	213.8	88.1	96.3
corporate farms	150.4	209.8	68.1	94.5
peasant farms	17.0	0.1	7.7	0.0
household plots	11.8	3.9	5.3	1.8
municipal meadows and pastures in household use	15.4	0	7.0	0
Reserve land	13.8	1.8	6.2	0.8
Other users	12.5	6.5	5.7	2.9

Source: Statistical Yearbook 2003, Goskomstat, Moscow.

Table 5.1 shows the structure of agricultural land resources by users in 1991 and 2003.[2] In addition to the features discussed above, it demonstrates the substantial transfer of land from corporate farms (former kolkhozes and sovkhozes) to the individual sector: corporate farms lost 59 million hectares, or nearly 30% of their total land endowment in 1991, while the individual sector gained 40 million hectares (the remainder was absorbed in the state reserve and by user reclassification). The distribution of land flows

[2] Note that the percentages for different categories of users in Table 5.1 are calculated on the basis of total agricultural land (about 220 million hectares). The structure of users is often calculated in the literature taking as 100% the agricultural land used by farms (about 195 million hectares). Thus, the share of land in individual use (peasant farms and household plots) is 13.0% in Table 5.1 and 14.8% when based on agricultural land used by producers.

between 1991 and 2003 is illustrated in Figure 5.1. While the agricultural land inventory remained fairly constant (see line 1 in Table 5.1), the individual sector absorbed nearly 70% of the area lost by the corporate farm sector. The state reserve also increased significantly, absorbing 20% of redistributed land. The rest went mainly to other users.

Fig. 5.1. Agricultural land flows 1991-2003

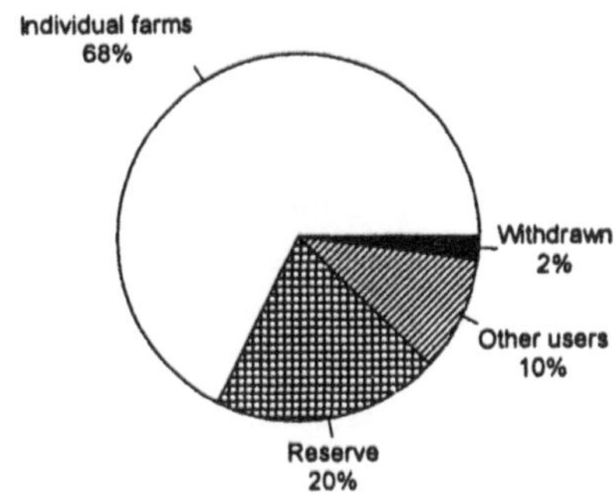

In percent of total redistributed land (59.4 mln ha)

Historical Background

Land in Russia (and in other parts of the former czarist empire) was nationalized within days of the Bolshevik revolution in October 1917, as Lenin's Decree on Land (*Dekret o zemle*) transferred all land to the state and prohibited private land ownership. The monopoly of the state on land ownership, although firmly established in Russia and the rest of the Soviet Union since 1917, was formally reflected in the Soviet constitution only twenty years later, when the Stalin constitution of 1936 articulated the classical Soviet formula for land ownership principles (Article 6):

> The land, the mineral resources, the water bodies, the forests . . . are the property of the state, that is, belong to the entire people.

"Land to the peasants" was the slogan that the Bolsheviks used to attract the largely conservative peasantry to the revolutionary cause. Indeed, after 1917, the land in large estates was distributed to the landless and the land-poor, and communal land was generally transferred to individual use (in a sense completing Stolypin's unfinished reform that began in 1905 and was interrupted by the outbreak of World War I in 1914). Although the Bolshevik platform advocated from the start collective farming as the road to socialism, individual farming dominated the Russian rural scene all through the 1920s. Thus, of the 20 million hectares distributed to the rural population during 1917-1919, fully 95% went to peasant families and only 5% to various collectives and state farms.[3] It was only in 1929 (after the severe grain procurement shortfalls of 1927-1928) that Stalin initiated a forceful and brutal collectivization program, which rapidly led to total elimination of individual farming and establishment of collective and state farms (kolkhozes and sovkhozes) as the monopolistic organizational form in Soviet agriculture. Russian peasants, who had received individual land plots only a decade earlier, were now forced to put their land and assets in collective farms. By the end of the 1930s, a relatively small number of socialized, or "public-sector" farms (about 30,000 in total) controlled 98% of agricultural land and produced most of the commercially marketed output.

However, despite pervasive collectivization and monopolistic state ownership of land, individual or private agriculture quickly reemerged in a different guise. Independent peasants had disappeared, but millions of rural households were now allowed to cultivate small plots of less than 0.5 hectare, producing mainly for subsistence (although part of the output always found its way to farmers' markets in nearby towns). The policy of allowing small-scale private farming on household plots within large collectives was implemented as a measure for improving rural family incomes and mitigating the resistance of the peasantry to socialized agriculture. Household plots operated by employees of collective farms and other village residents controlled in aggregate 2% of agricultural land, but still managed to produce 20% of gross agricultural output. During most of the Soviet period Russia's agriculture was thus characterized by a distinctly dual or bimodal farm structure, in which commercial production from collective and state farms was supplemented by subsistence-oriented individual agriculture based on small household plots within the large collectives.

Since the 1930s, collective and state farms were given state-owned land in permanent, perpetual use, without any payment. Transfers of land between these large corporate farms were initiated by orders from the authorities, as part of often changing government programs that aggregated or disaggregated collective farms and transformed kolkhozes into sovkhozes (or back). Rural households typically received their plots in use rights from the local collective farm and generally enjoyed security of tenure (subject to administrative whims of the local authorities).

---

[3] T. V. Osipova, *Rossiiskoe krestyanstvo v revolyutsii i grazhdanskoi voine*, Moscow: Izd. Strelets (2001), pp. 229-230.

Reform Legislation

After more than seven decades of state monopoly in land ownership, the first signs of readiness to reform the Soviet system were incorporated in the 1989 *Principles of Land Legislation of the USSR*. This Union-level law relaxed the traditional policy of giving state land in use rights only, and introduced a new category of land tenure for rural households—lifetime inheritable possession (*pozhiznennoe nasleduemoe vladenie*). Security of tenure was formally ensured, but land transactions (including subleasing) were absolutely prohibited. From this point on, Russia took over the land reform initiatives from the USSR legislators. In October 1990, more than a year before the dissolution of the Soviet Union, Russia (then still called RSFSR—the Russian Soviet Federated Socialist Republic) passed the Land Reform Law, which actually recognized the right of private ownership in agricultural land. This law required a constitutional amendment, which was duly adopted by the Russian legislature in December 1990. A single word, "ownership," was added to the amended article 12 of the Russian constitution:

> Land plots for agricultural production are given by the state in use rights, in lifetime inheritable possession, or in ownership.

This was truly the beginning of the latest revolution in land-tenure relations in Russia. As a compromise, however, the amended article 12 imposed a 10-year moratorium on buying and selling of land in private ownership and restricted alienation of land to the state (and not to other landowners).

The trail-blazing constitutional amendment was followed by the Russian Law on Peasant Farms (December 1990), which legalized private farming, allowed distribution of collective land in the form of paper shares to members, and provided the option of withdrawing land plots for the establishment of an independent peasant farm outside the collectivist framework. Russia's new Land Code passed in April 1991 formalized these various legal initiatives and laid the road for mass privatization of agricultural land (see Table 5.2 for a chronology of land reform legislation in Russia).

Land ownership is an emotionally charged issue in Russia. The passage of reform-oriented land laws was accompanied by dramatic political debates, and the sharply opposed positions of the reform-minded executive and the highly conservative legislature prevented the development of a full-fledged legal privatization mechanism for more than a decade. The main sticking point was the right to buy and sell privately owned land—a basic inalienable right associated with private land ownership in market economies. The two branches of government could not agree on this point. The legal deadlock was clearly reflected in the laws of that period. The 1993

Russian constitution simply reiterated the right of private ownership. The 1994 Civil Code suspended its land chapter until the passage of a new land code, while the 2001 Land Code skirted the issue by deferring all matters relating to transactions in agricultural land to a special law to be passed at some unspecified time in the future. All the legal advances since mid-1991 were realized in the form of presidential decrees and government resolutions—temporary instruments that required ultimate codification in permanent laws, but because of the persistent legislative opposition to buying and selling of agricultural land, these legal instruments mainly focused on mechanisms that progressively broadened the range of permissible transactions in land shares, not physical plots (see Table 5.2). Ownership rights in agricultural land (including buying and selling) were finally normalized in January 2003, when the Law on Agricultural Land Transactions came into force. Mortgage of agricultural land has been allowed since February 2004.

Table 5.2. Evolution of Russian legislation pertaining to agricultural land (Legend: PD—presidential decree, GR—government resolution)

Date	Title	Main rights
Feb1990	Principles of Land Legislation of the USSR (USSR Supreme Soviet)	Lifetime inheritable possession for individuals; leasing for individuals and agricultural enterprises; permanent or limited-term use for other organizations; principle of payment for land use
Apr1990	"On some issues of allocation and alienation of land" (RSFSR Supreme Soviet)	Underutilized land of kolkhozes and sovkhozes may be alienated and transferred to individual users in possession, use rights, and leasing for farming purposes and individual construction
Nov1990	Land Reform Law	Use, possession, non-state ownership of land
Dec1990	Amendment to Art. 12 of the Constitution of RSFSR	Use, lifetime inheritable possession, and non-state ownership of land. Alienation only to the state (except for inheritance). Buying and selling of plots prohibited for 10 years after acquiring ownership rights
Dec1990	Law on Peasant Farms	Plots for peasant farming allocated in use rights, including leasing, in lifetime inheritable possession, and in ownership; the land of kolkhozes and sovkhozes is divided in shares among workers (inheritable personal right), and workers may convert their shares to land plots for establishing a peasant farm; alienation to local government only
Apr1991	Land Code	Land ownership: state, individual, or collective; corporate farms get land in permanent use; corporate-farm workers get land in collective (joint or shared) ownership; individuals may hold land in ownership, lifetime inheritable possession, or leasing for farming purposes or housing construction; land is distributed free and for payment.

Table 5.2 continued

Date	Title	Main rights
Oct1991	Law on Payment for Land	Principle of payment for land use: private land owners pay land tax, lessees pay rent.
Dec1991	"Procedure for reorganization of kolkhozes and sovkhozes" (GR 86)	Definition of individuals entitled to received land shares; transactions in land shares: conversion to land plot for peasant farming, investing in the equity of a corporate farm, selling to other workers or to the corporate farm
Dec1991	"On immediate measures for implementation of land reform" (PD 323)	Mandatory reorganization of kolkhozes and sovkhozes in 1992; size limits for peasant farms and household plots; district quotas for free transfer of land to private ownership; creation of redistribution reserve for allocation of land to new owners; sale and leasing of reserve land to individuals and legal bodies; time limit for withdrawal of land by groups of workers and pensioners for group ownership; leasing of land shares; selling land shares to other workers of the same farm or to farm as legal body; mandatory conversion of land shares to plots for creation of new corporate farm; transactions with land plots allowed in case of retirement, inheritance, and relocation of peasant farm; transactions also allowed if proceeds are to be used for establishment of non-farm rural business; free allocation of land for household plots, gardening, and residential construction in the village
Mar1992	"On sale of land plots to individuals and legal bodies during privatization of state enterprises" (PD 301)	Land under privatized buildings may be bought or leased by new owner; land held in use rights for business activity may be bought from the state
Sep1992	"Procedure for privatization and reorganization of enterprises and agroindustrial complex" (GR 708)	Selling or leasing land shares to other joint land owners; restriction of the range of farm enterprises that may undergo privatization
May1993	"Procedure for buying and selling of land by individuals" (GR 503)	Buying and selling land for household plots and other individual uses (dachas, gardening, individual housing)
Oct1993	"Control of land relations and development of agrarian reform in Russia" (PD 1767)	Definition of a wider range of transactions involving land shares: land shares may be converted to plots for household farming, peasant farming, individual housing construction (without requiring the consent of joint land owners); land share may be sold to any buyer; land shares may be given as a gift, exchanged, or passed in inheritance; leasing of land shares; mortgage; titling of land ownership by the state; guarantees of inviolability and protection of ownership rights in transactions

Table 5.2 continued

Date	Title	Main rights
Dec1993	New Russian Constitution	Land may be privately owned; land rights of physical and legal bodies; land owners have freedom of disposition without violating the rights of other owners
Feb1995	"Procedure for realization of rights of owners of land and asset shares" (GR 96)	Details of land transaction mechanisms
Oct1995	Civil Code	Land chapter deferred until the passage of the Land Code
Mar1996	"Guarantees of constitutional rights of citizens in land" (PD 337)	Simplified conflict resolution procedure for conversion of land shares into plots
Jul1998	Mortgage Law	Mortgage of land plots, with the exception of agricultural land
Oct2001	Land Code	Free conversion of land previously received in permanent (inheritable) use to private ownership; mechanisms for transactions in land with the exception of land shares and agricultural land plots
Jan2003	Law on Transactions in Agricultural Land	Full legal framework for ownership rights and transactions in agricultural land
Feb2004	Amendments to Mortgage Law	Mortgage of agricultural land allowed

Although the 1990 constitutional amendment recognized private ownership of land, the Russian Constitution explicitly prohibited buying and selling of land for 10 years after the acquisition of the land plot (this moratorium provision was added to article 12 recognizing private landownership). In 1991, agricultural land held by collective and state farms began to be distributed in the form of land shares to individuals, who could then withdraw their land allocation for the establishment of a peasant farm. Alienation of land in peasant farms was allowed only to local authorities, not to other individuals. Starting in early 1992, land shares still held in the form of paper certificates could be sold to other members of the collective or to the collective farm as a legal body; physical land plots could be sold only under special circumstances (when the landowner retired, when the plot was passed in inheritance, when the peasant farmer relocated to another region, or when the seller undertook to use the proceeds from the sale of land for the establishment of a non-farm business). Procedures adopted in March 1993 allowed buying and selling of land for household plots and other individual uses (dachas, gardening, individual housing).

The new 1993 constitution declared that landowners had "freedom of disposition without violating the rights of other owners." However, the buying and selling of land (other than household plots) was neither allowed nor prohibited. Some regions (e.g., Saratov Oblast) adopted local legislation allowing transactions in land, but no federal law existed. This naturally created serious administrative barriers to the development of land markets, and until the adoption of the January 2003 law land transactions mainly involved conversion of land shares into physical plots.

In practice, the legal ban on buying and selling of agricultural land was bypassed even before January 2003 by presidential decrees and government resolutions. These decrees and resolutions allowed buying and selling of land shares (first to other members of the collective, and since October 1993 practically to any buyer). Having purchased a land share, the new shareowner could request its conversion into a land plot. The transferability of land shares has led to substantial redistribution of land ownership and land use in former collective farms. Presidential decrees and government resolutions also allowed (since October 1993) conversion of land shares into physical plots for household farming. Once the ownership of the new addition to the household plot was officially registered, it could be sold and bought in accordance with the different rules applicable (since May 1993) to land in household plots. A market has thus emerged for relatively small plots created through conversion of land shares by family members. The national average land share was about 10 hectares, and some regions actually allowed expansion of household plots to the combined size of the land shares held by all family members. In principle, the size of a household plot in the fields outside the village limits could thus reach several tens of hectares, which is comparable with the average size of a peasant farm.

Beyond the psychological opposition to buying and selling of land, other emotional issues included the concern about concentration of land ownership in the hands of few physical persons or corporations ("latifundiazation" of agricultural land), the fear of excessive fragmentation of land during privatization, loss of land holdings by former collective farms due to their weak financial situation and danger of bankruptcy, and that perennial bogey, the sale of Russian land to foreigners. The provisions of the 2003 Law of Agricultural Land Transactions were designed to address these concerns. While buying and selling of land plots (as well as land shares) was allowed, the state retained a preemptive right on land purchases; regional governments could impose limits on physical concentration of land by a single owner (typically 10% of the agricultural land in the district), on the one hand, and also limits on the minimum size of physical plots that could be surveyed and registered for farming purposes (household plots were exempt from this restriction); foreigner and companies with majority foreign capital could only lease agricultural land, not own it.

The introduction of preemptive rights for the oblast and the municipalities is perhaps most restrictive in this context. Negotiations for a piece of agricultural land (or even a land share) between private parties cannot be concluded without offering the authorities the option to buy the land on the same terms. The private deal can go through only if the authorities refuse or let the option lapse (within one month). Moreover, if the terms of sale are changed in the process of dynamic negotiations, the authorities again should be given the option to exercise their preemptive right, leading to a further delay of at least one month. Despite these and other reservations, the legal restrictions on the whole are regarded in Russia as striking a reasonable balance between the need to alleviate the original concerns and the goal of minimizing the restrictions on land markets (Table 5.3).

Table 5.3. Assessment of the impact of legal restrictions on transactions in agricultural land

Legal restriction	Impact assessment
Preemptive right of oblast government and municipal authorities to purchase land plots and land shares from joint shared ownership	Intended to prevent socially undesirable transactions, but significantly delays the completion of land transactions and complicates seller-buyer negotiations. Can be bypassed by resorting to alternative mechanisms: public auction, investment in the equity capital of a corporation, gift
Upper limit on concentration of land by physical persons or legal bodies created by physical persons	Intended to prevent "latifundiazation," but may be overcome by creating a legal body owned by another legal body, not a physical person (e.g., in Belgorod, whole raions are held by a single corporation created by another corporation)
Minimum size of agricultural plots	Intended to prevent excessive fragmentation, but some conservative regions abuse this provision by setting unacceptably large minimum sizes, e.g., 100 ha in Voronezh, 200 ha in Krasnodar.
Foreigners and foreign corporations may not own land	Intended to address social concerns; the restrictive impact of this provision has been alleviated by allowing long-term leasing for 49 years

Land Privatization

Contrary to the Baltic states and most countries in East Central Europe, Russia chose to privatize agricultural land by distribution to users, not restitution to former owners. The initial legislation in 1989-1991 focused on the principle of private ownership of land and the procedures for distribution of state land for individual farming (household plots, small-scale gardeners and vegetable growers, peasant farmers). Mass privatization was launched in 1991-1992, when large chunks of state land were privatized into joint ownership of the rural people who lived and worked in collective and state farms. This formal privatization affected most of the agricultural land in collective and state farms, while the rest of agricultural land and other rural land (including land under farm buildings, for instance) remained state property (creating the so-called redistribution reserve for future municipal and farming needs). The privatized land was then divided into equal shares, and each adult—collective farm worker, pensioner, or employee of rural social services—received one land share. The size of the share was determined by land availability in the district and was rigidly controlled (like the size of all land distributions to individuals; see Table 5.4).

A land share is a paper entitlement of fractional ownership in the agricultural land transferred by the state to the collective. This mechanism created a new ownership category that became known as "joint shared ownership." This was no longer state ownership (hence the use of the term "privatization"), but it was not individual ownership either. The reform laws typically allowed shareowners to withdraw physical land plots from joint shared ownership into individual ownership, but the requirement to survey and register the plot (with all the attendant costs and administrative complications) was deferred to the time in the future when the shareowner would actually decide to withdraw his or her land from the common pool of owners.

Table 5.4. Typical size of privatized land units in Russia

Privatization category	Size
Allocation for peasant farming	30-50 ha
Allocation to household plots	0.5 ha in the village; 1.0 ha in the fields outside the village
Allocation for gardening and vegetable growing	0.06-0.08 ha
Allocation for individual housing	0.10 ha
Land share	7-10 ha

Russian land privatization quickly produced 11.9 million shareowners with land shares covering 117.6 million hectares, or 9.9 hectares per share. By 1995 the state had privatized through land shares fully 56% of the original 209.8 million hectares controlled by former collective and state farms at the beginning of reform. The remaining land was transferred to the state redistribution reserve, which provided the pool of land for future creation of peasant farms, expansion of household plots, and various municipal needs.

The distribution of land shares immediately placed the shareowners in a decision node: they could choose to start an independent business by withdrawing their land from the collective; on the other hand, they did not have to do that, as they could simply leave their land shares in joint cultivation by the existing collective farm (which meanwhile had reorganized as a corporate farm in one of the standard organizational forms, such as shareholding company, limited liability company, partnership, or agricultural cooperative). It was clear from the outset that most shareowners would not start an independent farm and instead would prefer to keep their land shares in the collective. It suffices to mention that in 1992 half the shareowners were elderly pensioners. With time, rural people developed mixed conversion strategies, with several members of the same family or groups of relatives and neighbors pooling their paper shares to receive one contiguous land plot in return. This home-made consolidation strategy was a natural response to the concerns about excessive fragmentation of individual land plots. In retrospect, distribution of land shares proved more effective and less costly for land consolidation than distribution of physical plots. This is evident from any comparison with the experience in East Central Europe, where land fragmentation and consolidation efforts are still a major issue after nearly fifteen years of transition.

The privatization process itself provided the trigger for the first land market transactions in Russia. The former collective and state farms, now reorganized as corporate farms of various types, were formally left without any land for farming. They had to turn back to the newly created shareowners and lease their land shares or alternatively entice them to invest their land shares in the equity capital of the corporate farms. Initially, it was not too difficult to persuade the new shareowners not to withdraw their land and to let the corporate farm continue using it. However, as time went on and people began to get used to the new market mechanisms, some shareowners would withdraw their land from the former collective farm and lease it to another producer offering more competitive terms. A shareowner actually could avoid the bureaucratic hassle of withdrawing a land plot and simply lease out the land share. The lessor would then negotiate for a specific plot with the manager of the collective farm where the share was originally used. A two-tier leasing system has thus developed: leasing of land shares from individual shareowners (either by the original corporate farm or by other producers) and leasing of land plots from individuals who have independently converted their land share into a plot. All in all, it seems that the Russian land privatization strategy, based on distribution of land shares instead of the more conventional (to the Western observer) distribution of individual land plots, has had clear beneficial effects on the level of transaction costs, although possibly it delayed the productivity improvements normally associated with individual ownership and control of land.

The 2003 Law on Agricultural Land Transactions has interfered with this established process by prohibiting leasing of land shares and requiring that a land share be converted into a physical plot before it could be leased. All the existing lease transactions involving millions of land shares have thus been placed in a legally dubious position. According to the strict letter of the law, the existing lease agreements have to be rewritten for specific land plots or, failing that, the land shares have to be placed in a fiduciary trust. The trust manager would then supervise the distribution of revenues between the farmer-operator and the landowner. The option of converting land shares into plots is not simple to implement because of substantial costs and bureaucratic complications (see Table 5.28), whereas the option of putting the leased land in trust is alien to peasant mentality (not only in Russia, but anywhere in the world) and moreover suffers from various legal inconsistencies with the provisions of trust management in the Civil Code. Nevertheless, the new provision of the 2003 Law on Agricultural Land Transactions has definitely spurred the interest in surveying, registering, and titling individual land plots, and unintentionally has led to further cost increases as the demand for surveyor services outstripped the limited availability of these professionals.

In addition to transactions in private land, there are naturally transactions involving state-owned land. Original distributions of state land to peasant farmers and households (up to specific limits) were free. After that, state land has to be leased or purchased. Both types of transactions for state land legally require an auction or at least a bidding process, but in many cases this requirement is bypassed in practice.

Land Ownership Structure

Land use statistics are routinely collected by Goskomstat and consistent time series reported on the basis of agricultural land are available since 1990 (and much earlier). Land ownership statistics, on the other hand, are collected by Roszemkadastr—Russia's land authority[1]—and data are available only since 1997. Moreover, land ownership is reported on the basis of so-called "agriculturally zoned land" (*zemli sel'khoznaznacheniya* in Russian), not agricultural land. While agricultural land is an accepted international concept (consisting of arable land, land under perennials, hay meadows, and pastures), "agriculturally zoned land" is defined in Russia as "land intended for the needs of agriculture and located outside the limits of existing settlements." It includes agricultural land (but only the part outside the village limits!), and also roads, protective vegetation, closed water bodies, and land under agricultural buildings (i.e., buildings intended for agricultural production, storage, and primary processing of farm products). "Agriculturally zoned land" comprised 400 million hectares as of January 2003, of which 192 million hectares was agricultural land. Unfortunately for purposes of

---

[1] Roszemkadastr was the official name between 2000 and 2004. Previously the agency had been called Roskomzem (1990-1995) and Goskomzem (1995-2000). In 2004 its name was changed to Rosnedvizhimost' to reflect cadastral responsibility for all forms of real estate in Russia. The agency's web site is www.goscomzem.ru.

statistical estimation, agricultural land is not a proper subset of "agriculturally zoned land:" some 30 million hectares of agricultural land are not included in the latter category.

Fig. 5.2. Privatization of agriculturally zoned land vs. share of agricultural land by region (2003 data)

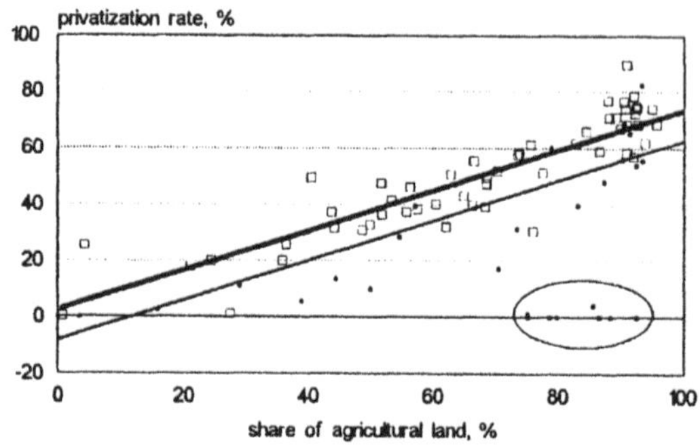

Legend: White squares with the thick regression line represent the mainstream "Slavic" regions. Black dots are the ethnic-minority regions; the ellipse in the bottom right-hand corner encloses the group of 8 ethnic-minority regions that do not recognize private land ownership. The thin regression line for the ethnic-minority regions has been fitted excluding the group of no-privatization regions. Source: *Zemel'nyi fond Rossii* 2003, Moscow: Roszemkadastr (2004).

Yet it is agricultural land that has always been the main subject of land reform and land privatization in Russia. Land shares represent fractional ownership of agricultural land; household plots are augmented by allotments of agricultural land; and peasant farms are created on agricultural land. When we examine Roszemkadastr's land privatization rates by region (based on "agriculturally zoned land"), we notice that the privatization rate increases with the increase in the share of agricultural land in the region. Low privatization rates are observed in regions with little agricultural land and, of course, also in a group of ethnic-minority regions that exercise their right under the Federal Treaty not to recognize private land ownership for cultural reasons.[2] If we exclude from analysis the eight ethnic-minority regions without private land ownership, we observe a correlation of 0.90 between the privatization rate and the share of agricultural land in the region. The relationship between the reported privatization rates and the share of agricultural land by region is presented in Figure 5.2, which incidentally

[2] Bashkiria, Dagestan, Kabardino-Balkaria, North Osetia, Chechnya, Tyva, Ingushetia, Kalmykia have zero privatization rates, although more than 70% of their agriculturally zoned land is agricultural land.

shows that even ethnic-minority regions that allow private land ownership (black dots outside the group of no-privatization regions enclosed in an ellipse at the bottom right-hand corner of the diagram) are on average characterized by privatization rates lower than the mainstream "Slavic" regions.

The findings presented in Figure 5.2 suggest that privately owned land is primarily agricultural land, and we can therefore approximately recalculate the structure of land ownership in Russia in percent of agricultural land. The results of this calculation are presented in Table 5.5 for 2001 and 2003. There have been no significant changes in the structure of land ownership between 2001 and 2003, and Figure 5.3 presents the structure of land ownership in 2003. The share of state-owned agricultural land dropped from 100% before 1990 to around 40% today. Nearly 60% of agricultural land is now privately owned, but most of this land (51%) is represented by land shares—abstract (though transferable) pieces of paper corresponding to virtual plots. Only 6% of agricultural land is in the form of physically demarcated plots, most of it owned by individuals (household plots and peasant farms).[3]

Table 5.5. Land ownership 2001-2003 (agricultural land, million hectares)

	2001	2003
Total agricultural land	221.1	220.9
Privately owned (as reported)	128.7	127.5
Individuals: land shares	113.4	112.7
Individuals: plots*	11.7	12.1
Corporations*	3.6	2.7
State owned (by difference)	92.4	93.4

*About 2.5 million hectares held in peasant farms registered as legal bodies has been moved from the category of corporations (where it is formally reported) to the category of individual plots.

Source: Total agricultural land from Statistical Yearbook 2003, Goskomstat, Moscow; breakdown by ownership from *Zemel'nyi fond Rossii* 2001, Moscow: Roszemkadastr (2002), pp. 49, 57, 100, 109, and *Zemel'nyi fond Rossii* 2003, Moscow: Roszemkadastr (2004), pp. 13, 253, 604, 616.

Examining the structure of land ownership by three main categories of users (corporate farms, peasant farms, and household plots), we see an astonishing similarity between corporate farms and household plots (Figure 5.4). For both types, about 30% of the land is state-owned and close to 70% is land owned by private individuals; land in corporate private owner-

[3] Due to the phenomenon described in Figure 5.2, the share of state ownership in agriculturally zoned land is much higher than in agricultural land. Thus, according to reported data for 2003, the state owned 276 million hectares of agriculturally zoned land, or 69% of the total 401 million hectares.

ship (leased from other corporate farms or invested by shareowners in the equity capital) accounts for a negligible 1%-2%. However, there is a huge difference in the kind of individual land ownership between corporate farms and household plots: virtually all individually owned land in corporate farms is in the form of land shares owned by the local rural population, while in household plots this is physically demarcated private land. Peasant farms have a smaller component of individually owned private land (about 40%, all of it in the form of physical plots) and they rely to a greater extent on land leased from the state and on land shares leased from outside owners who are not family members. This component of land leased in the form of shares from outsiders accounts for 18% of total holdings in peasant farms (shown as "other" in Figure 5.4).

Fig. 5.3. Structure of land ownership 2003
(percent of agricultural land)

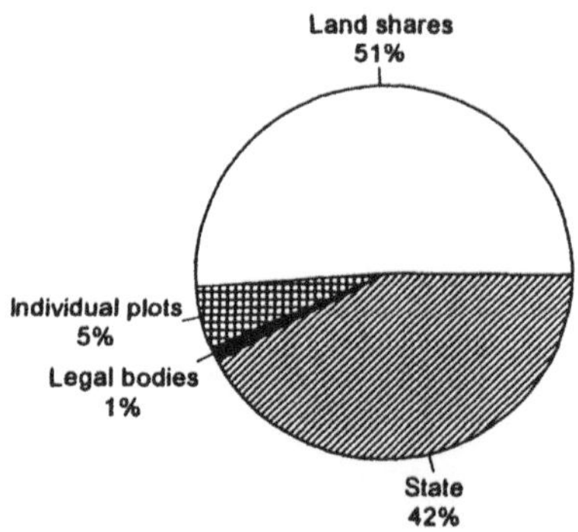

Fig. 5.4. Structure of land ownership by farm type 2001
(in percent of agricultural land)

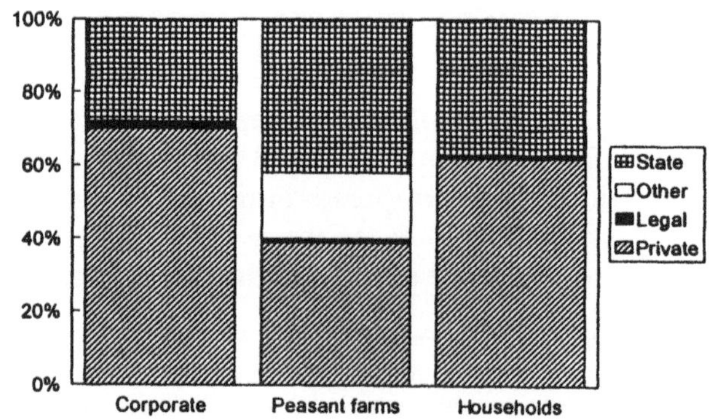

Some of the state land in individual farms is still reported in the old tenure forms (use rights or possession). This applies to two-thirds of the state land in household plots and one-third in peasant farms. Eventually this land will be transferred to private ownership of the users, increasing the component of individually owned land by as much as 25% in household plots and 15% in peasant farms.

Since 88% of privately owned land is represented by land shares (Table 5.5), it may be instructive to look at the disposition of land shares in corporate farms. According to Roszemkadastr data for 2004, 64% of the land shares are leased by corporate farms from individuals and 12% are given to corporate farms in use rights. The remaining 24% is represented by unclaimed land shares of beneficiaries who have died, left the village, or failed to exercise their right for other reasons (e.g., did not want to pay for the share certificate). The local corporate farm continues to use the unclaimed land shares by default.

Land Redistribution in the Process of Reorganization of Former Collective Farms

The basic land privatization mechanism—distribution of land shares accompanied by the option of individual or group withdrawal with land—has created the basis for redistribution of agricultural land among users. Although in reality this option has seldom been exercised among the 12 million shareholders, it always exists and in some specific instances has led to significant redistribution of privatized land resources.

One of the examples of significant land redistribution is provided by the Lodeinopl'skii Raion in Leningrad Oblast. In the pre-reform period, the farms in this district specialized in dairy production, largely relying on concentrated feed supplied by the state at subsidized prices. After 1991, state support to livestock production was curtailed. Farms did not have machinery to produce feed grain and had to reduce their dairy herd to one-third of the pre-reform level. The performance of all corporate farms in the district deteriorated and signs of impending bankruptcy began to appear. The local financial crisis provided a stimulus for a spate of reorganizations, which included creation of new corporate farms as well as expansion of peasant farms and household plots.

The process did not involve any buying and selling of land and it was enabled by a flow of land shares from owners to new users. Among the six former collective farms in the district, more than 50% of land resources were redistributed by land share transfers (Figure 5.5). Most of the land was absorbed by new corporate farms (primarily through leasing), but about 10% of the land shares were taken up by peasant farms and household plots with the purpose of expanding their holdings. The remaining 48% of the land shares did not find new users, in part because the shareowners had died or left without assigning their use rights (about 20% of the shares). By default, these shares continue to be used by the (greatly downsized) former collectives.

Fig. 5.5. Farm reorganization by transfer of land shares

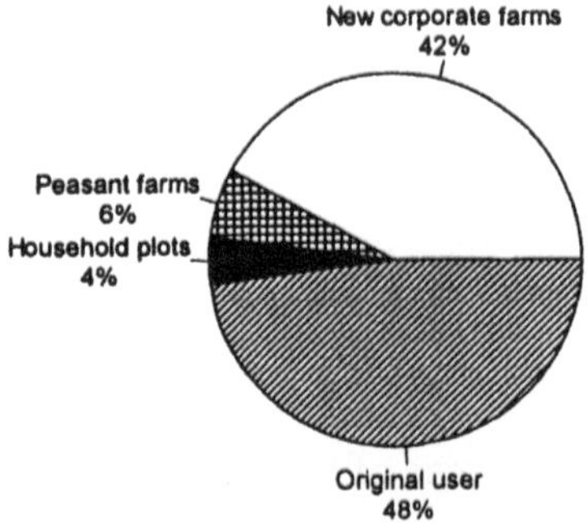

Lodeinopol'skii Raion, 2002-2003, percent of land shares

This example with half the land shares transferred to new users and the other half remaining by default with the original user—a former collective farm—is typical of agriculturally poor regions, where the demand for land is weak. In agriculturally fertile regions land is in greater demand and the process of redistribution proceeds in a different way. Thus in Belgorod and Orel oblasts in the highly productive central region the oblast authorities are actively trying to attract new investors to agriculture (see Chapter 4). The investors prefer not to acquire the failing, debt-ridden former collectives and instead create entirely new corporate farms by leasing or buying land shares and sometimes even persuading the shareowners to put their land in the corporate farm's equity capital. The former collective farms remain as empty shells with responsibility for most of the outstanding debt, to be eventually liquidated through bankruptcy proceedings.

This scenario is demonstrated in Figure 5.6. In the first phase, between 1995 and 2000, the entry of new investors was accompanied by creation of new corporate farms and decapitalization of some of the existing farms. The total number of corporate farms increased and the number of landless farms ("empty shells") also grew. In the second phase, between 2000 and 2002, the landless farms began to be liquidated, their number decreased, and so did the total number of corporate farms in the region.

The mechanism of privatization through land shares, and not physical plots, has an interesting side effect in that it actually reduces the fragmentation of privatized land. A priori the land shares do not correspond to specific plots of land: each share remains a paper entitlement to an unspecified land plot of a certain size until the moment when the shareowner applies to convert the land share into actual land. According to established procedures, the members of a family or another user group can pool their land shares before conversion and thus apply to receive a single contiguous plot the area of which is equal to the combined area of all their shares. This "consolidation" is illustrated in Table 5.6, where we see that 6 families holding jointly 17 land shares were allotted 6 consolidated plots. Each family received a single contiguous plot of appropriate size, regardless of whether its individual members had 2, 3, or 5 land shares. This particular example, illustrates conversion of land shares for the purpose of enlarging the household plot, but it applies to any instance when a group of shareowners applies to withdraw its land from joint shared ownership.

Fig. 5.6. Reorganization of corporate farms in fertile regions

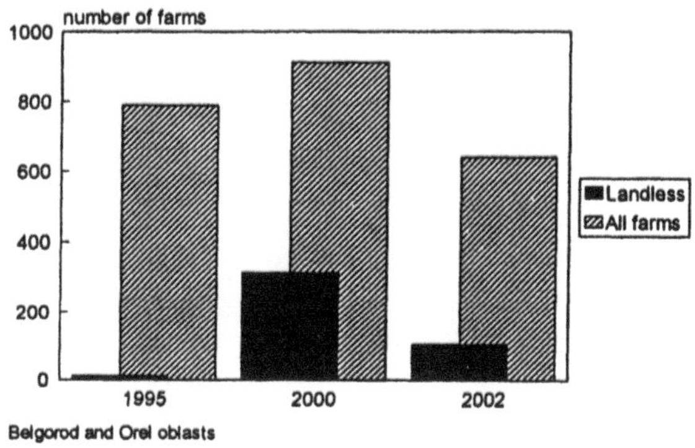

Table 5.6. Land consolidation through pooling of land shares before conversion

	Number of land shares	Number of plots received	Area received, ha
Family 1	2	1	11.13
Family 2	2	1	7.46
Family 3	3	1	12.83
Family 3	3	1	12.94
Family 5	5	1	16.27
Family 6	2	1	7.58
Total	17	6	68.21

Note: While all land shares are of equal size, the area of the actual plot is adjusted for land quality and the allotment per share therefore varies.

Emergence of Land Markets

Land market transactions are divided into two main categories: buy-and-sell transactions that involve transfer of legal ownership; and leasing transactions that involve transfer of use rights from owner to tenant without change of ownership. In this section, we review the information on transactions in land available from the federal land agency generally known as Roszemkadastr and recently (2004) renamed Rosnedvizhimost'. The discussion covers transactions in state land as well as transactions in privately owned land among individuals.

Land Transactions

Roszemkadastr publishes annual information on the number of all registered land transactions, including both agricultural and non-agricultural land. The existing registration procedures ensure a fairly complete record of transactions in state land (both leasing and buying). On the other hand, they do not capture any leasing transactions between individuals—whether plots or land shares. Buying and selling of land plots between private individuals and corporations are in principle covered by the statistics, but only to the extent that the buyers and sellers choose to go through the bureaucratic difficulties of registering the transaction (see below). The general trend of all registered transactions (including both agricultural and non-agricultural land is shown in Table 5.7.

Table 5.7. Registered land transactions 1997-2002

	Number of transactions, millions	Land area involved, million ha	Percent of transactions		
			Leasing of state or municipal land	Buy-sell by individuals and corporations	Other types of transactions*
1997	4.0	23.5	88	7	5
1998	4.4	24.6	91	5	4
1999	5.2	72.2	91	6	3
2000	5.3	60.3	90	6	4
2001	5.6	70.0	90	6	4
2002	5.3	43.9	89	6	5

*Sale of state and municipal land, sale of lease rights in state and municipal land, inheritance, mortgage, gifts.

Source: 1997-1999 from Goskomzem (private communication); 2000-2001 from *Svedeniya o sdelkakh s zemlei i platezhakh za zemlyu*, Mosow: Roszemkadastr (2002), p. 19; 2002 from *Gosudarstvennyi (national'nyi) doklad o sostoyanii i ispol'zovanii zemel' v Rossiskoi Federatsii v 2002 g.*, Moscow: Roszemkadastr (2003), p. 104; for 1999-2001 see also "Zemel'nyi rynok Rossii," *Otechestvennye zapiski* No. 1 (2004), www.strana-oz.ru.

Table 5.8. Transactions in agricultural land 2001

	Number of transactions, '000	Percent
Leasing of state land to households	1,695.6	93.0
Leasing of state land to agricultural producers	81.4	4.5
Sale of state land to households	2.6	0.1
Sale of private land to households	44.5	2.4
Total transactions in agricultural land*	1,824.1	100.0

*Transactions in agricultural land constitute 33% of the total number of 5.57 million registered land transactions in 2001.

Source: Estimated from official data on land transactions published in *Svedeniya o sdelkakh s zemlei i platezhakh za zemlyu*, Moscow: Roszemkadastr (2002), pp. 46, 52, 78, 84, 111, 113, 115. Data for later years not available.

The omission of transactions in land shares and leasing between individuals automatically biases the registered transactions toward predominance of land leasing from the state. In Table 5.8, over 90% of all registered transactions in agricultural land in 2001 involved transfer of state-owned land to individual use and another 4.5% involved leasing by the state to agricultural producers (peasant farms and corporate farms). Buy-and-sell transactions accounted for a mere 2.5% of the total, involving primarily activity between individuals.

The information on transactions in agricultural land is not published as a separate category, and we estimated the frequency and magnitude of these transactions by focusing on categories of registered transactions that are predominantly agricultural land. These included transactions by individuals for peasant farms, household plots, and related farming activities (both inside and outside the villages), as well as transactions by corporate farms (involving fields outside the villages).

Buying and Selling by Individuals

Buying and selling of agricultural land by individuals is miniscule compared to land leasing from the state and shareowners. Table 5.9 presents the available data in the form of three-year averages (1997-1999 and 2000-2002). Three main characteristics emerge from Table 5.9. First, the number of transactions shows a definite increase over time, while the transacted land area remains constant. As a result, there is a slight decrease in the average transaction size, which is more noticeable for household plots (0.24 ha per transaction in 1997-1999 compared with 0.21 ha in 2000-2002). Second, the number of land transactions for garden plots is an order of magnitude greater than the number of transactions for household plots, but the average transaction size is an order of magnitude smaller. Third, transactions by household plots involve primarily land inside village limits (the average transaction size inside village limits is naturally much smaller than in the open fields outside the village). The private land area transacted inside village limits exceeds by a factor of 6 the area transacted outside village limits (see Table 5.9).

Table 5.9. Sale of privately owned land for household plots and garden plots inside and outside village limits (averages for 1997-1999 and 2000-2002)

	Number of transactions		Area, ha		Average plot size, ha	
	HH plots	Garden plots	HH plots	Garden plots	HH plots	Garden plots
Inside village limits 1997-1999	91,054	5,879	21,216	643	0.23	0.11
2000-2002	106,626	5,847	21,069	587	0.20	0.10
Outside village limits 1997-1999	1,331	39,222	789	2,937	0.59	0.07
2000-2002	1,228	44,448	1,153	3,191	0.94	0.07
Total transactions 1997-1999	92,385	45,101	22,005	3,580	0.24	0.08
2000-2002	107,854	50,295	22,222	3,778	0.21	0.08

Calculated on the basis of yearly data from the following sources: 1997-1999 from *Svedeniya o sdelkakh s zemlei*, Moscow: Goskomzem (2000), p. 24; 2000-2001 from *Svedeniya o sdelkakh s zemlei*, Moscow: Roszemkadastr (2002), p. 31; 2002 from *Gosudarstvennyi (natsional'nyi) doklad o sostoyanii i ispol'zovanii zemel' v RF v 2002 g.*, Moscow: Roskomzem (2003), p. 115.

While statistics record 150,000 transactions annually between private landowners in villages, the amount of land transacted is about 0.5% of the total holdings in household plots and gardens. This information is presented in Table 5.10.

Table 5.10. Share of land transacted among private landowners 1997-2002

	Household plots			Garden plots		
	Total holdings, '000 ha	Trans-acted, '000 ha	Percent transacted	Total holdings, '000 ha	Trans-acted, '000 ha	Percent transacted
1997	5,782	24.7	0.43	1,263	7.0	0.56
1998	5,947	21.6	0.36	1,266	5.8	0.46
1999	6,217	24.7	0.40	1,265	6.3	0.50
2000	6,121	23.6	0.38	1,263	5.9	0.47
2001	6,432	24.8	0.38	1,261	5.7	0.45
2002	6,830	25.5	0.37	1,260	5.2	0.42

Source: Total holdings from Stat. Yearbook 2003, p. 399; transactions from previous table.

Land Prices Based on Official Data

About 90% of recorded land transactions involve leasing of state and municipal land. The share of these leasing transactions reaches 99.6%-99.8% in total area transacted. Therefore, the available land prices represent with any confidence only leasing from the state and the municipalities, not buying and selling of land and not transactions among individuals. Leasing rates charged by the state are shown in Table 5.11, which presents the data separately for different categories of lessees and for land leased in towns and inside or outside the village limits. Land in towns is naturally much more expensive than land in villages, while land inside the village is in turn more expensive than land on the periphery. In addition, there is strong differentiation by lessees: corporate farms pays less than rural households for land leased outside the village, whereas entrepreneurs leasing land for non-agricultural business pay the highest rate. In terms of the number of transactions, land for farming uses leads by an order magnitude. In the past, the authorities used to set lease payments as a multiple of the land tax. To this very day, corporate farms typically pay not more than 1-2 times the land tax, but for other users no clear pattern is discerned.

Contrary to leasing transactions, it is impossible to estimate statistically meaningful mean prices for buying and selling of land because of the extremely low frequency of these transactions (see the last row in Table 5.12). Official statistics thus provide only a very rough indication of the comparative level of prices for buying and selling of land by different groups of agents in different locations. Moreover, official statistical data are published only for buying and selling of state-owned land: transactions involving private land are not monitored at all. The available price data for buying and selling of state land are summarized in Table 5.12. To the extent that any relationships can be inferred from these data, it seems that land prices run at between 10 and 100 annual lease payments.

Table 5.11. Average lease rates charged for leasing state and municipal land, rubles per hectare

	Individual farming activities			Individual non-farm activities			Corporate farming
	In towns	Inside village	Outside village	In towns	Inside village	Outside village	Outside village
1997	640	35	60	303,800	5,920	53	10
1998	1,400	100	90	172,100	182,200	500	50
1999	1,300	160	90	386,600	179,800	38,900	40
2000	1,700	500	80	325,400	196,400	17,300	50
2001	1,400	500	90	230,900	230,900	7,900	50
2002	2,500	1,200	200	570,100	210,100	129,300	60
Transactions	1,018,100	963,600	1,535,600	122,800	n.a.	n.a.	85,200

Source: 1997-1999 from *Svedeniya o sdelkakh s zemlei*, Moscow: Goskomzem (2000), p. 17; 2000-2002 from *Gosudarstvennye (natsional'nyi) doklad o sostoyanii i ispol'zovanii zemel'*, Moscow: Roszemkadastr (2003), p. 109.

Although the Land Code stipulates that state-owned land should be sold in an auction, this practice is not enforced for individual farmers. Thus, of the 3,500 plots sold for farming uses outside the villages, only 84 were auctioned (land inside the villages and in towns is sold by the municipality, not the state). Corporate farmers are more bound by auction procedures (14 auctions among 19 transactions in Table 5.12), but the number of transactions in this category is too small to reach any meaningful conclusion about prices.

Table 5.12. Average prices for buying and selling of state and municipal land, rubles per hectare

	Individual farming activities			Individual non-farm activities			Corporate farming
	In towns	Inside village	Outside village	In towns	Inside village	Outside village	Outside village
1997	51,000	32,000	19,000	525,000	61,000	21,000	600
1998	174,000	21,000	117,000	705,000	120,000	22,000	600
1999	184,800	47,800	104,200	957,400	177,400	580,100	710
2000	162,600	58,100	36,800	2,639,400	904,400	28,100	1800
2001	158,400	61,400	24,100	2,379,200	88,900	47,600	650
2002	102,900	85,000	73,600	427,400	41,500	47,800	200
Transactions	18,300	15,200	3,500	1,900	700	77	19

Source: 1997-1999 from *Svedeniya o sdelkakh s zemlei*, Moscow: Goskomzem (2000), p. 21; 2000-2002 from *Gosudarstvennye (natsional'nyi) doklad o sostoyanii i ispol'zovanii zemel'*, Moscow: Roszemkadastr (2003), p. 112.

Factors Determining Land Transactions

Since Roszemkadastr reports land transactions by region, we have tried using regional data to model land transactions in a cross section of 71 regions (omitting those where private land ownership is prohibited and where data are suspect). The registered transactions are a mixture of different types, but the main category includes transactions that involve leasing of state land (outside village limits) by corporate and peasant farms (see Table 5.8). Household plots are not included in this category, since they generally lease land from municipalities inside the village limits.

Our modeling exercise was subject to severe restrictions due to lack of regional data. We have tried a model in which the total number of leasing transactions is explained by the following independent variables:

- The number of potential lessees, i.e., the number of corporate and peasant farms in the region. The expectation is that a higher number of potential lessees will have a positive impact on land leasing transactions. To allow for the possibility of differential effects of farms from the two categories, the number of corporate farms and the number of peasant farms were introduced separately.

- The quality of agricultural land, the expectation being that better land will increase the scope of leasing transactions. Land quality was represented by two alternative measures: the so-called "cadastral price," which is a value calculated on the basis of local fertility of land (Model 1), and partial productivity of land calculated as regional agricultural product per hectare of agricultural land (Model 2).
- Availability of agricultural land in the region: more agricultural land should lead to more leasing transactions.

Other a priori relevant factors, such as the financial situation of farms (farms that are better off would tend to engage more in leasing transactions), could not be used for lack of data.

The regression results are presented in Table 5.13, which also gives the mean values of the variables across 71 regions. On the supply side, both availability of agricultural land and land quality (in both models) have a significant positive effect on the number of lease transactions. On the demand side, the number of peasant farms has a positive effect on the number of transactions, while the effect of the number of corporate farms is not statistically significant. This result is understandable because peasant farms exist in much larger numbers (nearly 3,500 peasant farms in the average region) and accordingly engage in more numerous transactions. The few hundred corporate farms in each region (350 on average) cannot produce a noticeable impact on the total number of transactions.

Table 5.13. Regression modeling of land lease transactions in a cross-section of 71 regions (2001)

	Model 1		Model 2		Mean
	Coefficient	p-level	Coefficient	p-level	value
Dependent variable: number of transactions					1,131
Number of corporate farms	-1.35	0.214	0.25	0.827	347
Number of peasant farms	0.11	0.032	0.21	0.000	3,374
Agricultural land in farms, '000 ha	0.48	0.001	0.33	0.031	2,081
Cadastral price, ruble/ha	0.13	0.000	—	—	10,950
Land productivity, ruble/ha	—	—	0.07	0.080	7,360
R^2	0.545		0.42		

Source: Transactions from *Svedeniya o sdelkakh s zemlei i platezhakh za zemlyu*, Moscow: Roszemkadastr (2002), pp. 46-47; agricultural land in farms from *Zemel'nyi fond RF na 1.1.2001g.*, Moscow: Roszemkadastr (2002), pp. 113-129, 199; number of farms and productivity from Statistical Yearbook of Russia 2001.

The Effect of Land Transactions on Landowner Composition

Land transactions are naturally driven by additional factors that could not be formalized for analysis due to lack of data. We will briefly consider three groups of factors that seem to be conducive to the development of land transactions in rural areas. These factors include the general poverty of the rural population, which often drives families to sell their property for cash; the inability of current land users to pay competitive rates, which encourages rural landowners to look for new clients for their land; and the emergence of cash-rich non-agricultural companies looking for new investment opportunities in agriculture.

Rural Poverty and the Composition of Landowners

Agricultural land is mainly owned by the rural population, i.e., workers and pensioners of farm enterprises created by reorganization of kolkhozes and sovkhozes. This is a low-income segment of the Russian population. The agricultural wages remain very low (40% of the national average) and every third worker in agriculture suffers from salary arrears (compared with every eighth worker in manufacturing industries).[4] In April 2002, 51% of agricultural workers earned less than 1,000 rubles per month and about 40% earned less than 800 rubles per month. The per capita income of rural households was 1,971 rubles (exceeding the minimum cost of living by only 4%).[5] Naturally, if anybody will suggest these workers to sell their land shares, they will sell it. For comparison, the minimum price of land (in the form of land shares, before conversion into plots) in Moscow Oblast is 10,000 rubles per hectare, and the average land share is 3-4 hectares. These people would gladly sell their land if anyone was willing to buy it.

The practice shows that when willing buyers turn up in a village, only 15% of land-share owners refuse to sell. The buyers offer cash on the table, and the price is comparable with wages or pensions that rural people receive during 1-3 years. Land shares have become a commodity in impoverished areas, which was not the case during the initial stages of land privatization. Thus, from the beginning of land privatization to 2003, heirs never bothered to register the rights to their inherited land, because registration would only lead to a real estate tax without any increase of income. According to calculations for four farm enterprises in Moscow Oblast, the dead "owned" nearly 30% of land shares as of the beginning of 2004. When the demand for land shares materialized, most of the heirs began to register their rights to the inherited land.

Because of the absence of statistical data on transactions in land shares and registration delays, this trend will be indirectly confirmed by official statistics only one or two years later. It will be reflected, for instance, in the

---

[4] Calculated from *Sotsial'noe polozhenie u uroven' zhizni naseleniya Rossii*, Moscow: Gostkomstat (2003), pp. 84, 114, 120, 138, 142.

[5] Ibid., p. 227.

decrease of the share of agricultural land used by farm enterprises and the increase of land for gardening, dacha building, and peasant farming.

Uncompetitiveness of Traditional Agricultural Producers in the Land Market

To compete against those who buy land for non-agricultural purposes, lease payments from farm enterprises and peasant farmers should at least be equal to the bank interest rate on the market value of a land share, i.e., on average 1,000 rubles/ha. Agricultural producers cannot afford such payments, and as a result they will not be able to retain in the long term the privately owned land that they are currently using, mainly on trust. Rural people will therefore continue to accept the offers of speculative buyers in the future.

Cash Accumulation in Other Industries May be Fueling the Land Market

Surplus cash accumulated in other sectors of the economy may encourage buying of land from rural landowners by non-agricultural business. This hypothesis has been prompted by the analysis of the high demand for land in some regions close to Moscow Oblast. The available statistical data point to the following factors: surplus profits that are not used for investment in assets are increasing, deposits in Russian banks are becoming less attractive, and the decrease of the dollar/ruble exchange rate drives depositors away from foreign-currency accounts. Real-estate prices begun to grow in 1999, and they are now at their peak, with experts predicting an imminent fall in prices. All these facts have driven businesses to search for new investment opportunities. The demand for suburban homes, the poverty of rural land-share owners, and the end of the debate about buying and selling of agricultural land—all these factors have made it possible for non-agricultural businesses to buy agricultural land. These developments are discussed in more details in the chapter on agroholdings (Chapter 4).

Implications of Rezoning

Although in theory agricultural land is protected by law, there is nothing in reality that prevents its rezoning into non-agricultural land. This naturally stimulates the interest in land of individuals and organizations whose aim is not agricultural production. The cost of buying cheap agricultural land is recovered with a substantial profit by selling plots for country cottages near Moscow or in other desirable locations. For example, 0.01 ha of agricultural land can be bought for $3-$10, but after rezoning the same plot is sold for residential construction for at least $500.

Rezoning of agricultural land for residential construction is governed by the Land Code. The Land Code explicitly allows building a summer house on agricultural plots of certain types (a vegetable garden or a so-called dacha plot for a secondary country house), but the rules for allowing general construction on agricultural land are not clear. All this may stimulate transactions in agricultural land in the short term, but ultimately it may severely deplete the stock of agricultural land in some regions.

Agricultural land can be protected from non-agricultural uses by establishing a clear legal procedure for rezoning of agricultural land; setting zoning rules that specify where construction is allowed and not allowed in rural areas; introducing standard criteria to control the designated use of agricultural land and identify improperly used of unused land.

Evidence of Land Transactions in the 2003 Basis Survey

Land transactions and land market constraints cannot be studied only on the basis of official sources because no statistical data are available on transactions in agricultural land and land shares and there is absolutely no information on the terms of land transactions, on the composition of buyers and sellers, or on supply and demand. The required data can be obtained only through specially designed questionnaire-based surveys. We carried out such surveys in three regions—Rostov, Ivanovo, and Nizhnii Novgorod—that differed by natural conditions, economic development, and the level of policy reforms.[6] The surveys covered agricultural producers of three basic organizational forms—farm enterprises (a corporate form of organization), peasant farms, and household plots (two individual forms of organization). A detailed discussion of these organizational forms is presented in Chapter 2. One of the significant behavioral differences between peasant farms and household plots was eliminated from the survey by restricting the sample to household plots with declared commercial orientation (see the description of survey design in Chapter 1).

Land Market Activity

The survey shows that leasing of land is widespread among corporate farms and peasant farms. Only household plots rely mainly on owned land (Figure 5.7). In corporate farms and peasant farms, the share of leased land is on average 60% of the total area of agricultural land used (Table 5.14).

---

[6] The level of policy reforms as proxied by administrative intervention scores is discussed in Chapter 8.

Fig. 5.7. Structure of land by tenure

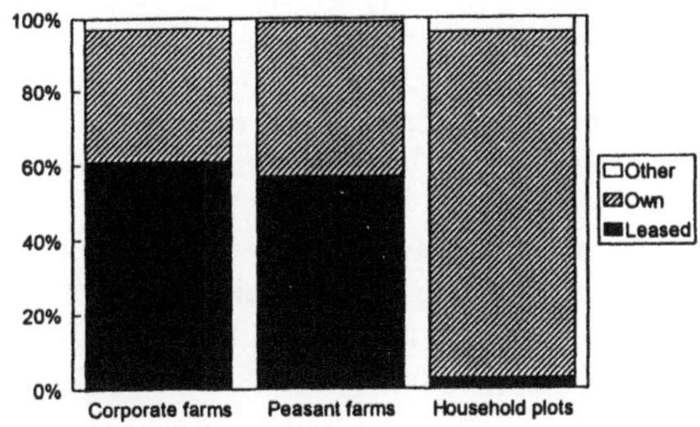

Source: 2003 BASIS survey

In corporate farms, three-quarters of the leased land is in the form of land shares, and only one-quarter is leased as land plots. Peasant farms, on the other hand, tend to lease land plots to a greater extent (more than 40% of all leased land). The owned land in peasant farms consists of land shares and land plots allotted without payment to the members of the farm during land privatization (land in joint shared ownership), as well as land shares and land plots purchased in the market. The owned land of corporate farms consists of land shares invested by members in corporate equity capital. The share of owned land in corporate farms (36% in Table 5.14) is overestimated in part due to the incorrect interpretation of survey questions by the managers, who improperly regard leased land shares as land shares invested in their corporate farm's equity. According to official national statistics, the share of owned land in corporate farms is only 1.3%.[7]

Table 5.14. Sources of land in the survey (percent of the total area)*

	Corporate farms (*n*=136)		Peasant farms (*n*=222)		Household plots (*n*=190)
Average farm size	4100 ha		270 ha		2.6 ha
Leased land	61		57		3
Leased land shares		(46)		(32)	—
Leased land plots		(15)		(25)	—
Owned land	36		42		93
Purchased land shares		(2)		(30)	—
Purchased land plots		(1)		(11)	—
Land shares invested in equity		(33)		(1)	—
Other	3		1		4
Total	100		100		100

*The numbers in parentheses are rough estimates based on part of the respondents.

[7] See *Zemel'nyi fond Rossiiskoi Federatsii na 1.01.03 g.*, Moscow: Roszemkadastr (2003), p. 13.

Respondents in 553 farms of various organizational forms in three regions reported 97 land transactions during one year. The frequency of transactions was virtually the same in farms of different organizational forms. There were only three cases of buying and selling of land. All other transactions involved land leasing. Strengthening the data in Table 5.14, this shows that land leasing is a dominant form of transaction in land markets across Russia, but unfortunately most of these transactions remain outside the scope of official statistics.

The survey did not detect any dependence of the frequency of land transactions on the distance from the regional center. Yet we are witnessing a particularly active land market in areas close to Moscow and in Moscow Oblast, where land is bought for non-agricultural purposes. This subject requires a special study.

Reported Land Transactions

The incidence of land transactions in the sample is not very pronounced and basically only leasing transactions are reported. Of some 550 respondents in the survey, one respondent (a corporate farm) reports selling land in 2001 and none reports buying land. A total of 96 respondents (17% of the sample) report engaging in land lease transactions of some kind in 2001. Of these, 57 respondents (10%) report that they lease in additional land, 34 respondents (6%) report that they lease out land, and 4 respondents (1%) report both leasing in and leasing out of land.

Leasing In

A total of 61 respondents (11% of the sample) reported acquiring additional land by leasing in 2001. The distribution of the frequency of transactions across the three organizational forms was not significantly different from uniform (Figure 5.8). Data were collected for 56 instances involving leasing of physical plots and 13 instances involving leasing of land shares (land shares can be leased from private individuals only).

Table 5.15 shows the distribution of transactions by main sources of land and the estimated prices per hectare per year (mean and median). The prices reported for land leased from the raion administration (40-50 rubles per hectare) were significantly lower than the prices paid to corporate farms and private individuals (400-500 rubles per hectare). There were no statistically significant differences in prices paid to private individuals for land leased in the form of plots or paper shares.

The average amount of land per farm acquired through land shares is significantly greater than the amount of land acquired by leasing of physical plots. On the other hand, there are no statistically significant differences in the size of physical plots leased from different sources. The median plot areas in Table 5.15 seem to suggest that smallest areas are acquired from corporate farms and largest from the raion administration and other sources. However, these differences are not statistically significant.

Fig. 5.8. Leasing transactions by farm type

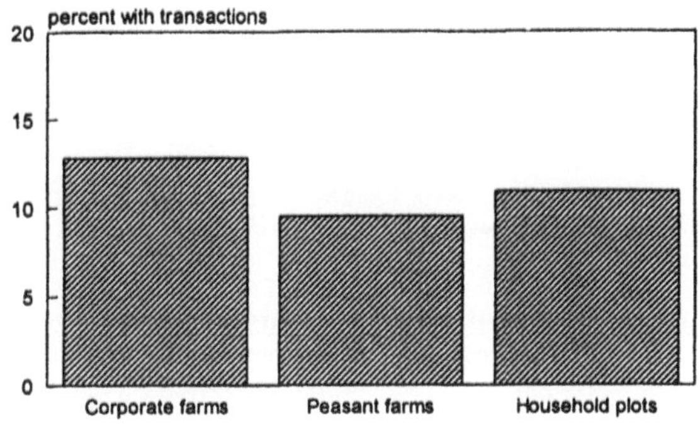

Source: 2003 BASIS survey

Table 5.15. Lease-in transactions and prices by source of land (including leasing of plots and land shares)

Source	Number of transactions	Ha, mean	Ha, median	Number of price data	Price/ ha, mean	Price/ ha, median
Plots:						
Corporate farms	22	112	5	17	480	324
Private individuals	18	298	29	13	576	362
Raion administration	10	163	142	8	42*	48
Other	6	1,102	425	2	100-143	
Total plots	56	287	28	40	405	212
Plots from corporate farms and private individuals	40	196	16	30	522	343
Land shares from private individuals	13	1,198	524	13	607	420

In general, the amount of new land acquired in 2001 is strongly correlated with total land holdings of the acquiring farms (R=0.94 between logged variables, Figure 5.9). The correlation with land holdings remains statistically significant even when we control for organizational form (corporate farms, peasant farms, household plots), which in itself accounts for a lot of variability in land transactions (R^2=0.83).

Fig. 5.9. Acquired land vs land holdings

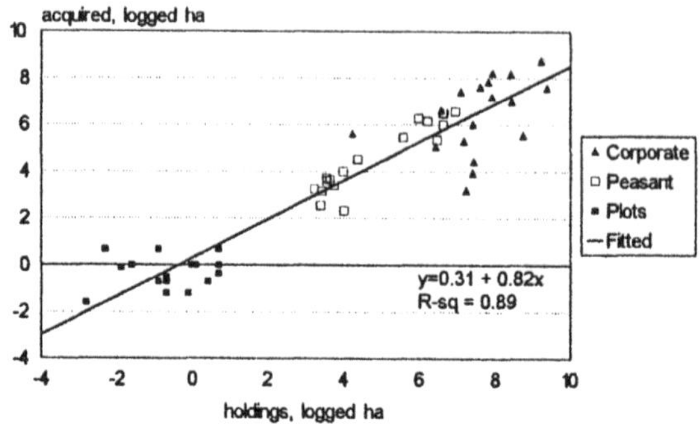

Source: 2003 BASIS survey

Table 5.16. Distribution of farms of different organizational forms by sources of land

	Corporate farms (*n*=23)	Peasant farms (*n*=24)	Household plots (*n*=22)
Corporate farms	22	21	55
Private individuals			
plots	22	21	36
land shares	26	29	0
Raion administration	17	21	5
Other	13	8	4
Total	100	100	100

There are no significant differences in access to different sources of land for corporate farms and peasant farms (Table 5.16). Unlike corporate and peasant farms, household plots do not resort any leasing of land shares: they only lease land plots—mainly from corporate farms and to a certain extent also from private individuals. Their transactions with the raion administration are minimal. This is also in contrast to corporate and peasant farms, for which more than 25% of lease transactions are with the raion administration. These differences in the pattern of leasing sources, and specifically the preference of household plots for leasing land from corporate farms, may explain the observation in Table 5.15 according to which the smallest plots are acquired from corporate farms.

Leasing Out

In addition to 61 respondents who lease in land, 38 respondents (7% of the sample) report leasing out land and one respondent reports sale of land. It is mostly corporate farms that lease out land (the one seller is also a corporate farm), presumably because of lack of profitability and inadequate business opportunities. At the other extreme, very few peasant farms lease out land—for exactly the same reasons, but in reverse (Figure 5.10). Household plots fall in between, with 9% leasing out land. A working hypothesis suggests that these are probably plots of older families, although lack of household demographic data in the survey instruments makes it impossible to check this hypothesis. The survey only reveals that farms lease out land predominantly because they are unable to cultivate it themselves. This is the reason provided by 32 out of 38 lessor farms. Yet these responses are equally distributed among corporate farms and household plots and we cannot learn anything about the specific reasons for leasing out land by household plots.

Fig. 5.10. Leasing out of land by fann type

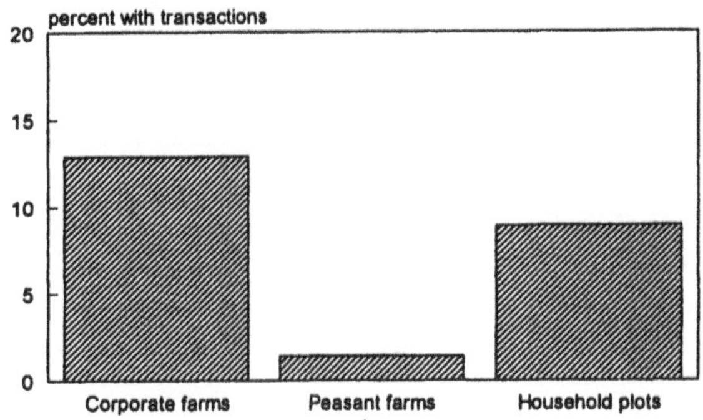

Source: 2003 BASIS survey

The average size and the average price received in leasing-out transactions are given in Table 5.17 (means and medians in the sample). The general pattern is essentially the same as for leasing-in transactions (see Table 5.15), except that raion administration is not included. Household plot operators lease out land exclusively to the local corporate farm. Corporate and peasant farms lease out land to other corporate farms and private individuals in roughly equal proportions (in such a small sample percentage frequencies are meaningless). In this sense, the leasing-out and leasing-in patterns are identical.

Table 5.17. Lease-out transactions and prices by source of land

Source	Number of transactions	Ha, mean	Ha, median	Number of price data	Price/ha, mean	Price/ha, median
Corporate farms	29	163	13	15	467	350
Private individuals	10	262	139	8	361	238
Other	2	145	145	2	24-158	
Total	41	186	28	25	403	310
Corporate farms and private individuals	39	188	18	23	430	325

There is no clear linear relationship between the amount of land that farms lease out and their land holdings (unlike the sharply linear relationship for the amount of land leased in, see Figure 5.9). Yet larger farms (i.e., corporate farms) clearly lease out more land than smaller farms (i.e., household plots). This association between farm size and the amount of land leased out is demonstrated in Table 5.18. An interesting fact clearly emerges from Table 5.18: the large corporate farms lease out but a small fraction of their holdings; the small household plots and peasant farms, once they decide to lease out land, release most of their initial holdings and are sometimes left with less than the amount leased out to other users. Of the 21 small farms that act as lessors, 14 farms leased out about half their initial holdings while the remaining seven farms leased out 94% of their initial holdings. These "superaggressive" lessors cultivate on average a small residual of 0.7 hectares after leasing out more than 10 hectares.

Table 5.18. Amount of land leased out and farm size

	Number of lessor farms	Average farm size, ha	Average leased out, ha
Corporate farms	16	7,400	463
Plots and peasant farms	21	8.6	11.4
"Aggressive" lessors	14	12.5	11.9
"Superaggressive" lessors	7	0.7	10.3

Payment for Land

The survey enabled us to estimate the level and structure of lease payments. The data were analyzed separately for agricultural producers leasing in land (56 lessees) and those leasing out land (23 lessors).

Prices of leased-in and leased-out land were compared for transactions involving corporate farms and private individuals as lessees and lessors (see last line in Tables 5.15 and 5.17, respectively). The differences in prices are not statistically significant. The median price in the sample for all leasing transactions in these channels is 330-340 rubles per hectare. Supplementary data were obtained by analyzing the lease payments for land shares (median 420 rubles) and separate partial responses of lessees and lessors on structure of lease payments (which give 450 rubles/ha for leasing in and 440 rubles/ha for leasing out; the difference is not statistically significant). The various numbers suggest median lease payments of 350-450 rubles per hectare in the sample (excluding transactions with the raion administration, which command much lower prices).

Most lessees made lease payments in kind; payment in cash was less common. Many lessees assumed the responsibility for the land tax. Lessors also indicated that lease payments were typically a share of the output. The mean lease payment was 440-450 rubles/ha (Table 5.19). Lease payments estimated separately for land shares gave a median of 420 rubles/ha, while the median of all leasing transactions in the survey was 340 rubles/ha. Thus, the lease payments range between 350-450 rubles/ha, excluding the transactions in state land. The lease payments to the district administration for state land are much lower (about 50 rubles/ha, see Table 5.15). The differences in lease payments across farms of different organizational forms are not statistically significant.

Table 5.19. Lease payments estimated from the survey

Types of lease payments	Lessees, %	Structure of payments for leased-in land, %	Lessors, %	Structure of payments for leased-out land, %
Land tax	45	2	13	1
Fixed, in cash	25	20	22	17
Fixed, in-kind	52	57	17	22
Share of output	9	18	43	59
In services	9	3	13	1
Total	*	100	*	100
	56 respondents	451 rubles/ha	23 respondents	441 rubles/ha

* Adds to more than 100% because of multiple answers.

Contrary to the consistency in lease payments, there seems to be a significant difference in the lease term structure between lessees and lessors (Table 5.20). Lessees report leasing in land for terms up to 10 years; nobody reports leasing in land for more than 10 years. Among lessors, on the other hand, more than 40% report leasing out land for a term longer than 10 years. Moreover, it is only household plots that lease out land for more than 10 years—not corporate farms or peasant farms.

Table 5.20. Lease term as reported by lessees and lessors

A. Lease in

Lease term	Percent of lessees (n=66)	Corporate farms (n=22)	Peasant farms (n=25)	Household plots (n=19)
One year	24	18	4	58
2-3 years	30	36	32	21
4-10 years	47	50	64	21
More than 10 years	2	5	—	—

B. Lease out

Lease term	Percent of lessors (n=36)	Corporate farms (n=16)	Peasant farms (n=3)	Household plots (n=17)
One year	22	38	33	6
2-3 years	22	38	67	0
4-10 years	14	25	0	6
More than 10 years	42	0	0	88

Potential for Land Transactions

The potential for land transactions was examined by exploring the perceived need in farms for additional land (Table 5.21). Nearly 30% of respondents desired to increase their landholdings. This potential for future land transactions should be compared with the actual frequency of land leasing in 2001, which covered 11% of respondents (uniformly distributed over the three organizational forms, see Figure 5.8). The greatest need for additional land is expressed by peasant farmers: 50% of respondents in this category desire more land, compared with less than 20% among household plots and corporate farms. Peasant farmers who would like to increase their holdings typically have less land than the rest, although the difference is not dramatic (225 hectares compared with 314 hectares for those who do not need more land). A similar situation is observed for corporate farms (3,350 hectares compared with 4,320 hectares). Among household plot operators, on the other hand, the difference in land holdings between those who say they need more land and the rest is not significant.

Table 5.21. Potential and actual land transactions (percent of respondents)

	Potential: desire additional land	Actual: acquired additional land in 2001
All sample	29	11
Corporate farms	18	13
Farms	49	9
Plots	13	11

In principle, we can expect the demand for land to depend on the financial situation of farms. However, the only indicators of financial performance in the survey were sales revenue and surplus—a very crude profit-like measure of financial sources in excess of uses (see Chapter 9). Neither of these financial indicators showed a clear association with the perceived demand for land.

Table 5.22. Perceived sources for acquiring additional land (percent of respondents)*

	All respondents (n=161)	Corporate farms (n=25)	Peasant farms (n=111)	Household plots (n=25)
Lease plot from individual	44	32	52	16
Lease from the state	35	28	40	20
Buy land shares	18	8	23	4
Buy plot from individual	17	0	19	24
Lease from corporate farm	16	28	9	32

*Percentages in each column add up to more than 100% because multiple answers are allowed; "lease" includes also taking land in use rights.

The most common option for acquiring additional land is by leasing a plot from a private individual (40% of respondents with perceived need for additional land). Other accessible options (in multiple-choice answers) include getting a plot from the state in leasehold or use rights (35%), buying land shares (18%), and even buying a land plot from a private individual (17%). There are clear differences in potential access patterns of different organizational forms to various sources of land (Table 5.22). While corporate farms and peasant farms envisage mainly leasing from private individuals and the state, household plots primarily intend leasing from the corporate farm and buying from individuals. Buying of land shares is envisaged as a viable option only by peasant farms.

Buying land is thus not perceived as impossibility in the current environment. Indeed, fully 30% of respondents indicate that they would be able to get additional land in private ownership as needed (Table 5.23). However, leasing is clearly perceived as the most accessible option, with more than 60% indicating that they would be able to lease additional land as needed.

Table 5.23. How can you acquire additional land if needed? (percent of respondents)

	Acquire in private ownership	Lease
Yes	31	61
Undecided	48	34
No	20	5

As there are no observations of buy-and-sell transactions in the survey, it is unfortunately impossible to analyze the role of access to credit and other farm-related factors as potential constraints in land market development. Still, it is encouraging to note the emergence of land leasing transactions as a first stage and the positive perception of buying and selling of land as the second stage of land market development in some undefined future.

Estimating the Demand for Agricultural Land

The holdings of existing agricultural producers, including the leased component, formed a long time ago, so that the observed situation does not necessarily reflect a true satisfied demand for land. The survey has shown that the demand for land and thus the potential for land transactions vary by region and by type of farm.

In all three regions surveyed, peasant farms revealed a greater demand for land than corporate farms. Thus, 30%-70% of peasant farmers indicated a demand for land (Table 5.24). Among corporate farms, on the other hand, the maximum demand for land (30% of respondents) was reported in Rostov (a region with a highly developed agriculture), whereas in the less developed Ivanovo and Nizhnii Novgorod oblasts corporate farms showed a much lower demand for land.

Table 5.24. Land users' intentions to reduce or enlarge their land (percent of responses)

	Corporate farms	Peasant farms	Household plots
Ivanovo			
Reduce	43	10	1
Enlarge	3	30	14
Total agricultural land, ha	98,000	420	80
Nizhnii Novgorod			
Reduce	26	0	7
Enlarge	11	67	12
Total agricultural land, ha	99,000	460	141
Rostov			
Reduce	3	1	—
Enlarge	30	50	—
Total agricultural land, ha	271,000	57,000	—

In Ivanovo (an example of an agriculturally depressed region), 43% of corporate farms plan to reduce the use of land and only 3% plan to enlarge it (Table 5.24). A similar trend is observed in Nizhnii Novgorod (an example of a region with medium agricultural development). This suggests that corporate farms in agriculturally less developed regions will probably shed some of their land when they begin re-registering their use rights in state land and lease contracts for land shares, as mandated by the new law. Some of the released land will be absorbed by peasant farmers, who in general seek to enlarge their holdings, but the growth potential of this segment is not particularly large: while corporate farms control tens of thousands of hectares in each oblast, all peasant farms combined have less than 1,000 hectares in Ivanovo and Nizhnii Novgorod. Due to the absence of other interested parties, much of the land released by corporate farms may remain unused. The opposite situation will probably occur in Rostov with its highly developed agriculture and a different specialization (more crops, less livestock). There will be no unused agricultural land in this oblast (Table 5.24), and unsatisfied demand for land can actually arise.

To conclude, we see that there is a demand for land for large-scale agricultural production in some regions (the Rostov example). In all regions, however, peasant farmers and household-plot owners show less intention to reduce their holdings and more willingness to enlarge their land than managers of large corporate farms. This tendency does not depend on natural and climatic conditions. In all three regions surveyed, peasant farmers demonstrate a greater demand for land even than household-plot operators. Nevertheless, the physical potential of peasant farms and household plots in

Ivanovo, Nizhnii Novgorod, and similar less developed regions is not sufficient to absorb the surplus land that will be released by corporate farms.

Constraints on Land Transactions

Analyzing the 2003 BASIS survey and Roszemkadastr data on sources of land used by agricultural producers, we conclude that farms of all types heavily rely on leased land and some even purchase land from individual and corporate owners. Yet the state land registry contains records of relatively few transactions that represent a very small portion of agricultural land. Two main groups of reasons may be responsible for this curious state of affairs. First, there is a general lack of market information pertinent to land transactions. The agents do not have sufficient knowledge of mechanisms and procedures necessary for registration of land transactions. Many rural people still do not know that land transaction are allowed and prefer to deal informally; many do not know how to draw up a contract or where to get standard forms for this purpose. Second, the legal registration procedures are very cumbersome, costly, and time-consuming. People may be avoiding land registration because of such bureaucratic barriers.

Table 5.25. Main constraints to registration of land transactions (percent of respondents)

	All respondents (n=558)	Corporate farms (n=142)	Peasant farms (n=214)	Household plots (n=202)
No need to register	42	44	17	69
Lack of information	18	23	11	23
High costs	19	10	34	7
Complex procedures	16	15	25	6
Clear procedures, no problems	23	23	36	9

Note: Numbers do not add up to 100% because multiple answers were allowed.

These groups of obstacles have been suggested by the analysis of the survey responses as summarized in Table 5.25. It may be instructive to note some differences across farms of different organizational types. Thus, the large corporate farms and the small household plots both feel that they can disregard the registration requirements. This is much less so for peasant farmers, who are apparently much more sensitive to the protection they get through land registration. At the same time, peasant farmers complain much more frequently of high registration costs and complex procedures. Somewhat paradoxically, however, more than one-third of the peasant farmers report that the registration procedures are clear and they have no problems in that respect (last line in Table 5.25).

Lack of Market Information

The respondents in the 2003 BASIS survey were asked if land transactions were permitted, if they knew the land prices, if they knew where to find a standard lease contract form and how to register a transaction. With the exception of Rostov peasant farmers (12%), 22%-32% of respondents indicated that lack of information on these matters was a problem for engaging in land transactions. Land price information and transaction registration procedures were mentioned as the most important obstacles.

The respondents did not know the prevailing land prices. The survey showed that many (though not all) knew the land tax rate: 33%-50% of respondents in different groups knew what the land tax was because they paid it once or twice a year. However, most respondents could not answer the other questions. The response rate was highest among the peasant farmers in Rostov, where land transactions were more frequent and the interest in the enlargement of holdings greater (Table 5.26). The fact that most respondents did not know the prevailing land prices apparently means that there are no established prices for land. There is no benchmark that could help rural people with the decision to sell or lease land.

Table 5.26. Frequency of responses to land-price questions (percent)

	Ivanovo		
	Corporate farms	Peasant farms	Household plots
Lease payment for state land	20	20	3
Lease payment for land share	17	50	2
Land tax	40	70	34
Price of land share	3	0	1
Price of state land	0	0	0
Price of land in corporate farms	0	0	0
Number of respondents	35	10	94

	Nizhnii Novgorod		Rostov	
	Corporate farms	Household plots	Corporate farms	Peasant farms
Lease payment for state land	8	11	14	46
Lease payment for land share	5	12	26	49
Land tax	49	45	53	47
Price of land share	3	1	9	27
Price of state land	0	2	0	2
Price of land in corporate farms	3	2	1	1
Number of respondents	39	93	70	209

The issue of market price of land is of special importance because the 2003 Federal Law on Agricultural Land Transactions mandates the use of market prices in the process of the partition of a land in joint shared ownership (i.e., when converting land shares into plots). According to this law, disputes between withdrawing and remaining owners are resolved by applying the market price of specific land plots (on a per share basis). If the agricultural land market is not developed and there are no consistent land prices in the district, it is impossible to speak of the market price of specific fields and plots. This is a severe barrier to the partition of joint shared land and to the withdrawal of share owners with land plots for individual farming. A similar problem arises with the buyout of land that was formerly given in permanent use. According to the law, state land should be bought on the basis of market prices, but none exists. The state is therefore obliged to set calculated (normative) prices, but these normative prices and the rules of their calculation are not readily available.[8] There are no statistical reports on the number and characteristics of transactions in agricultural land. While some information is partially collected for land plots, there are absolutely no data on transactions in land shares, which as we have seen are the most common form of land transactions today (Table 5.14).

High Registration Costs and Complex Procedures

Expert judgments suggest that high registration costs and complex procedures are an obstacle to land transactions. This view is confirmed by the responses of the Rostov peasant farmers, who have the strongest tendency toward land enlargement. Most of the peasant farmers in Rostov (84%) indeed regard these two issues as a major problem (Table 5.27; in other regions, where land markets are less developed, a much smaller percentage of respondents addressed these issues).

Table 5.27. Are high registration costs and complex procedures an obstacle to land transactions? (percent of yes responses)

Regions	Peasant farms and household plots	Corporate farms
Ivanovo	10	12
Nizhnii Novgorod	15	5
Rostov	84	58

The analysis of registration procedures shows that government bodies have created numerous administrative and organizational restrictions to land registration. In the land registration system that was in force until 1998, all the procedures were carried out by a single administrative level—the district committees for land resources and land planning. These committees kept the land redistribution maps, the lists of shared landowners,

[8] According to the recommendations of the Ministry of Real Estate, the market price of an agricultural plot is determined by comparing with past sales, and in the absence of sales—by discounting of lease payments (similarly to the calculation of cadastral prices).

copies of land ownership certificates, and other documents confirming land rights. The district committees also registered the transactions with land shares and land plots.

This system was changed in 1998. At present, three organizational levels are involved in registration of land transactions: (1) district committees for land resources and land planning; (2) district divisions of regional registration chambers (subordinated to the Ministry of Justice); (3) district divisions of regional cadastral chambers.

The cadastral chambers are responsible for the register of land plots and the registration chambers manage the register of use rights and land transactions. Previously issued certificates of land ownership rights theoretically have the same validity as new entries in these registers. In practice, however, each new transaction requires full registration of the previous rights. This is not done automatically, and it requires a full set of documents issued ten or twelve years ago at the time of original reorganization, including old cadastral plans of the land plots that became joint shared ownership. One does not need to bother with all this as long as there are no new transactions, but the first transaction (be it leasing or sale) immediately triggers the need to go through the entire process of formal registration of previous rights.

The new system suffers from at least two serious problems. The first problem is the refusal of the cadastral chambers to issue registry extracts for land plots in joint shared ownership. As a result, the whole area in joint shared ownership (often several thousand hectares) has to be surveyed. This is not only a very expensive operation (500 rubles per hectare), but it also takes a long time to complete (at least two months). The second problem is the multi-step and absolutely opaque operation of the registration and cadastral chambers, especially regarding the requirements for documents. These bodies develop internal instructions that are not always compatible with the relevant law and require additional documents that were not envisaged by the law. These administrative barriers involve additional expenses for the applicants and lead to a sharp increase of transaction costs.

Table 5.28, based on a case study for Moscow Oblast in 2004, lists the various organizations that are involved in the process of selling a land share or requesting conversion of land shares into a physical plot. The numbers in the table show the sequence of actions for a landowner in his dealings with these organizations. The sequence of actions to register previously existing rights is shown separately in column 2. Registration of rights associated with a new transaction is identical with the registration of previously existing rights. The case study in Table 5.28 assumes that the share owner has a land ownership certificate in his possession, i.e., a document that certifies his ownership rights and theoretically has the same validity as a record in the State Register of Rights.

Table 5.28. List of organizations involved in the registration of a transaction with land shares (column 1) and the sequence of relevant actions (columns 2-4): Moscow Oblast, 2004

	Registration of previous rights	Selling a land share	Converting 10 land shares into a 42 ha plot
(1)	(2)	(3)	(4)
Registration chamber	1, 8	3	7
Land committee	5	—	5
Cadastral chamber	3, 6*	4	6, 8
District archive	7	—	—
Corporate farm	2	—	—
Surveyor	4*	—	5
District newspaper	—	1	1, 4
Notary	—	—	3
General assembly of land-plot owners	—	—	2
Ministry of Real Estate	—	2	—
Total elapsed time	from 2 weeks to 6 months	not less than 3 months	from 3 to 6 months
Cost, '000 rubles	0.3-1.5*	not less than 2.4	not less than 88.1

* In case of refusal to issue a cadastral extract.

As Table 5.28 shows, the costs of registering previously existing ownership rights vary between wide limits. Theoretically the total shared area is already registered in the cadastre and the applicant only has to pay 100 rubles to get the cadastral extract for his land share. In practice, however, officials often require a full survey for the identification of the individual plot, which is too expensive for the private landowner. This requirement is apparently based on the provisions of the 2001 Law on Agricultural Land Transactions, which stipulates that land in joint shared ownership should be physically divided into plots, a very costly proposition for the small landowners.

The calculations in Table 5.28 were carried out assuming that each action requires only one visit to the administrative office. Usually each action requires at least three visits, and the time delays are substantially longer than shown. In the author's own experience, the withdrawal of a single land plot from joint shared ownership requires up to one year of constant occupation.

The list of documents that should be presented in each office is not strictly defined. So each office should be visited again and again. For example, as soon as the joint share owners have reached an agreement regarding the withdrawal of a specific plot, the plot should be marked on the ground and registered in the cadastre. The registration procedure is determined by law, but the law does not specify the precise requirements for documents. That is why officials at the local level set their own demands. In Moscow Oblast, nine out of the ten steps that an applicant has to complete are not prescribed by the law, require submission of additional documents, or are part of a list of verbal requirements that are not listed in any official document.

In this situation, corporations that can afford to hire advisors and have specialized staff responsible for transaction registration are in an advantageous position. This is a typical example of market asymmetry, where some agents have more information than others by virtue of their official position, greater financial possibilities, or the ability to hire experts. Land-share owners, peasant farmers, and traditional corporate farms are weak players in this process: they have to spend so much time and money on registration that they often simply give up their rights or use land that is not legally registered.

In order to simplify the land purchase procedure, buyers resort to general power of attorney or give the land away as a gift. With general power of attorney, the seller gets the money and empowers a third person to sell the land share and complete all the necessary arrangements. With a gift of land, there is no need to offer the share to other pre-emptive buyers (the joint owners, the oblast government, or the municipality). These "under-registration" mechanisms are risky for the buyer, as a power of attorney can be revoked before the registration of rights transfer to the buyer is completed, and a gift can be annulled as a fictitious transaction. Still, buyers are willing to take the risk because the prices of land grow so fast (the price of land in Mozhaisk Raion, Moscow Oblast increased by a factor of 20 between January 2003 and June 2004).

The cost of the entire procedure of converting a land share into a plot of land (according to the actions in column 3 of Table 5.28) can be estimated by comparing the market price of a land share with the market price of a registered plot in the same area. In Volokolamsk Raion, Moscow Oblast, the price of a plot is double the price of a land share before conversion.

The state is gradually toughening the rules for the registration of rights to a land plot. Following the provisions of the 2001 Law on Agricultural Land Transactions, which mandated that land in joint shared ownership should be physically divided into plots regardless of the cost to the small landowners, the land authorities issued instructions in September 2003 according to which the rights to a land plot cannot be registered until the full cadastral plan has been prepared.

Expected Effect of Mortgage on Transactions in Agricultural Land

The inability to mortgage land or, more generally, to use it as collateral for obtaining credit is often cited in the literature as a major obstacle to buying and selling of land. Even the general mortgage law passed in 1998 excluded agricultural land. The mortgage of agricultural land was allowed only in January 2004 by a special amendment of the 1998 mortgage law. The language of the original mortgage law extended to agricultural land imposes certain conditionalities that are highly restrictive in the new context:

- Only a land plot can be mortgaged, while most agricultural land is in joint shared ownership.
- Only owned land can be mortgaged (which accounts for a mere 1.3% of agricultural land used by corporate farms).
- The land owned by legal entities is registered in the cadastre in large massifs. To mortgage a specific plot, the entire area has to be surveyed, the plot has to be registered as a separate object, and the ownership rights to the plot have to be registered. These procedures are costly and take time.
- The mortgaged property must be insured by the borrower, even if this requirement is not explicitly stated in the mortgage contract. If the property is not insured, the lender can demand immediate repayment of the loan. Insurance companies still do not have such a service in their portfolio because they do not know how to assess the risks.

These legal provisions suggest that although the 1994 amendment formally allows agricultural land to be mortgaged, the technical barriers to meeting the basic requirements are practically insurmountable. At the present stage land mortgage will have a limited potential in Russia and will be unable to fulfill its theoretical role as a facilitator of transactions in land.

Conclusions and Policy Recommendations

Land markets act as a medium for the transfer of agricultural land from passive to active landowners, or more generally from less efficient to more efficient producers. Through this basic economic function land markets contribute to the improvement of productivity and efficiency in agriculture, while at the same time facilitating adjustment (or optimization) of farm sizes by mechanisms that do not require government intervention. Smooth functioning of land markets is taken for granted in market economies, where farmland is traditionally in private ownership, landowners are free to exercise their property rights through market transactions, and farmers are generally independent to make decisions on disposition and use of land. These prerequisites for normal functioning of land markets were definitely not a feature of the former socialist countries up to the early 1990s. These countries, with Russia as the largest representative of the group, had virtually no land markets during the Soviet era, when land use rights were allocated by the state and farm sizes were determined by fiat. Land markets began to emerge, quite hesitantly, after 1990, as the various transition countries gradually implemented reforms that corrected the previous distortions in property rights.

As of today, Russia has met the necessary pre-conditions for the development of agricultural land markets: agricultural land has been largely privatized, individual landowners have legal rights to most agricultural land in the country, and previous prohibitions on buying and selling of land have been removed by recent laws. Land markets have responded positively to these changes and we are beginning to witness transactions that involve individual landowners, and not only the state. While the Russian media, politicians, and scholars generally argue that market development is restricted by the low demand for agricultural land, our survey results seem to indicate that this is not really so: a substantial proportion of farms in some regions are actually interested in expanding their holdings.

Table 5.29. Measures to improve market information availability and reduce transaction costs for land market development

Measures to improve market information systems:

- introduce standard reporting requirements for organizations that deal with registration of ownership rights and transactions in land shares and land plots
- specify in law the sources of public data on market prices that will be available on the web sites of the relevant bodies
- establish an authority for conducting market-price surveys and collecting price information needed for re-registration of previous rights for permanent use of land

Measures to reduce transaction costs:

- introduce a simplified cadastral registration procedure that allows varying degrees of detail in the description of land plots by owner's consent, utilizes to the maximum extent the available information in district committees for land resources and land planning, cadastral chambers, and land-surveying institutes, and enables land users and landowners to access the cartographic data banks created at the taxpayers' expense
- establish a procedure for registration in the Central State Register of Rights on the basis of documents issued to landowners before 1998, without any additional demands from landowners and land users
- develop a single and complete set of clear requirements for documents from applicants, publish a full binding list of required actions, and prevent the officials from making arbitrary demands on applicants during registration of plots and rights

However, further development of the embryonic land market is severely circumscribed by the inadequacy of the administrative and technical infrastructure. There is no public registry of plans and maps that can be used to complete the transactions, the bureaucracy has created numerous procedural obstacles that complicate land transactions, and the agents effectively do not have access to market information about land prices or demand and supply of land. All these factors contribute to very high transaction costs in land markets. In the absence of competitive demand for agricultural land in many regions, landowners have no motivation to complete the required procedures for registration of their property rights, be it registration of land shares or physical plots. The actual costs are simply not justified by the expected benefits from making their property "ready for the market."

Restrictions on transferability of land and general nontransparency are often characteristic of the legal and administrative framework for land markets throughout the world (and not only in Russia). Unfortunately, these are serious obstacles to achieving economic efficiency—one of the goals of transition. If the registration procedures are made more transparent and less costly, we will probably see a growing number of buy-and-sell transactions alongside the burgeoning lease market. Measures to create adequate market information systems and to achieve significant reduction of transaction costs will alleviate the main barriers to further development of land transactions in this huge, land-rich country. Table 5.29 summarizes some policy recommendations that can be expected to alleviate the main barriers to land transactions in Russia.

Chapter 6

Completing Agricultural Land Reform in Russia: A Western View of Outstanding Policy and Legal Issues

Leonard Rolfes, Jr.

Collective agriculture was an economically unsound method of farming. Yields were much lower than those in the market economies, more inputs were needed to produce a unit of output, and much of the output was left in the fields, lost, or otherwise wasted before reaching the end consumer. Since 1991 Russia has moved away from collective farming through a variety of agrarian reform initiatives, with land reform being a key component. Russia has made progress on the land question, but if land reform is to be "completed" several issues still require attention.

Examining these issues begins with identifying the main objectives of land reform. What ultimate benefits should a "completed" land reform agenda provide for Russia and its people?

1. Land reform should deliver meaningful economic benefits directly to rural households, most of whom are poor;
2. Land reform should help improve the efficiency of agricultural production;
3. Land reform should help generate higher levels of agricultural production.

The legal and policy changes needed to reach the objectives of land reform, to deliver its benefits, can be grouped into three broad topics: completing land privatization; establishing a land market; and completing the law and policy framework for farm restructuring as it relates to land. For each topic, the discussion below reviews the progress to date and then examines the reforms still needed.

Completing Land Privatization

Farmers must have secure rights to land. Those who have a secure relationship to the land they work, and get to keep the production from the land, have a natural incentive to work hard and invest in the land. As a result production will increase, as will the efficiency by which that increased production is gained. This will mean more income for the farmer, and more agricultural output and economic development for the country at large.

In the post-Soviet environment, the best method to provide farmers with secure rights to land is to make them owners. Since the state owned all of Russia's agricultural land during Soviet times, land privatization is required to make farmers into owners. Therefore, the transfer of land ownership rights from the state to the private individual is an important reform needed to boost agricultural productivity and raise rural household incomes.

Progress to date

Russia has made substantial progress in two areas toward the goal of privatizing its agricultural land base. First, 12 million rural people have received land shares that represent ownership of 118 million hectares of arable land. Second, the vast majority of Russian families—both rural and urban—have become owners of small-scale landholdings, including dacha plots, vegetable patches, and household plots. These small landholdings generate over half of Russia's total agricultural output on less than 10% of the agricultural land base. They are an important source of food for both rural and urban households and provide supplemental income for many families (Rolfes, 2002).[1]

Reforms still needed

The major remaining privatization task concerns the 12 million land shares. As a practical matter, land share owners do not possess land rights they can actively use, largely because their rights do not correspond to an identifiable plot of land. Land shares are linked to the entire arable land base of the former collective or state farm, which is usually several thousand hectares in size. The typical land share owner holds these rights in common with several hundred other land share owners, and the location of his individual share is not identified. This greatly limits the influence an ordinary land share owner has over the land he formally owns.

---

[1] The legal rules governing small plots as set out in the Law on Household Plots (2003) are sound and do not require major changes.

Land share owners have long had the right to claim a land parcel in kind under the law, but the right has been difficult to exercise for a variety of reasons. These include opposition from agricultural enterprise managers and government officials, the high cost of boundary surveys needed to describe the parcel, and a registration system that is not functioning well. As a result, very few land share owners have claimed land parcels and become real land owners. By and large this is not a reflection of their lack of interest in becoming land owners,[2] but evidence of the land share's fundamental inadequacy as a vehicle for people to use individually or in small groups to claim land parcels.

To complete the privatization effort begun with land shares, the Russian government should implement a program for mass conversion of land shares into physical land parcels. Such a program would carry out boundary surveys, assign land parcels to individual land share owners, register the parcels in the names of the land share owners, and issue them certificates on state registration (Law on Land Registration (1997), art. 14).

Under this mass conversion program an individual land share owner could receive a land plot in one of two forms. The share owner could receive an individual land parcel, which would be farmed directly or leased to a commercial agricultural producer. Alternatively, the share owner could join with a small group of other share owners to receive a larger land parcel in common ownership. This group would acquire a specific field or portion of a field, and would continue to be guided by the common ownership rules.[3] For example, 10 relatives who own land shares of 5 hectares each could claim a single land plot of 50 hectares. The land plot would be surveyed and then registered in the common ownership of each relative. Each relative would receive a certificate on state registration reflecting his common ownership right to a one-tenth portion (in this example) of the 50-hectare land plot. The group could farm the land themselves, or lease it to a commercial agricultural producer.

If such a program could be carried out, land share owners would know which land they own, and therefore could make use of the land to benefit themselves and their families. For the program to be successful, though, it

---

[2] On the contrary, most land share owners probably would like to become owners of specific land parcels. Ukraine provides compelling evidence for this proposition. Ukraine currently is in the midst of a mass conversion of land shares into individual land parcels. A survey in 2003 of 800 people in rural Ukraine (Rolfes, 2003) showed that 78% of the new land owners had wanted their land shares converted into physical land parcels. This is not a self-selected group of especially motivated people, but rather ordinary people who live in villages where mass conversion has occurred. Both Ukraine and Russia come from the same tradition of collectivized agriculture, and both have utilized the land share in their reform processes.

[3] See Civil Code Part I (1994), chapter 16, and Law on Agricultural Land Transactions (2002), arts. 12-14.

should allocate land only to individual land share owners or small ownership groups numbering no more than 10 people. Permitting large groups with large parcels would keep the individual's land rights too diffuse, as they are currently in the land share system, and thus would perpetuate the fundamental problem with the land share.

Apart from the land share question, the other major privatization issue relates to the tens of millions of hectares of arable land that remain in state ownership (Deininger, 2003; see also Chapter 5). The Land Code, the Law on Agricultural Land Transactions, and the Law on Peasant Farms together set forth adequate general procedures for privatization, but could be improved in two ways. First, the procedures leave most of the initiative for privatization to private parties, rather than the state. Increasing the state's obligation to actively offer land for privatization would help to move more land into the private sector. Second, the land of many enterprises that carry out special activities such as agricultural research or elite seed breeding was exempt from privatization during the 1990s. Much of this land was not used for the special activities, though, and should have been subject to privatization (Prosterman et al., 1999). These enterprises should receive careful scrutiny to determine how much of their arable land should remain in state ownership, and how much should be transferred into private ownership.[4]

Establishing a Market in Land

Russia needs a free, active market for agricultural land. A market will help to put land in the hands of the most effective users, since lessees and buyers who can pay the highest prices generally are those who most efficiently use land. A market will provide an environment where parties can find each other efficiently, negotiate over pricing and other important issues, and conclude a transaction that satisfies each party. In addition, a market will encourage investment in land because a land owner will know that, when the time comes to sell the land, the sales price will reflect the value of the investment made.

An active land market is also important to the millions of land share holders who will become real land owners if the mass land share privatization recommended above is implemented. While these people can make money by farming their land, most are likely to allocate their land to private farmers and large agricultural enterprises in order to generate income. Allo-

[4] The Russian Government's 2004-2006 privatization plan envisioned the privatization of agricultural enterprises still owned by the state, but did not state clearly to what extent the underlying land would also be privatized (Privatization Plan, 2003).

cating land by leasing or selling it on a free market is the best method of transfer for the owner, for the land is offered to a wide universe of lessees and buyers who must compete with each other through price to gain access to the land.

Progress to date

To a large degree, Russian law provides for a land market and secure private rights to land. The security of rights to small land holdings (dachas, household plots, etc.), and the ability to buy and sell these holdings, were well-established in the law over 10 years ago, and have been further bolstered since then.[5] The 2001 Land Code (art. 27) and the 2002 Law on Agricultural Land Transactions settled the question of private parties having the legal right to buy and sell commercial agricultural land. The right to lease land is also established in the law. Agricultural land mortgage, the key to financing land purchases in many market economies, is now legal in Russia after amendments to the Law on Mortgage (Law on Mortgage Amendments, 2004). Finally, inheritance is not a market transaction strictly speaking, but as the second most prevalent method of land transfer (after leasing) is very important nonetheless. Part III of Russia's Civil Code (arts. 1110-1185) provides adequate inheritance rules.

Reforms still needed

Eliminating the state's authority to terminate private rights to unused land. The Law on Agricultural Land Transactions provides that, if a portion of a commonly owned agricultural land parcel has not been used for a two-year period, Russia's provincial governments (or municipalities if authorized) can terminate the ownership right to the unused portion, with no clear provision made for compensating the owner (art. 13(3)). Presumably the reasoning behind this provision is to enable the state to sell the land through a tender process to another private party who will use it (art. 10).

Giving the state the power to take a person's land away because it is not being used represents a gross violation of ownership principles. Non-use of land should rarely, if ever, be a basis for state action, and certainly not in a society such as Russia, which has no significant population pressure on land (the non-use question is distinct from the question of active use of land in

---

[5] See Law on Household Plot Sales (1992); Land Code (2001), arts. 27(1), 27(7); Civil Code Part I (1994), art. 260; Law on Household Plots (2003), art. 4(6). The Law on Agricultural Land Transactions (2002) does not apply to household plots, see art. 1(1).

violation of land use norms). This reason alone justifies eliminating the provision from the law.

The state's authority to terminate private rights to unused land is unsound for several other reasons as well. It short-circuits the development of the rural land market by injecting the state into land use decisions that are properly the domain of the land owners. It penalizes land share owners even though land shares by their nature are very hard to actively use because they do not correspond to a particular parcel of land. It assumes that the state can facilitate economic use of the land where a private party cannot, although as demonstrated throughout the capitalist world it is private parties, not the state, that are best equipped to deliver an economic result. The provision also opens the door for land to be taken by overzealous local officials or corrupt bureaucrats for transfer to bribe-paying corporate farms.

The likely outcome of this provision is that hundreds of thousands of mostly poor people will lose the land rights given to them during the 1990's land reform, and will have no compensation to show for it.

For all of these reasons the provision on terminating rights to unused commonly owned land parcels should be eliminated from the law. Rather than expending energy on a campaign to terminate private land rights, the state instead should work to make these rights real by converting land shares into physical land parcels as recommended above.

Eliminating the preferential right of provincial governments to purchase private land. The Law on Agricultural Land Transactions gives Russia's provincial governments a priority right to buy private land plots and land shares offered for sale on the market (art. 8).[6] The procedure for using this right makes it difficult for a seller of land to obtain the highest price possible. The seller must first offer the land to the provincial government for a specific price. If the provincial government declines to buy the land, then the seller can offer the land on the open market, but only at a price not lower than that offered to the provincial government.

A land seller should be entitled to receive the highest price possible, and the only way to determine this price is to offer the land for sale on the open market. The law as it stands will result in sellers either receiving less than the market price (because they underestimate the market price when offering the land to the provincial government), or having to re-offer the land to provincial governments if it turns out that the market price is lower than the price offered to the provincial government in the first place. These are major obstacles to an orderly, efficient land sale process, and the harm they will cause outweighs any legitimate reason to give the provincial government a priority purchase right, if any exist.

---

[6] The provincial government can transfer this priority right to local governments.

For these reasons, the priority right of provincial governments to buy private land plots and land shares should be eliminated from the law. Eliminating this right will not stop provincial governments from acquiring land, but they will have to do so through the market like everyone else.

Also on this topic, provincial governments can challenge a private land sale for violating the preferential-purchase procedure within one year from the date the sale is registered. This provision has the potential to cause significant economic damage as buyers, fearful of challenges to transactions, refrain from investing in their new land until the one-year period has lapsed. This provision should be eliminated from the law as well.

Eliminating the preferential right of lessees to renew leases. The Law on Agricultural Land Transactions gives the lessee of land who has performed his obligations under the lease agreement the preferential right to sign a new lease agreement when the existing lease expires (art. 9(5)). It is the fundamental right of a landowner—the lessor—to control how his land is used, who uses it, and for how long. This right should be narrowed only to serve some higher important purpose. Granting preferential lease renewal rights to lessees does not meet this test for the following reasons.

First, while other countries have laws that grant lessees preferential renewal rights, the policy reason behind these laws reflects circumstances that are the reverse of those in Russia: namely, a weaker lessee with little bargaining power who leases land from a powerful land owner. By contrast, in Russia the land owners are typically low-income rural workers and pensioners, while the lessees typically are much more powerful large agricultural enterprises. Thus, lessees in Russia do not need special legal protection in their relationships with land owners. On the contrary, if any party requires protection it is the land owners.

Second, supporters of preferential lease renewal rights justify them on the basis of production efficiency. If the lessees know they can renew their leases, so the argument goes, they will work harder and invest in the land. But Russia's large-scale agriculture leases most of the land and remains inefficient,[7] so to give them preferential lease renewal rights is to reward their inefficiency and thus stifle the improvement of agriculture.

Third, preferential rights are likely to reduce the interest of other potential lessees in the leased land parcels. Potential lessees must make a significant effort to gather information about land that may be available for lease

---

[7] Low grain yields in Russia demonstrate this point. For the three-year period 2001-2003 Russia produced 1.9 tons of grain per hectare. By comparison, the ex-Soviet Baltic countries produced 2.4 tons (26% more), northerly Canada produced 2.5 tons (32% more), neighboring Poland produced 3.1 tons (63% more), far-north Finland produced 3.2 tons (68% more), Europe as a whole produced 3.4 tons (79% more), and the United States produced 5.8 tons (305% more) (FAO, 2004).

and then negotiate lease contracts. But entrepreneurial family farmers and agricultural enterprises may not be willing to make this effort since they know that existing lessees have the right to renew. As a result demand for land will be stifled artificially, thus suppressing lease rates, reducing the movement of land to more efficient lessees, and decreasing the income to the land owners.

Fourth, the preferential lease right in the law has no time limit, therefore once a lessee concludes a lease he can use the land into perpetuity as long as he complies with the lease contract conditions. The land owner loses effective control of his land, and with it much of his own incentive to invest or think creatively about how to best use his land.

For these reasons, the provision granting lessees the preferential right to conclude new lease agreements should be reversed, instead prohibiting the inclusion of preferential rights in land lease agreements. If prohibiting preferential rights is not possible, then at a minimum preferential rights should be excluded from the law and left to negotiation between the parties.

Elimination or reform of trust management of land. As part of the effort to bring the agricultural land share into compliance with civil law property norms, the Law on Agricultural Land Transactions states that any land share lease agreements not brought into conformity with the Civil Code by January 2007 will be governed by rules on trust management (art. 16(2)). The law also provides that common owners of a land parcel may, at their discretion, transfer their share into trust management (art. 12(1)). These legal rules effectively prohibited the long-standing practice of land share owners leasing their land shares to agricultural producers.

The goal of bringing land shares into compliance with civil law norms is acceptable in principle, but using the trust management rules formulated in the Civil Code (see arts. 209(4), 1012-1026) for this purpose is fraught with risks to the land share owner and the common owners of a land parcel. (Hereinafter in this section these parties are referred to as "land owners," and land shares and land parcels are referred to as "land rights.") Trust management is a relatively untested concept in Russian law, and how it will work in practice if applied to land rights is difficult to predict. Furthermore, trust management is a rather difficult and sophisticated concept to understand, unlike a sale or lease, and one that interjects an additional party into the transaction. This difficulty creates opportunities to cheat land owners out of rightful income from their land rights, or potentially out of their land ownership rights altogether.

Specific problems with the trust management concept as applied to land rights are described below:

- Land in trust management is to be managed for the benefit of the land owner (or designated beneficiary) by a trustee (Civil Code Part I (1994), art. 1012(1)). It is difficult to imagine who is fit to serve as a trustee for an owner of agricultural land. Neither the agricultural enterprise using the land nor its managers are fit, because they have a major conflict of interest with the land owner. Their interest as producers is to utilize the necessary means of production—in this case land—for the lowest possible price, while their responsibility as trustees would be to maximize income for the land owner.
- A trustee must be compensated for his services and necessary expenses (art. 1023). Would this compensation consume a significant portion of the proceeds from the land rights under trust management? The answer is likely to be yes, thus calling into question the benefits that trust management would deliver to a land owner.
- Even though a land owner can transfer his land rights to trust management without agreement from the other common owners of the land parcel, the parcel itself is still subject to common management. In other words, the trustee does not have more freedom of action than did the land owner in deciding how to use the land. And yet, one of the stated justifications for introducing trust management has been to retain some of the freedom of action that was a characteristic of land share leasing.
- Trust management could lead to land owners losing their land rights. While transfer of land into trust management is not a transfer of ownership (arts. 209(4), 1012(1)), a trustee may sell or otherwise transfer the land parcel as part of his trust management authority (art. 1012(2), 1020(1)). While this transfer authority is limited to situations foreseen in the trust management contract concluded between the land owner and the trustee (art. 1020(1)), most land owners are not sophisticated with regard to contract negotiation and are unlikely to fully comprehend the contract terms. Clearly they would not expect to lose their land in a trust arrangement. In addition, while a transfer, especially a transfer for inadequate compensation, would be a clear violation of the trustee's fiduciary duty to the land owner, the aggrieved owner will probably not be able to successfully pursue the violation and receive an appropriate remedy.

The best solution to the potential problems caused by trust management is to eliminate it as an option for land owners. Returning to the system of land share leasing that has functioned decently for a decade is preferable to the flawed trust-management approach.

If trust management is not eliminated as an option for landowners, a variety of incremental reforms are needed to minimize the potential problems of the trust management regime and to preserve the land owners' long-term rights along with their flexibility to use the land in new ways as future opportunities arise. First, the law should prohibit agricultural enterprises, as well as managers and administrative personnel of such enterprises (chief accountant, chief agronomist, etc.), from serving as trustees over agricultural land. Second, the law should prohibit permanent transfers of agricultural land under trust management and restrict leases of such land by trustees to no longer than 10 years in length. Also, the law should automatically void trust management contracts that violate these provisions.

Improving the registration system. Russia adopted a land registration law in 1997 and put a network of registration chambers into place over the following three years (United Nations, 2003). Unfortunately the registration system is not working well with regard to agricultural land. Most agricultural land has not been initially registered following its transfer from state to private ownership, costs are high, and the bureaucracy is stifling. Because of these factors, transactions either are not being concluded or are taking place outside of the registration system (see Chapter 5 for a more extensive discussion of problems with registration).

A variety of measures would improve the functioning of the registration system. First, the state should ensure that all land rights transferred from state to private ownership are registered. This especially pertains to the desirability of physically demarcating and registering agricultural land shares as proposed above. Second, the high cost of preparing survey and other technical documentation needed for registration should be rigorously examined. Third, the bureaucratic process required for registration should be significantly shortened and simplified. Currently it can take several work days, spread over several months, for a person to successfully register his rights to land (Pulin, 2004).

Completing the Law and Policy Framework for Farm Restructuring As It Relates to Land

The issue of farm restructuring is integrally related to the completion of land reform, since the farm types and sizes resulting from restructuring play an important role in determining who will control the land and how it will be used. The reverse is also true to a significant degree: those who control land will influence the types and sizes of farms that are formed.

The state's ability to command effective farm restructuring through the law is limited. This is shown historically. In the early 1990s a set of presidential decrees and government resolutions ordered the conversion of state and collective farms into private agricultural enterprises. The resulting partnerships, joint-stock companies, agricultural cooperatives, and other enterprise forms represented largely cosmetic reorganization rather than real reform of management and operations. During the same time period, the state's effort to facilitate creation of family-style peasant farms foundered once large-scale financial support from the state was no longer available. This support primarily took the form of subsidized credit (see Chapter 3; subsidized credit was even more attractive in the early 1990s because of high inflation, which resulted in repayment of the credit with devalued rubles).

Farm managers continue to have a relatively powerful position in the post-Soviet countryside. These managers have persuasive influence over the rural population, are more knowledgeable and experienced business people than the population at large, and frequently control services and facilities upon which the population depends. By contrast, most of the local people have low incomes, and do not have sufficient knowledge and experience to fully protect their economic interests and effectively operate in a market environment. But these local people now own much of the land used by the enterprises under the managers' control. As such, the economic interests of these two groups will often diverge, creating a natural incentive for farm managers to use their relative power to degrade the rights of new landowners.

The limited ability of the state to order meaningful restructuring, and the oft-divergent interests of farm managers and landowners, lead to a two-part paradigm for farm restructuring as it relates to land reform. First, law and policy should seek to minimize the obstacles to the creation and operation of all types of agricultural operations, ranging from household plots to peasant farms to agricultural cooperatives and other types of large-scale corporate farms. Second, farm restructuring law and policy should not provide for inequitable or deceptive practices which result in the rights of new landowners being significantly impinged upon or lost outright. This may require restrictions on some types of transactions until new land owners are ready to make knowledgeable and informed decisions about their land.

Progress to date

Russia's progress on farm restructuring has been disappointing overall. On the bright side the collective and state farms have been converted into private agricultural enterprises, some smaller corporate and peasant farms have developed, and the system of small household plots gained an increasingly prominent role as an agricultural producer.

Unfortunately, though, large-scale producers, the effective successors to the Soviet collective and state farms, still use 80% of the agricultural land, and are the main recipient of subsidies from the government (see Chapter 2). Some of these enterprises are profitable, but overall world experience suggests that their gargantuan scale and unwieldy management make them inferior to smaller-scale enterprises as agricultural producers.[8]

Reforms still needed

Changing the law can do little to alter the fact that large-scale producers still use 80% of Russia's agricultural land, but some modest changes can help protect the rights of landowners whose land is used by these producers.

Eliminating contribution of land rights to charter capital of enterprises. Russia's large agricultural enterprises are frequently incorporated as joint-stock companies (open and closed) and limited-liability companies or partnerships (see Chapter 2). The legislation describing and regulating these forms of legal entity provides for contributions of money and property into their equity capital.[9] Land-specific legislation authorizes contribution of commonly held ownership rights to land and land lease rights to the equity capital of a legal entity.[10]

As with trust management discussed above, contribution of land rights to the equity capital of agricultural enterprises is not likely to provide benefits for the land owner. Contribution requires giving up ownership of the land in exchange for an ownership (or partnership) interest in the enterprise, an interest that will theoretically return dividends or profits to the contributing land owner. But the reality in rural Russia is that enterprise management holds the power and controls information, thus the contributing land owner will rarely be able to verify independently that fairly measured dividends or profits are being paid. Contribution is a way for enter-

---

[8] Large-scale agriculture is usually justified on the basis of purported economies of scale. However, a large body of literature has demonstrated that many agricultural activities do not exhibit true economies of scale in production. In addition, several studies show an inverse relationship between farm size and productivity for all but the smallest farm size classes. Finally, when other markets function reasonably well, optimal farm sizes often do not exceed the scale at which family labor is fully occupied (using seasonal hired labor for specific tasks). See Deininger (2003), p. 83.

[9] Law on Joint Stock Companies (1995), art. 25; and Law on Limited Liability Companies (1998), art. 15.

[10] See Law on Agricultural Land Transactions (2002), art. 12(1), and Land Code (2001), art. 22(5).

prise managers to gain permanent control of land at low cost, at the expense of ordinary rural people.

And these people have a much better option. They can lease their land rights to the agricultural enterprise and receive rent in return. They will have a contract specifying the rent, and thus can tell if they are being paid the agreed amount. In the vast majority of situations, the land owner will receive more income by leasing the land to the enterprise than by contributing it to the enterprise's equity capital. Moreover, these people will retain their land ownership rights, and thus remain able to take advantage of future opportunities to use their land profitably.

For all of these reasons, the ability to contribute commonly held rights to land to the equity capital of enterprises should be eliminated from the law. These reasons apply to contribution of lease rights as well, thus the option to contribute should also be eliminated from the law.

Reforming contribution of land rights to the share fund of agricultural production cooperatives. In addition to joint-stock companies and limited-liability companies, many of Russia's large agricultural enterprises are organized as agricultural production cooperatives, though this number is declining (see Chapter 2). This organizational form is defined and regulated by the Law on Agricultural Cooperation (1995). Cooperative members make contributions of property, including land, to the cooperative's share fund—the equivalent of equity capital in other corporate forms (arts. 1, 11(1)(6), 14).

For the reasons discussed above regarding contribution of land rights to the equity capital of agricultural enterprises, contribution of land rights to the share capital of agricultural production cooperatives brings with it many potential problems for the contributor. These include inadequate profit sharing and possible permanent loss of land rights. To minimize these problems, the cooperative law requires three amendments:

- Land rights contributed to the cooperative's share fund cannot be designated as part of the cooperative's indivisible property fund (arts. 1, 11(1)(8)).
- Land rights contributed to the cooperative's share fund cannot be sold (art. 6).
- When a cooperative member withdraws from the cooperative, he should receive at least as much land in kind as he contributed to the share fund when he joined the cooperative (art. 13(3)).

Conclusion

Land reform has been an ongoing process in Russia since 1991. Resolving the policy and legal issues discussed in this chapter would largely complete needed state efforts at land reform and represent a major achievement. Millions of ordinary people would become landowners, and the market would be a functioning mechanism for buying, selling, and leasing land for real value.

The most important reform still needed is the mass conversion of land shares into physical land parcels, which will help to improve farming efficiency and deliver economic benefits directly to average rural households. Improving the registration system, and eliminating the state's authority to terminate ownership rights to unused, commonly owned agricultural land parcels are other changes that will lead to benefits for Russia and its people.

References

Deininger, K. (2003). *Land Policies for Growth and Poverty Reduction.* A World Bank Policy Research Report. World Bank and Oxford University Press.

FAO (2004). Production Yearbook 2003.

Privatization Plan (2003). "Forecast Plan (Program) on Privatization of Federal Property for 2004 and the Basic Directions for Privatization of Federal Property through 2006." Approved by Order of the Government of the Russian Federation No. 1165-r (15 August 2003).

Prosterman, R., Rolfes, L., and Duncan, J. (1999). *A Vision for Agricultural Land Reform in Russia.* Rural Development Institute Reports on Foreign Aid and Development #100. Seattle, WA.

Pulin, A. (2004). Center for Land-Reform Support of Vladimir Oblast, personal correspondence, March 5.

Rolfes, L. (2002). "Making the Legal Basis for Private Land Rights Operational and Effective." Paper presented at the World Bank Regional Workshop on Land Issues in Central and Eastern Europe, Budapest, Hungary, April 3-6.

Rolfes, L. (2003). *The Impact of Land Titling in Ukraine: An Examination of the Results from an 800 Person Random Sample Survey.* Rural Development Institute Reports on Foreign Aid and Development #119. Seattle, WA.

United Nations (2003). *Land Administration Review: Russian Federation.* United Nations Economic Commission for Europe, Committee on Human Settlements, Working Party on Land Administration, reports HBP/2003/7-HBP/WP.7/2003/7 [http://www.unece.org > Human Settlement > Working Party on Land Administration > Workshops and Operational Activities].

Chapter 7
Rural and Agricultural Labor Markets

Vladimir Bogdanovskii

The institutional reforms of the 1990s have produced significant changes in the level and structure of rural employment. Gone is the colossus of collective and state farms with their 8 million workers that traditionally dominated the rural labor scene. The individual sector—independent family farms and household plots—has emerged as a strong alternative employer drawing labor from the shrinking corporate farm sector. Yet agricultural employment has demonstrated considerable stickiness compared with other factors of production. This is apparently attributable to insufficiently developed non-farm employment opportunities in rural areas, shrinking urban demand for labor, housing difficulties that limit rural-to-urban migration, and inadequate training of rural labor for alternative occupations. Contrary to the situation in some transition countries, such as Estonia, the Czech Republic, Hungary, and most strikingly former East Germany, the downward adjustment of agricultural employment in response to decreasing agricultural production has been limited. Instead of swelling ranks of rural unemployed, we are mainly witnessing adjustment through reduction of working time and reduction of wages, which allow relatively large numbers of rural people to remain nominally employed. The decreasing contribution of wages to rural incomes is offset by growing reliance on food production from the household plot.

In this chapter we use national statistics and the results of several rural surveys (including the 2003 BASIS survey) to describe and analyze the ongoing changes in rural and agricultural employment during Russia's transition to a market economy. We distinguish between rural and agricultural employment for two conventional reasons: first, although agriculture is the dominant employer in rural areas, the rural population is actually engaged in a wide variety of occupations; second, the contingent of employed in agriculture includes both rural and urban residents, although it is the rural people that constitute a large majority. The chapter is according divided into two parts, which are separately devoted to the non-overlapping rural workforce and agricultural workforce. The chapter starts with a discussion of the dynamics of the rural population, including natural mobility and cross-country migration. It then moves on to issues of rural employment, including a discussion of rural unemployed and a comparison to the urban situation. Since nearly 40% of the rural employed work in agriculture, we then proceed to examine agricultural employment in more detail, including the breakdown into corporate and individual farms. As part of our discussion of the agricultural workforce, we deal with the quality of human capital, the length of the working week, agricultural wages, and finally productivity of labor in agriculture.

Rural Workforce

Agriculture is the main employer in rural areas, although its role is declining over time. The rural potential for agricultural employment is also shrinking, as the population declines and life expectancy decreases.

Population Growth

Russia's rural population is just below 39 million in a total population of 144 million, or 27%. The total population has shown a slight but steady decline since 1992, which is primarily due to the decline of the urban population by about 3% during the last decade (Table 7.1). The rural population, on the other hand, showed a slight increase up to 1995-1996 (by less than 3% since 1989), and then gradually dropped back to the 1989 level. As a result of these slightly divergent trends between town and village, the share of rural population on the whole has remained stable, nudging up by just half a percentage point from 26.5% in 1989-1990 to 27% in the years since 1994.

Although the rural population has remained fairly stable, the birth rate has been steadily declining and the death rate increasing since the late 1980s. The rural birth rate dropped from 16.4 per thousand to 9.9 per thousand between 1989-2001, while the death rate rose from 12.7 per thousand to 17.2 per thousand. Since 1992 the death rate has consistently exceeded the birth rate, leading to a sharply negative natural increase of the rural population (Figure 7.1; a similar pattern is observed in urban areas).

In recent years, the rural natural increase has been between −6 and −7 per thousand. It is significant that a negative natural increase is a very recent phenomenon in Russia. During the 1980s the births exceeded the deaths by a healthy margin, producing a natural increase of 4-5 per thousand in rural areas, whereas two decades earlier, during the 1960s, the rural natural increase was around 10 per thousand. The switchover occurred in 1992, in the early years of transition, although the birth rate began to exhibit a downward trend somewhat earlier. Although decreasing birth rates and, surprisingly, increasing death rates appear to be part of a long-term trend, there is no escaping the conclusion that the switch to negative natural increase may have been precipitated by the overall economic and social stresses of the early transition period.

Table 7.1. Russia's population 1989-2002
A. Yearly average population, millions

Year	Total population	Urban	Rural	Share of rural population, %
1989	147.33	108.34	38.99	26.5
1990	147.91	109.05	38.86	26.3
1991	148.24	***109.27***	38.97	26.3
1992	***148.31***	108.83	39.48	26.6
1993	148.15	108.24	39.91	26.9
1994	147.97	107.95	***40.02***	27.0
1995	147.77	107.78	39.99	27.1
1996	147.37	107.51	39.86	27.0
1997	146.94	107.21	39.73	27.0
1998	146.53	106.97	39.56	27/0
1999	145.94	106.48	39.46	27.0
2000	145.19	105.85	39.34	27.1
2001	144.39	105.34	39.05	27.0
2002	143.53	104.82	38.71	27.0

B. Population in percent of 1989

Year	Total population	Urban	Rural
1989	100.0	100.0	100.0
1990	100.4	100.7	99.7
1991	100.6	100.9	99.9
1992	100.7	100.5	101.3
1993	100.6	99.9	102.4
1994	100.4	99.6	102.6
1995	100.3	99.5	102.6
1996	100.0	99.2	102.2
1997	99.7	99.0	101.9
1998	99.5	98.7	101.5
1999	99.1	98.3	101.2
2000	98.5	97.7	100.9
2001	98.0	97.2	100.2
2002	97.4	96.8	99.3

Source: Statistical Yearbook 2003, pp. 87-90.

Fig. 7.1. Natural increase of the rural population 1989-2001

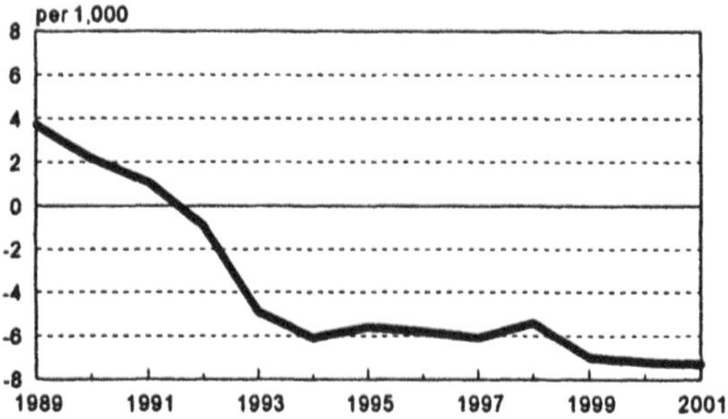

The negative natural increase has been generally offset by in-migration (and residually also by reclassification of urban into rural areas). The observed stability of the rural population during the last decade is thus the net outcome of these two opposing forces—negative natural increase and positive migration. Rural in-migration (Table 7.2) has been mainly driven by new arrivals from outside Russia—a phenomenon that emerged after the dissolution of the Soviet Union, when Russian minorities began leaving their established places of work and residence in other former Soviet republics. Migration within Russia continued to follow the traditional rural-to-urban channel (with the exception of only one year, 1992, which was characterized by pronounced urban-to-rural migration).

During the 12-year period 1990-2001, rural in-migration from former Soviet republics (CIS countries and the Baltic states) averaged 150,000 per year, whereas rural-to-urban migration within Russia averaged 40,000 per year (Table 7.2). The net contribution of the migration component thus averaged 110,000 per year, or 1.3 million during 1990-2001. This is equal to almost 3.5% of the actual rural population in 2001. Without the substantial inflow of migrants from former Soviet republics, Russia's rural population would have dropped to 37.8 million in 2001 (instead of the actual figure of 39.1 million). Table 7.2 shows, however, that rural in-migration from former Soviet republics has probably exhausted itself: the net inflow decreased sharply after 1998, and in 2001 it did not even offset the rural-to-urban migration within Russia (not to mention the negative natural increase).

The rural areas held particular attraction to immigrants from former Soviet republics. Thus, of the 5.5 million people who returned to Russia from other CIS countries and the Baltic states during 1992-1998, 2 million or 36% were absorbed in rural areas. This is substantially higher than the share of the rural population, which runs at 27%. The migration intensity to rural areas was thus 7.2 immigrants per thousand rural people compared to only 4.6 per thousand urban people. Immigrants probably found it easier to get housing in rural areas; the option of receiving free land for a household plot in the village was apparently also an important consideration for immigrants facing absorption difficulties.

Table 7.2. Rural migration flows within Russia and from former Soviet republics (thousands)

Year	Within Russia	From CIS and Baltics	Net rural in-migration
1990	-40	134	94
1991	-86	80	-6
1992	201	198	399
1993	20	225	245
1994	-66	324	258
1995	-66	200	134
1996	-62	142	80
1997	-92	160	68
1998	-72	128	56
1999	-9	74	65
2000	-68	79	11
2001	-91	17	-74
Average , '000 per year	-35.92	146.75	110.83
Cumulative, '000	-431	1761	1330

Source: Demographic Yearbook of Russia 2000, pp. 317-319; Demographic Yearbook of Russia 2002, pp. 318-320. The migration numbers do not match arithmetically the numbers in Table 7.1 because they are taken from residential registration and border control records.

Working-Age Population

The traditional Soviet definition of "working age" covered individuals from the age of 16 to 59 for men and 54 for women, but the alternative definition of working age as the 15-72 age group is becoming more widely used in Russian statistics. This alternative definition is clearly more appropriate for rural areas, where people start farming on the household plot at an earlier age and continue to farm their private land even after retiring from paid employment at the age of 59 (or 54). We generally follow the definition of working age as the 15-72 group, which is larger by 6-7 million people than the 16-59(54) group. Thus, the working-age rural population in the 15-72 group was 28.3 million in January 2003, compared with 22.1 million in the 16-59(54) group.

Table 7.3 shows that the potential rural workforce is around 70% of the rural population (it has actually increased over time from 68% in 1992 to 73% in 2002). The table also shows the dependency ratio for the rural population, which is calculated as the ratio of the number of people younger than 15 and older than 72 to the number of working-age people in the 15-72 group. While the dependency ratio is around 400 per 1000, it shows a definite declining trend over time. Both trends—the declining dependency ratio and the increasing share of the working-age group—are due in part to the declining birth rate, which steadily reduces the number of rural children, and in part to the increasing death rate, which reduces the number of rural people above 72 years of age.

Table 7.3. Rural population of working age 1992-2002

Year	Working-age population (15-72), millions	Dependency ratio (per 1000)	Rural population, millions	Share of working-age population, %
1992	26.94	465	39.48	68.2
1993	26.83	487	39.91	67.2
1994	27.39	461	40.02	68.4
1995	27.83	437	39.99	69.6
1996	27.82	433	39.86	69.8
1997	27.81	357	39.73	70.0
1998	27.98	414	39.56	70.7
1999	27.98	410	39.46	70.9
2000	28.07	401	39.34	71.4
2001	28.24	383	39.05	72.3
2002	28.43	362	38.71	73.4

Source: Author's calculations based on *Ekonomicheskaya aktivnost' naseleniya Rossii 2002*, Goskomstat, Moscow (2002), pp. 13, 16; *Obledovaniya naseleniya po problemam zanyatosti 2003*, Goskomstat, Moscow (2003), pp. 30, 36.

Table 7.4. Life expectancy at birth (years)

	All population		Urban		Rural		OECD countries	
	Males	Females	Males	Females	Males	Females	Males	Females
1990	63.79	74.27	64.39	74.35	62.00	73.89	71.7	78.3
1991	63.46	74.27	64.06	74.33	61.70	73.87		
1992	62.02	73.75	62.48	73.80	60.67	73.45		
1993	58.91	71.88	59.25	71.97	57.94	71.51		
1994	57.59	71.18	57.88	71.29	56.75	70.82		
1995	58.27	71.70	58.48	71.76	57.70	71.50	72.7	79.1
1996	59.75	72.49	60.22	72.70	58.44	71.85		
1997	60.75	72.89	61.43	73.10	58.94	72.29		
1998	61.30	72.93	61.82	73.13	59.90	72.32		
1999	59.93	72.38	60.41	72.65	58.63	71.55		
2000	59.00	72.20	59.38	72.40	57.99	71.55	74.3	80.2
2001	58.96	72.34	59.34	72.50	57.90	71.79	74.5	80.4
2002	58.47	72.04	58.99	72.26	57.07	71.31	74.6	80.5

Source: Statistical Yearbook of Russia 2003, p. 117; OECD countries (36) from www.oecd.org (>Statistics>Health> Frequently requested data>Table 1: Life expectancy).

Another trend that will ultimately affect the size of the working-age cohort is the steadily declining life expectancy at birth of the Russian rural population. The life expectancy at birth for Russian rural males today is 57 years, down from 62 in 1990; for rural females the life expectancy is 71, down from 74 in 1990 (Table 7.4). This is slightly lower than the urban life expectancy in Russia, but dramatically lower than in the OECD countries, where life expectancy has been steadily rising (and not falling!) over time, reaching 75 years for males and 80 years for females in 2002. The decline in life expectancy is probably attributable to the high levels of psycho-emotional, social, and economic stress that have persisted during the transition period since 1991. Deteriorating health services and increasing alcoholism and crime rate have contributed to this situation. While cardiovascular diseases remain the leading cause of death in Russia, nearly 38% of deaths are classified as "accidents," of which one-third are suicides, murders, and traumas. The government must devise policies that will counteract the declining life expectancy and thus help maintain the country's working-force potential.

Rural Employment

A cursory look at the rural employment time series as reported by Goskomstat (Table 7.5, column 1) shows that a fairly steep decline up to 1998 gave way to an impressive recovery since 1999, which brought rural employment in 2003 back to the level of 1993-1994 (16 million).

Table 7.5. Rural employment 1992-2003

Year	Rural employment as reported	Commercial household plots	Subsistence household plots	Old series (excl. commercial HH plots)	New series (incl. commercial HH plots)
	(1)	(2)	(3)	(4)	(5)
1992	17.02			17.02	
1993	16.51			16.51	
1994	15.61			15.61	
1995	15.25			15.25	
1996	14.84			14.84	
1997	13.76			13.76	
1998	14.31			14.31	
1999	***15.55***	1.69		13.86	15.55
2000	***15.93***	1.86		14.07	15.93
2001	***15.33***	1.30	5.91	14.03	15.33
2002	***15.57***	1.36	5.83	14.21	15.57
2003	***15.95***	1.78	5.82	14.17	15.95

The reported data, however, are a mixture of two entirely different time series: 1992-1998 and 1999-2003, a fact highlighted by the use of different fonts in column 1. Until 1998 the rural employed did not include individuals who reported the household plot as their main or only occupation (it did include, however, individuals for whom the household plot was a source of secondary activity). In 1999, Goskomstat began to include in rural employment statistics those individuals who reported *commercial* farming on the household plot as their main or only activity (column 2). Commercial farming is a term used to characterize household plots that sell some of their output to consumers, as opposed to subsistence plots, where the total output is consumed by the family. This change in statistical methodology immediately added more than 1.5 million people to rural employed and produced the illusory growth in rural employment after 1998.

Once the reported rural employment series for 1999-2003 is adjusted to the same basis as before 1998 by taking out the cohort of commercial household plots (column 4), the decline in rural employment between 1992-1998 is seen to change to relative stability (not an increase!) at the level of about 14 million between 1999-2003. The number of employed in commercial household plots also remained stable during this period (at about 1.5 million). The total number of rural employed after 1998 was thus constant at about 15.5-16 million (column 5), and it could not be compared to the period 1992-1998 because of the change in methodology. The two time series are shown separately in Figure 7.2.

Fig. 7.2. Rural employment with and without household plots

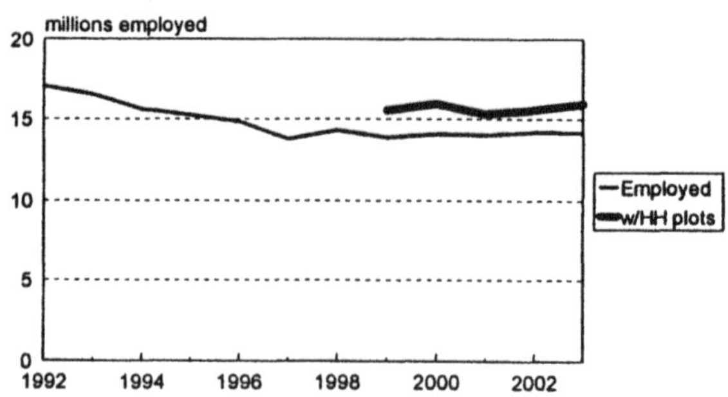

The inclusion of commercial household plots in rural employment statistics constituted a significant improvement in the reporting of rural employed, bringing the statistical time series closer to the real situation on the ground. Yet in addition to 1.5 million people whose main or sole occupation is a commercial household plot, there are close to 6 million rural people who report their main or sole occupation as *subsistence* farming on the household plot. These people work on their household plot, but the total output is consumed within the family and nothing is sold. These household plots do not qualify as commercial plots, and the corresponding cohort of rural people is not counted among the rural employed, although they definitely fall within the definition of employed in the Russian Employment Law and even in the Goskomstat employment survey guidelines. Thus, according to the Employment Law, "employment is a legal activity associated with satisfaction of personal or social needs that typically produces earnings or income" (ch. 1, article 1). This clearly covers even subsistence household plots. According to Goskomstat employment criteria, adults who worked (i.e., engaged in an income-generating activity) at least one hour during the week before the survey should be classified as employed, which also clearly covers work on subsistence household plots.

Cognizant of the actual role of household plots for the rural population and of the implications of these legal definitions, Goskomstat is gradually expanding the coverage of rural household farming. Thus, the information on commercial household plots began to be collected as of 1999, and the information on subsistence household plots is collected since 2001 as part of Goskomstat's quarterly employment surveys. The available numbers are presented in Table 7.5 column 3 and their impact on rural and agricultural employment is discussed below.

There is a slight negative correlation between rural employment and total rural population, but it is not statistically significant. Thus, the total rural population increased slightly during 1992-1998, while rural employment decreased during this period; conversely, rural population was declining (slightly) after 1998, while rural employment (without employment on the household plot) remained stable. Over the entire period 1992-2003, the coefficient of correlation between rural employment (without household plots) and rural population is -0.05 ($p = 0.88$).

The rate of rural employment (including those employed on commercial household plots) is around 55% of the economically active rural population in the 15-72 age group (Table 7.6). This is lower than for the urban population, where employment rates have been around 61% of the economically active population. This gap is certainly attributable to greater availability of conventional employment opportunities in urban areas and the tendency of the rural population to shift to "hidden" employment in subsistence farming from statistically recognized formal employment. The rural employment rates increase to over 90% if the nearly 6 million people engaged in subsistence farming on household plots (Table 7.5, column 3) are added to the economically active population and the rural employed. No such estimates are possible for the urban population, but here also there is a substantial stratum of "hidden" employment, so that the actual employment is probably around the same level for both urban and rural population.

Table 7.6 presents some additional comparative characteristics of rural and urban employed. We see that the rural workforce is more "masculine" than urban: 53% of the rural employed are men, compared with 51% in urban areas. A substantially larger proportion of the rural youth (the 15-20 age group) are employed: 3.8% of rural employment compared with 1.7% in urban areas. On the other hand, the rural employed have a lower educational attainment: the percentage of university graduates is 12% among the rural employed compared with 29% in urban areas. The last two factors are probably interconnected: today (as opposed to the Soviet period) the access to higher education is more restricted for the rural youth, primarily for economic and financial reasons, and as a result they start working earlier than their contemporaries in urban areas. The education gap will have negative long-term effects on the skills and preparedness of the rural workforce.

Table 7.6. Selected comparative characteristics of rural and urban employed, 2003 (percent)

	Rural	Urban
Gender structure of employed: males	53.1	50.6
Age structure of employed (percent of all employed):		
15-20	**3.8**	**1.7**
20-29	20.0	22.4
30-39	24.5	25.1
40-49	**32.9**	**29.8**
50-59	14.2	16.3
60-72	4.6	4.6
all employed	100.0	100.0
Share of employed by age (percent of each age group):		
15-20	**17.0**	**9.8**
20-29	63.9	68.5
30-39	76.0	84.3
40-49	77.8	85.4
50-59	63.8	71.7
60-72	14.6	17.7
average	54.9	61.1
Educational attainment of employed:		
Higher	12.1	28.6
Secondary	55.5	49.3
Primary	32.4	22.1

Source: *Obsledovanie naseleniya po problemam zanyatosti, Noyabr' 2003*, Moscow: Goskomstat (2004), pp. 56 (gender), 54, 58 (age structure), 62, 64 (share of employed by age), 72, 76 (education).

A partial response to the rural/urban education gap is probably manifested in the fairly pronounced shift from general to vocational education, which so far is occurring only at the primary school level (Figure 7.3). Thus, the share of employed with vocational primary education increased from 9% to 16% between 1999 and 2003, whereas the share of employed with general primary education dropped from 24% to 17% during the same period (the total percentage of employed with higher and secondary education remained virtually unchanged). The rural labor market under the new conditions apparently values vocational training more than basic general education.

Fig. 7.3. Changes in education structure of rural employed 1999-2003

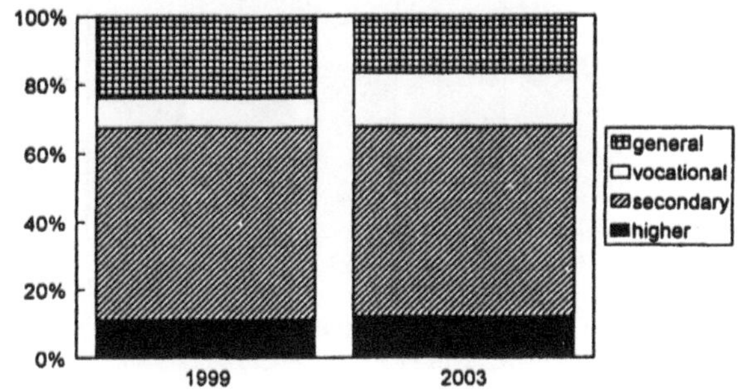

Structure of Rural Employment

Comparable data on the structure of rural employment by occupation are available only since 1999. Yet despite the myopic view these data provide, we observe fairly dramatic changes in the employment mix of the rural population between 1999 and 2003. The most notable change is the shift of rural employed from agriculture to other sectors—mainly to trade and consumer services (Figure 7.4). Thus, the share of agriculture in rural employment decreased from 46% in 1999 to 37% in 2003, while the share of trade and consumer services increased from 8% to 13% (Table 7.7). The remaining 4 percentage points lost by agriculture were absorbed in equal shares by the industrial sector (including manufacturing, construction, transport, and communication) and by the social services sector (health, education, rural administration, etc.).

In absolute numbers, agriculture lost 1.6 million employed, or more than one-fifth of its 1999 base (Table 7.7). Of these, 0.8 million shifted to trade and consumer services—the fastest growing sector that increased its workforce by 60%. Another 0.3 million shifted to the industrial sector and 0.2 million to social services. The remaining 0.3 million people (or 2% of the 1999 rural employed) simply dropped out of the workforce. This was due not to increasing unemployment, but primarily to increasing "disemployment," i.e., an increase in the number of inactive people in the working age group: the number of inactive people in the 15-72 age group increased by 0.7 million between 1999-2003, which was partially offset by a substantial decrease of 0.4 million in the number of unemployed to produce the observed net drop of 0.3 million in rural employment.

Fig. 7.4. Structure of rural employment 1999-2003

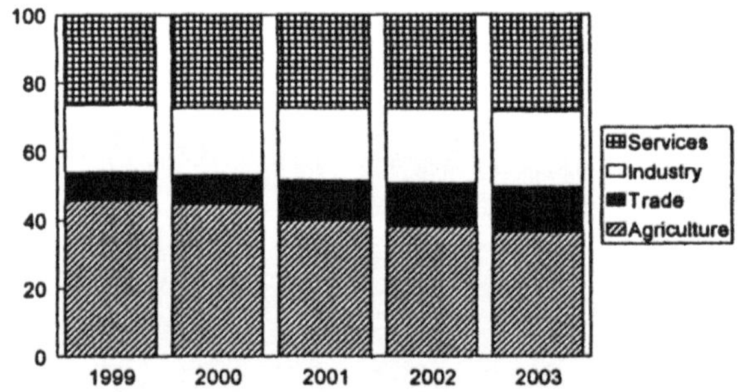

Fig. 7.5. Structure of rural employment 1999-2003
(adjusted for "household workers")

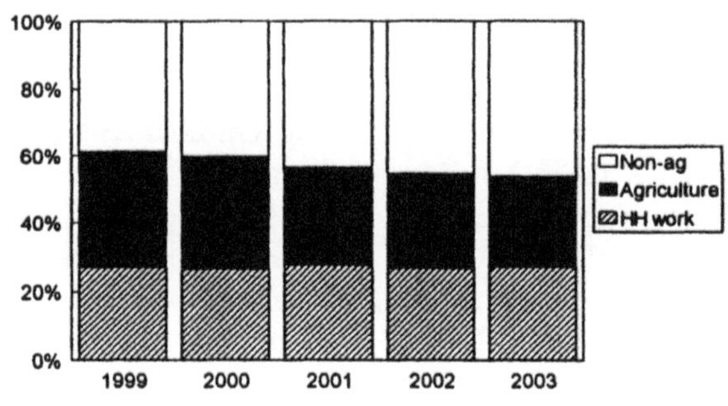

The rural employment statistics (15.6 million employed) do not include people for who "household work" (*domashnee khozyaistvo*) is the main or only occupation. This is the category of rural people who farm their household plot for pure subsistence needs and do not work anywhere else. These people are not even classified as self-employed by statistical organs, and they all fall in the group of inactive working-age population, although according to the Russian Employment Law they are clearly employed (see the previous section). Because of the importance of the household plot to rural family incomes, we have looked separately at this cohort of "household workers" and reconstructed the overall employment picture by transferring them from inactive population to active population employed in agriculture.

Table 7.8 summarizes the structure of rural population according to economically active (employed and unemployed) and economically inactive cohorts after this correction. The published statistical data are given in Table 7.8A, whereas Table 7.8B presents the results of the suggested adjustment, with pure-subsistence "household workers" moved to economically active population.

Table 7.7. Rural employment by sectors of the economy 1999-2003

	1999	2000	2001	2002	2003	Absolute change 1999-2003	2003 in percent of 1999
Number of employed in millions							
Total rural employed	15.89	16.16	15.25	15.9	15.57	-0.32	98.0
Agriculture	7.27	7.19	6.09	6.04	5.69	-1.58	78.3
Industrial sectors	3.17	3.19	3.24	3.48	3.46	0.29	109.1
Trade and consumer services	1.27	1.38	1.76	2	2.02	0.75	159.1
Social services	4.18	4.4	4.16	4.38	4.4	0.22	105.3
	1999	2000	2001	2002	2003		
Employment structure in percent							
Total rural employed	100.0	100.0	100.0	100.0	100.0		
Agriculture	45.8	44.5	39.9	38.0	36.5		
Industrial sectors	19.9	19.7	21.2	21.9	22.2		
Trade and consumer services	8.0	8.5	11.5	12.6	13.0		
Social services	26.3	27.2	27.3	27.5	28.3		

Source: author's calculations based on 2002 and 2003 employment surveys (*Obsledovanie naseleniya po problemam zanyatosti, Noyabr' 2002*, Moscow: Goskomstat (2003), pp. 110, 117; *Obsledovanie naseleniya po problemam zanyatosti, Noyabr' 2003*, Moscow: Goskomstat (2004), pp. 106, 111) and the 2003 labor statistical yearbook *Trud i zanyatost' v Rossii 2003*, Moscow: Goskomstat (2004), p. 187.

The "household workers" have numbered 5.8-5.9 million in recent years. This is 21% of the working-age population, and by reclassifying this cohort as economically active, we change the structure of the rural working age population to 82% active and 18% inactive (compared to 61% active and 39% inactive in the official statistics). Within the active rural population, the adjustment affects only agricultural employment, which is practically doubled by the inclusion of "household workers." The adjustment increases agricultural employment in 2003 from 5.7 million to 11.5 million, and its share rises from 36% to 54% of the employed.

Note, however, that the technical increase in agriculturally employed due to the inclusion of "household workers" does not correct the general trend of declining agricultural employment since 1999: the component of "household workers" has remained fairly constant, so that the adjusted number of agriculturally employed still declines by 1.6 million, as discussed previously.

The trend of decreasing agricultural employment and increasing non-agricultural occupations is again demonstrated in Figure 7.5. In this figure, the total workforce includes the "household workers," as shown in Table 7.8B. The percentages are therefore different from those in Figure 7.4, where "household workers" are excluded from the workforce.

Table 7.8. Rural population by occupation and age group (millions, beginning of year)

A. As reported in official statistics

	1999	2000	2001	2002	2003	Change 1999 to 2003
Rural population	39.46	39.45	39.22	38.87	38.55	-0.91
Below 15 and over 72	11.48	11.27	10.78	10.45	10.2	-1.28
Working age 15-72	27.98	28.18	28.44	28.42	28.35	0.37
Non-agricultural occupations	8.28	8.74	9.24	9.86	9.88	1.60
Agriculture	7.27	7.19	6.09	6.04	5.69	-1.58
Employed	15.55	15.92	15.34	15.9	15.57	0.02
Unemployed	2.22	1.86	1.84	1.76	1.82	-0.40
Economically active	17.77	17.78	17.18	17.66	17.39	-0.38
Economically inactive	10.21	10.4	11.26	10.76	10.96	0.75
Of which: "household workers"	5.8	5.8	5.91	5.83	5.82	0.02

B. After adjustment for "household workers"

	1999	2000	2001	2002	2003	Change 1999 to 2003
Rural population	39.46	39.45	39.22	38.87	38.55	-0.91
Below 15 and over 72	11.48	11.27	10.78	10.45	10.2	-1.28
Working age 15-72	27.98	28.18	28.44	28.42	28.35	0.37
Non-agricultural occupations	8.28	8.74	9.24	9.86	9.88	1.60
Agriculture	7.27	7.19	6.09	6.04	5.69	-1.58
"Household workers" (pure subsistence farming)	5.8	5.8	5.91	5.83	5.82	0.02
Employed (including "household workers")	21.35	21.72	21.25	21.73	21.39	0.04
Unemployed	2.22	1.86	1.84	1.76	1.82	-0.40
Active (including "household workers")	23.57	23.58	23.09	23.49	23.21	-0.36
Inactive (excluding "household workers")	4.41	4.6	5.35	4.93	5.14	0.73

Source: see Table 7.7.

Rural Unemployment

Russian unemployment statistics are based on two measures—total unemployment and registered unemployment. The two measures are radically different and describe essentially different concepts of unemployment.

Total unemployment is measured according to ILO guidelines on the basis of employment surveys regularly conducted by Goskomstat since 1992. According to these surveys, a person of working age (15-72) is classified as unemployed if in the week of the survey he or she satisfied the following three conditions simultaneously:

– did not work or did not engage in an income-generating activity;

– was actively looking for work by any means during the four weeks preceding the week of the survey;

– was ready and willing to start work.

Registered unemployment is based on registration at local labor exchanges (employment centers). According to current legislation registered unemployed are Russian citizens (in the 16-59 age group for men, 16-54 for women) who have no work and no labor income and register at the labor exchange at the place of their residence in an effort to find work. The status of registered unemployed is conferred by a special Ministry of Labor certificate issued at the local labor exchange. Only registered unemployed are entitled to receive unemployment benefits from the state.

Given the Ministry of Labor definition of registered unemployed as a person without labor income, it is not surprising that the prevailing policy at the local level is not to register as unemployed people who have no formal job but do have a household plot or a land share. The argument is that these people earn income from farming their own land and from leasing their land shares, and should not be eligible for unemployment benefits even if they do not have "proper" work. Thus, according to a survey of managers of 58 district-level labor exchanges in 18 provinces, an unemployment certificate is not issued if the person has a land share (26% of the districts), a household plot (4%), or rental income from land (43%) and property (30%).[1] It is ironic that subsistence farmers simply fall between two stools: they are registered neither as employed in Goskomstat employment surveys nor as unemployed in Ministry of Labor statistics. As a result, all these people (some 5-6 million of them) fall in the category of economically inactive population, although in practice they are engaged in active farming of their household plots.

---

[1] *Sostoyanie sotsial'no-trudovoi sfery sela i predlozheniya po regulirovaniyu*, Moscow: MinAg, VNIIESKh (2003), p. 34

Because of these differing definitions of unemployment, it should not surprise us that there is a wide gap between the total unemployment figures published by Goskomstat and the (much lower) registered unemployment figures reported by Ministry of Labor (Figure 7.6). The gap increased quite sharply after 1995, when the Employment Law was amended to restrict even further the eligibility for registration of unemployed and thus alleviate some of the budgetary pressures through reduction of unemployment benefits (these pressures were in part connected with the inability of Russian businesses to make timely contributions to the unemployment benefits fund). As a result, registered unemployment remained at the level of about 500,000 people between 1996 and 2002, while total unemployment according to Goskomstat data soared from 1.4 million in 1995-1996 to a peak of 2.2 million in 1999 and then subsided slightly to 1.8 million in 2002. The gap between the two unemployment measures thus broadened from about 700,000, or 50% of total rural unemployment, in 1995-1996 to 1.3 million, or 74% of total rural unemployment, in 2002.

Fig. 7.6. Total and registered unemployment rural population of working age (16-59)

Despite the fairly sharp increase in total unemployment over time, the rural unemployment rate since 2000 has ranged around 6% of the working-age population or 4.5% of the total rural population (see Table 7.8). These rates are by no means negligible, but they are far from the catastrophe that experts predicted in the wake of the economic collapse during the early years of transition. It seems that unemployment (and especially rural unemployment) in Russia is kept in check by two fairly unconventional mechanisms: reduction of effective working time per person and reduction of wages. Both these mechanisms make it possible to keep more people employed, although at an obvious cost of deteriorating standard of living. In rural areas, reduced working time and reduced wages in formal workplaces have contributed to gradual squeezing out of workers from formal employment to a kind of subsistence-oriented self-employment on their household plots (see the sections on *Working Time* and *Wages* below).

Agricultural Workforce

Agriculturally employed are individuals—mainly rural, but also urban—who report in Goskomstat employment surveys that agriculture is their main occupation. According to established Goskomstat criteria, agriculturally employed include (a) workers in farm enterprises and agricultural service organizations; (b) members and hired workers of private peasant farms; (c) people engaged in individual work activity in agriculture without forming a legal entity (i.e., self-employed physical persons); (d) individuals who farm their household plot with the intention of selling their output ("commercial household plots"). At this stage, the statistically reported cohort of agriculturally employed (like rurally employed) does not include individuals whose main or sole occupation is farming the household plot for subsistence purposes, without intention to sell.

Agricultural Employment

The statistical methodology suggests that agricultural employment is made up of three components: farm enterprises, household plots, and peasant farms (Table 7.9). Household plots and peasant farms constitute the individual sector of Russian agricultural, and alternatively agricultural employment can be viewed as consisting of two sectors: the corporate sector (farm enterprises) and the individual sector (household plots and peasant farms combined).

The majority of agriculturally employed are rural residents and are thus included in the agricultural component of rural employment (see the previous part *Rural Workforce*). Yet there is no one-to-one correspondence: some agriculturally employed (especially hired workers of agricultural service organizations) are in fact urban residents and are therefore not part of rural employment. The differences for the period 1999-2003 (when data on the structure of rural employment by occupation are available) clearly emerge from the comparison of Tables 7.10 and 7.8.

Agricultural employment slightly increased between 1990 and 1994 and then declined fairly sharply (by 2.6 million people, or 25% of the 1994 pea-level in 2002). The share of agriculture in total employment also dropped from a peak of 15% in 1994 to below 12% in 2002. During the early 1990s (definitely up to 1994-1995) agricultural employment exhibited a considerable measure of stickiness, generally remaining at the 1990 level while total employment dropped by more than 10%. After 1994, agricultural employment decreased, like the general trend in total employment, but much more rapidly, as if to compensate for the earlier stickiness. As a result, by 2002, agricultural employment had dropped to 80% of the 1990 level, compared to 87% for total employment in Russian economy. It seems that, in the early transition years, people (and especially immigrants from other former Soviet republics—see Table 7.2) were attracted to agriculture as a source of land and food that would help alleviate the unusual economic hardships of that time. This understandable response had worked itself out by 1995, and agricultural employment began to decline in parallel with the rest of the shrinking economy.

Table 7.9. Schema of agricultural employment statistics

Agricultural employment category	Main sources of primary data	Contingent covered by primary data	Main employment identification criterion	Registered employment status; available data
Corporate farms (farm enterprises)	Population employment surveys	Population in 15-72 age group regardless of occupation	Paid hired work (min. one hour during survey week)	Employed; historical time series
	Balance of labor resources	Workers in corporate farms	Inclusion in corporate-farm payroll	Ditto
	Consolidated annual report of corporate farms (form No.5-APK)	Ditto	Ditto	Ditto
	MinAg data on labor sufficiency of corporate farms	Ditto	Ditto	Ditto
Household plots	Population employment surveys	Population in 15-72 age group regardless of occupation	Production of farm commodities for sale	Employed; available since 1999
			Production of farm commodities for own consumption	Not employed (inactive); available since 2001
Peasant farms	Population labor surveys; survey of agricultural activity (peasant farms)	Population in 15-72 age group regardless of occupation	Work on family farm for family members; paid work for hired help	Employed; available only for 1998-1999

The decline in agricultural employment was clearly caused by the decline in the number of employed in the corporate sector (Table 7.10, Figure 7.7). The employment in farm enterprises dropped by 55% between 1990 and 2002 (much more than, for instance, in industrial enterprises, where employment dropped by 36% in this period).

Fig. 7.7. Total and agricultural employment 1990-2002

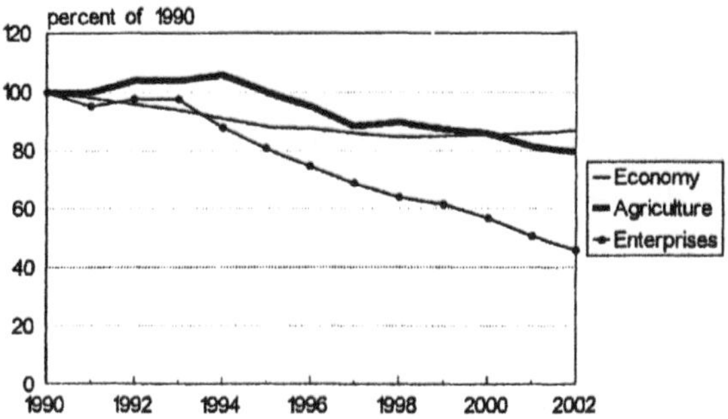

Table 7.10. Agricultural employment 1990-2002 (million full time equivalents)

Year	Agriculture	Corporate farms	Individual sector*	Total employed, % of 1990	Agriculture, % of 1990	Enterprises, % of 1990	Agriculture, % of total employed
1990	9.7	8.3	1.4	100.0	100.0	100.0	12.9
1991	9.7	7.9	1.8	98.0	100.0	95.2	13.1
1992	10.1	8.1	2.0	95.8	103.9	97.6	14.0
1993	10.1	8.1	2.0	94.0	103.9	97.6	14.3
1994	10.3	7.4	3.0	91.0	105.7	88.0	15.0
1995	9.7	6.7	3.0	88.2	100.0	80.7	14.6
1996	9.3	6.2	3.1	87.6	95.2	74.7	14.1
1997	8.6	5.7	2.9	85.9	88.3	68.7	13.3
1998	8.7	5.3	3.4	84.7	89.7	63.9	13.6
1999	8.5	5.1	3.4	85.0	87.3	61.4	13.3
2000	8.4	4.7	3.7	85.4	86.0	56.6	13.1
2001	7.9	4.2	3.7	85.9	81.4	50.6	12.2
2002	7.7	3.8	3.9	86.8	79.4	45.8	11.8

Sources: Statistical Yearbook of Russia 2001and 2003.

*Calculated as the difference between agricultural employment and employment in farm enterprises.

Since agricultural employment as a whole dropped by only 20%, the catastrophic decline of the number of employed in farm enterprises was partially compensated by a sharp increase of employment in the individual sector—household plots and peasant farms combined. While farm enterprises lost 4.5 million workers, the number of individually employed increased by

2.5 million—from 1.4 million in 1990 to 3.9 million in 2002. We are thus witnessing a significant shift of labor from corporate farms to individual employment. In 1990, farm enterprises were the dominant agricultural employer, accounting for 8.3 million workers out of total 9.7 million. In 2002, individual employment in household plots and peasant farms is practically equal to that in corporate farms, with each sector employing 3.8-3.9 million people (Figure 7.8; see Box 7.1 for details of estimation of labor in peasant farms).

Fig. 7.8. Agricultural employment 1990-2002

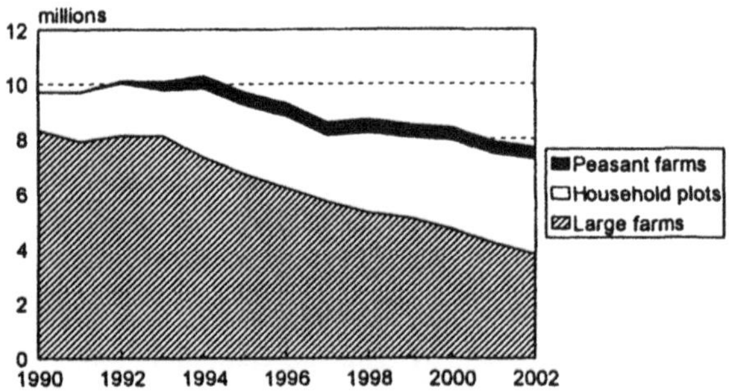

Box 7.1. How Many People Work in Peasant Farms?

The individual sector in agriculture consists of two components—household plots and peasant farms, but while detailed land and production data are published for each subsector separately, no disaggregated employment data exist. The statistical yearbooks for Russia provide information on the number of peasant farms, but there are no data on employment, i.e., the number of people working in peasant farms. The recent survey of peasant farms conducted by Goskomstat with professional assistance from USDA statistical experts shows that in 1998-1999 there were on average 4.3 workers per peasant farm, including 1.2 hired workers. These national results closely match the findings of the 2003 BASIS survey in three oblasts (Rostov, Nizhnii Novgorod, and Ivanovo), which gave an average of 3.1 permanent workers plus 1 seasonal worker per peasant farm. Hired workers primarily provide seasonal labor as their time input is between 1/3 and 1/4 of the time input of the family members (40 days annually compared with 140 days for family members).

Converting the workers in peasant farms to full-time equivalents based on 220 workdays annually, we obtain 2.2 FTE workers per farm. The total work-time input in peasant farms was 85-90 million man-days in 1998-1999, which is equivalent to about 400,000 workers. This is slightly over 10% of the total number of employed in individual agriculture in recent years (see Table 7.10).

The comparable averages obtained for three points in time (1992-1993, 1998-1999, and 2003) suggest that the 1998-1999 figures can be extended to the entire period of study in proportion to the actual number of peasant farms. It is in this way that we have estimated the number of employed in peasant farms for 1990-2002 shown in Figure 7.8.

Yet the individual sector did not absorb the entire slack created by the exit of labor from farm enterprises: 2 million people appear to have left agriculture altogether. They may have moved to other non-agricultural occupations or become inactive. Another possibility is that at least some of them simply dropped out from official statistics because they had moved to that blind area of individual employment characterizing subsistence-oriented household plots.

There are strong indications that the number of agriculturally employed in Table 7.10 (7.7 million or 13% of the total labor force) is understated. We have discussed the methodological problems in the employment surveys conducted since 1999: among people for whom the main or sole occupation is the household plot the employment surveys include only those who farm for "commercial" purposes, i.e., sell some of their output. The entire cohort of subsistence-oriented household plot operators—about 5.8 million people according to Table 7.8—remains outside the current definition of agriculturally employed.

Table 7.11. Household plot statistics 2003

	Total	Rural	Urban	Sole occupation		
				Total	Rural	Urban
All household plots	27.20	14.94	12.26	12.45	7.40	5.05
"Commercial" plots	3.47	2.99	0.48	1.86	1.58	0.28
Subsistence plots	23.73	11.95	11.78	10.59	5.82	4.77

Source: *Obsledovanie naseleniya po problemam zanyatosti, Noyabr' 2003*, Moscow: Goskomstat (2004).

Table 7.11 summarizes the available information on household plot operators (both rural and urban) in 2003. In total, 27.2 million people farm on household plots, but only 12.5 million (less than one-half) report this as their sole or main occupation and thus qualify to be treated as agriculturally employed. Among these 12.5 million household plot workers, 1.9 million engage in "commercial" farming and are presumably included in agricultural employment statistics. The remaining 10.6 million engage in subsistence farming, i.e., the entire output of their household plots is consumed by the family. It is remarkable that among the 1.9 million "commercial" household plot operators fully 85% are rural residents and only 15% are from urban areas; on the other hand, among the 10.6 million subsistence household plots, 55% are rural and fully 45% are urban (Figure 7.9). For urban people the household plot is thus mainly a supplementary source of food for the family, and it is primarily rural people who use household plots as a supplementary source of cash income from sales (in addition to food for family consumption).

Fig. 7.9. People with subsistence farming as sole employment

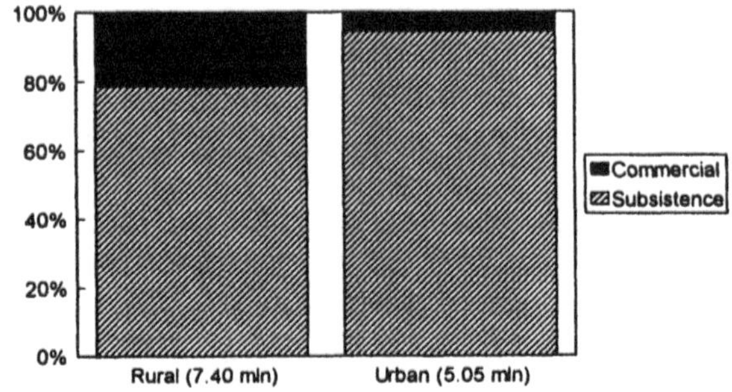

There are no statistical data on the actual work-time input of the 10.6 million subsistence farmers and as a result there are no estimates of the additional number of agriculturally employed that this cohort represents. The most conservative crude estimates suggest that agricultural employment should be increased by about 2.5 million employed to reflect "subsistence" farmers. Correcting for this cohort, who so far are not included in agricultural employment statistics, we raise agricultural employment from 7.7 million to about 10-10.5 million. This adjustment raises the share of the individual sector in agricultural employment to at least 60% (compared with 50% based on the official numbers in Table 7.10).

Factors Determining Changes in Agricultural Employment

It is reasonable to assume that agricultural labor should generally grow with the total population in the country and decrease as non-agricultural sectors of the economy grow faster than agriculture, thus creating alternative employment opportunities outside the agricultural sector. Using data on agricultural employment, population, and economic activity (GDP and GAO) for 74 Russian regions at two time points (1995 and 2000), we calculated the percentage changes from 1995 to 2000 and ran the regression

$$dAgEmp = dPopulation + d(GDP/GAO)$$

Here d(GDP/GAO) represents the nonagricultural GDP growth, calculated the percentage increase in GDP divided by the percentage increase in agricultural output. For instance, if GDP increased to 200% of the 1995 level and GAO increased to 150% of the 1995 level, then non-agricultural growth was 2.0/1.5=1.33, i.e., non-agricultural sectors grew 33% more than agriculture.

Regression results confirm that the change in agricultural employment is positively related to population growth and negatively related to non-agricultural GDP growth (Table 7.12). The coefficient of non-agricultural GDP growth has the correct sign (greater job opportunities outside agriculture reduce employment in agriculture), but unfortunately it is not statistically significant.

Table 7.12. Determinants of the changes in agricultural employment

Dependent variable: change in regional agricultural labor 1995-2000 (n=74)

Independent variables	Coefficient	t-value	Pr > t
Change in population	1.81	2.7	0.01
Relative non-agricultural growth	−3.92	−0.8	0.45
R^2	0.18		

Source: data from *Regiony Rossii 2002*, CD-ROM database, Moscow: Goskomstat (2003).

These regression findings are confirmed on a more qualitative level by the observed pattern of changes in agricultural labor versus non-agricultural growth. We see in Table 7.13 that on average positive non-agricultural growth reduces agricultural employment, whereas negative non-agricultural growth increases agricultural employment.

Table 7.13. Pattern of changes in agricultural employment versus non-agricultural growth

	Positive non-agricultural growth (n=36)	Negative non-agricultural growth (n=38)
Mean change in agricultural labor	−0.06	7.6
Mean non-agricultural growth index	1.44	0.85

Source: see Table 7.12.

Table 7.14. Clustering of 74 regions by changes in agricultural labor, population, and non-agricultural growth

	Cluster 1 (n=39)	Cluster 2 (n=34)	Cluster 3 (n=1: Chukotka)
Change in agricultural labor	17.8	−11.8	−24.3
Change in population	−1.0	−3.6	−25.5
Non-agricultural growth index	0.97	1.25	3.81

Source: see Table 7.12.

A similar result is produced by cluster analysis that groups the 74 regions by three variables: the change in agricultural labor between 1995 and 2000, the change in population during the same period, and the non-agricultural growth index as defined above (Table 7.14). The clusters with strong non-agricultural growth (clusters 2 and 3) are characterized by declining agricultural employment, whereas the cluster with negative non-agricultural growth (cluster 1) is characterized by increasing agricultural employment.

The fact that in the regression in Table 7.12 the sign of the non-agricultural growth factor is in the right direction (negative), but there is no statistical significance of this factor with the 1995-2000 data for all Russian regions suggests that there are barriers to free flow of labor out of agriculture. These barriers include insufficient skills of the rural population for non-farm employment, general shortage of alternative employment opportunities in rural areas, and persistent housing difficulties in urban areas that prevent migration.

Fig. 7.10. Agricultural employment in corporate farms 1990-2002

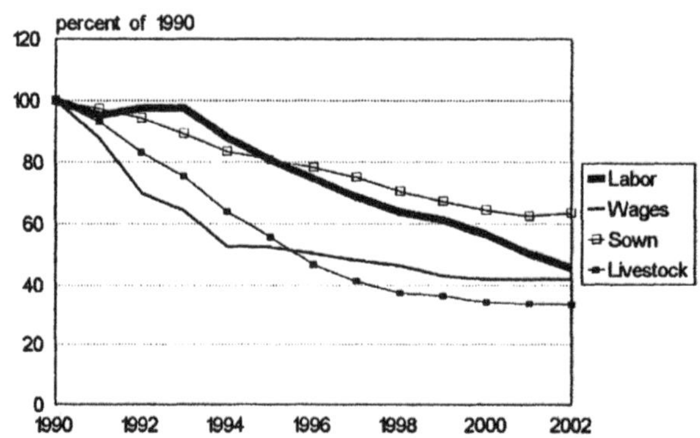

The conclusion that alternative opportunities increase labor outflow from agriculture is also supported by a different analysis of agricultural employment in the subsector of corporate farms, where more accurate statistical data are available. In general, the number of employed in corporate farms decreases with the shrinking activity of this sector, i.e., with the decrease in sown areas, the livestock herd, the pool of farm machinery, and gross agricultural output in general. It is, of course, impossible to determine the direction of causality in this case: does the labor force decrease due to shrinking activity of corporate farms, or does agricultural activity shrink due to the reduction of the labor force? A factor which is relatively free from such simultaneity concerns is the ratio of agricultural wages in corporate farms to the average wage in the economy. This ratio is a rough measure of attraction of other occupations for corporate-farm employees. It persistently decreased during the transition, which may be taken as a signal of diminishing attractiveness of employment in corporate and increasing attractiveness of other sectors of the economy. By 2002, the relative wage index had dropped to 42% of the 1990 level, generally paralleling the decline in corporate-farm employment (46% of the 1990 level). The coefficient of correlation between the relative wage index and the number of employed in corporate farms was 0.82 (data for 1990-2002). Unfortunately, an econometric analysis combining the relative wage index with activity-based factors did not produce significant coefficients. This is probably understandable because of high correlations among all the variables in the analysis and the short time series. We are thus forced to present our conclusion regarding the possible impact of agricultural wages on agricultural employment in corporate farms only in the form of a diagram (Figure 7.10).

Demand for Labor

The prevailing opinion in the media, among politicians, and even among some scholars is that agriculture suffers from excess of labor, which depresses productivity. MinAg, on the other hand, regularly publishes the results of normative calculations demonstrating that agriculture (or more specifically, the subsector of corporate farms) suffers from a shortage of qualified labor. Thus, the number of managers and skilled livestock workers is about 5% below the normative requirement over the years, while the shortage of machinery operators and mechanics reaches 10% of the norm. These results cannot be accepted at their face value, as they are based on normative calculations that reflect old central-planning practices. We accordingly turn to the 2003 BASIS survey to examine the sufficiency of labor in Russian agriculture.

Table 7.15. Assessment of labor sufficiency in farms of different types (percent of respondents)

	Corporate farms	Peasant farms	Household plots
How do you evaluate the size of your permanent labor force?			
Adequate	55.0	92.4	92.9
Redundant labor	10.6	—	—
Shortage of labor	34.5	5.4	5.1

Source: 2003 BASIS survey (Ivanovo, Nizhnii Novgorod, and Rostov oblasts).

There is essentially neither shortage nor redundancy of labor in the individual sector: over 90% of peasant farmers and commercial household-plot operators report that they have adequate labor for their needs (Table 7.15). The majority of corporate-farm managers (55%) are also satisfied with the size of their labor force, but fully one-third indicate that they experience shortages. The main shortages are identified for skilled and professional workers, such as machinery operators (27% of managers), livestock workers (23%), and agricultural specialists (15%). Only 10% of corporate farms (and none of the individual farms) report excess labor.

Comparison of corporate farms in two labor sufficiency categories—farms with adequate labor force and farms experiencing shortage of labor—does not detect striking differences in the production and resource profiles (Table 7.16). It may be significant that farms experiencing shortage of labor have more livestock in their product mix (a notoriously labor intensive enterprise), a higher proportion of women (typical unskilled labor), and a lower proportion of skilled workers. Yet these farms actually employ more workers than farms characterizing their labor force as adequate and the differences in other resource endowments (not shown in the table) are negligible. The only striking difference concerns factors related to remuneration of labor. In farms reporting adequate labor, the annual wages are higher, the wage arrears are substantially lower, and the share of wages paid in kind is much higher. All the incentives combine to ensure that these farms are more successful in attracting labor than the other farms in the survey. Workers seem to particularly appreciate the option of receiving a part of their wages in kind, as payments in kind are valued at cost, and not at the higher market prices, adding an extra benefit (estimated at 3-4% of the reported wages).

Table 7.16. Profiles of corporate farms from different labor sufficiency categories

	Adequate labor (*n*=82)	Shortage of labor (*n*=61)
Specialization: share of livestock production	37%	47%
Labor force (annual average)	114	134
Women	31%	40%
Workers younger than 30	15%	14%
Seasonal and temporary	6%	5%
Managerial	15%	14%
Non-agricultural	12%	10%
Skilled workers (top two categories)	35%	31%
Remuneration:		
annual wage, rubles	12,100	10,600
share of wages in kind	20%	10%
wage arrears, months	1.7	2.7

Source: 2003 BASIS survey (Ivanovo, Nizhnii Novgorod, and Rostov oblasts).

Managers of corporate farms complaining of labor shortages were asked about possible reasons for their failure to recruit an adequate workforce. The main reasons were shortage of housing (49% of respondents), uncompetitive wages offered in agriculture (38%), and difficulty finding workers with a good behavioral record (31%). Managers of farms reporting excess labor did not lay off the redundant workers because they did not want to be responsible for reducing the standard of living of the workers' families (70% of respondents), because they needed the redundant workers as a buffer for seasonal peaks (49%), and because they were concerned that layoffs would only increase theft, drunkenness, and the general criminal behavior in the village (35%).

Human Capital in Agriculture

Human capital includes issues of gender and age, educational attainment, and professional qualifications of the employed. Our discussion of the human capital in agriculture examines separately the cohort of agriculturally employed in official statistics (i.e., people with formal employment in agricultural organizations as well as peasant farmers and operators of commercial household plots) and the cohort of agricultural "household workers" farming their household plots for subsistence purposes (Table 7.17). This separate analysis has become possible with the publication of survey data on "household workers" since 2001.

Table 7.17. Gender and age structure of employment 2002 (by main occupation, percent)

	National economy as a whole	Formally employed in agriculture	Subsistence farming*
Men	51.3	59.9	47.3
Women	48.7	40.1	52.7
Age groups:			
Up to 30 years	24.3	19.9	23.1
30-59 years	71.3	71.5	63.1
60 years and older	4.4	8.6	22.8

Source: *Trud i zanyatost' v Rossii 2003*, Moscow: Goskomstat (2003), pp. 75, 76.
*Calculated from employment survey, November 2003, pp. 152, 157.

Formal agricultural employment is a male-dominated occupation: agriculturally employed in official statistics are 60% men and only 40% women. Subsistence farming, on the other hand, is dominated by women: 53% compared with 47% men. Men more readily go out to work in farm enterprises and agricultural service organizations; men generally organize peasant farms or push for commercial farming on household plots. Women dominate subsistence farming because it is a residual activity: women are left behind to take care of the household plot while men move on to outside occupations.

The age structure of the two cohorts is also sharply different. Formal agricultural employment is not particularly attractive to young people, and their share in agriculture is less than nationally (20% compared with 24% in the entire economy). Subsistence farming, being a residual activity, strongly relies on youngsters, just as it relies on women. The proportion of young people in subsistence farming is 23%, higher than in formal agricultural employment. The striking feature of the age structure of subsistence farming, however, is the exceptionally high share of people older than 60 years of age: this age group accounts for 23% of subsistence farmers, compared with only 9% in formal agricultural employment and 4% nationally. Subsistence farming is thus the responsibility of women and elderly people, who are strengthened and reinforced by youngsters. People in their prime (the 30-59 age group) tend to take on formal employment if they stay in agriculture.

The educational endowment of agriculturally employed (Figure 7.11) repeats the rural/urban education gaps that we have discussed previously. The proportion of agriculturalists with higher education is very low: 7% compared with 26% nationally. Primary education, on the other hand, is the highest attainment for 36% of employed in agriculture, compared with 19% nationally. Secondary education is evenly matched among agriculturally employed and nationally (slightly over 55%).

Fig. 7.11. Educational attainment: agriculture and nationally, 2002

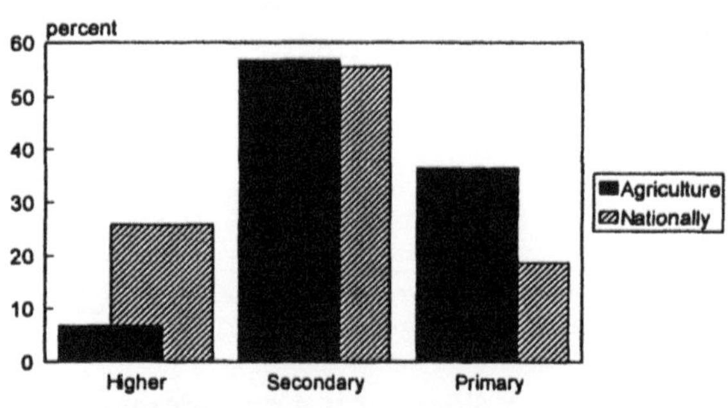

There is some evidence that the quality of human capital among agriculturally employed is slowly increasing. For instance, the proportion of people with vocational training (both primary and secondary) increased from 27% in 1998 to 34% in 2002, and the percentage of those with higher education moved up (albeit very slightly) from 6.5% in 1998 to 6.9% in 2002. However, these positive changes in human capital are not reflected in the quality of managerial personnel in farm enterprises, where the share of people with higher education decreased markedly between 1991-2001 and continued to decline in 2002 (Table 7.18). Moreover, the percentage of young managers (up to 30 years of age) decreased, while the percentage of managers in the retirement-age bracket increased during this period. Managerial positions in farm enterprises do not attract people with higher education, apparently because the continuing decline of this subsector has created a general impression of failure and deterioration that deters qualified candidates. The decline in managerial cadres with higher education is accompanied by a certain increase in the proportion of practitioners in managerial positions. A deterioration of human capital is observed not only among managers, but also among rank workers in farm enterprises. Thus, the percentage of livestock personnel in the top two professional-qualification categories dropped from 25% to 22% of all livestock workers between 1996 and 2001.

Peasant farms can boast a much greater endowment of human capital than farm enterprises. Among heads of peasant farms, 20% are people with higher education (compared with only 7% for agriculture as a whole) and 37% have vocational secondary training (compared with 21% for agriculture as a whole). Most farmers have attained at least secondary education, whereas among agriculturally employed in general 36% have only primary schooling. Moreover, the same pattern of relatively high educational attainments is observed not only for the farmer, but actually for all family members in peasant farms (Table 7.19). The high quality of human capital in peasant farms is not surprising if we recall that peasant farms were initially established by former managers or specialists of farm enterprise who decided to go into individual farming.

Table 7.18. Structure of managerial human capital in farm enterprises 1991-2002 (percent)

	Managers			Chief specialists		
	1991	2001	2002	1991	2001	2002
People with higher education	85.6	70.1	68.5	62.6	53.5	52.7
Practitioners	1.4	6.4	6.7	2.9	4.5	4.7
Of which: trained by correspondence	85.3	13.0	19.3	42.0	19.3	20.3
Age structure:						
30 years and younger	3.5	1.7	2.0	16.2	7.1	7.4
retirement age	3.3	5.3	5.5	2.3	4.3	4.5

Source: *Chislennost', sostav i dvizhenie kadrov agropromyshlennogo kompleksa*, Moscow: MinAg (2001), 1, p. 4.

Table 7.19. Educational attainment in peasant farms and in agriculture as a whole

	Peasant farmers (nationally)[1]	Peasant farmers (survey)[2]	Other family members (survey)[2]	All agriculturally employed[3]
Higher education	20.1	29	24	6.9
Vocational secondary	36.6	39	38	20.9
General secondary	43.3	24	32	35.9
Primary	—	8	6	36.4

Source:s [1]*Sel'skokhozyaistvennaya deyatel'nost' krest'yanskikh (fermerskikh) khozyaistv Rossii*, Moscow: Goskomstat (2000), p. 39; [2] Survey of 300 peasant farms conducted by the author in 1992-1993 in Sverdlovsk, Orel, Tambov, and Leningrad oblasts; [3] *Trud i zanyatost' v Rossii 2002*, Moscow: Goskomstat (2003), p. 78.

Working Time

People employed in agriculture work fewer hours than the national average. Thus, according to national labor surveys, the actual number of work hours per week for the entire economy is 38.6, whereas those employed in agriculture work only 34.5 hours per week (Table 7.20). In agriculture, 22% work less than 30 hours a week, compared with 7% in the economy as a whole and 2% in industry. However, agriculture consists of two distinct sectors: corporate farms (farm enterprises) and individual farms (peasant farms and household plots). Supplementary data from other sources can be used to disaggregate the average work week for agriculture into its components and to compare the work load in the corporate and the individual farm sector.

The consolidated annual report of farm enterprises prepared by MinAg and Goskomstat show that the average farm-enterprise employee worked 1,893 hours in 2002, which allowing for 4 weeks' leave gives 39.4 hours per week—the same as in industry. This is substantially higher than the average for agriculture, which obviously implies that the average work week in the individual sector is substantially less than 34.5 hours. Household employment surveys conducted in recent years by Goskomstat show that individuals reporting informal agricultural employment (this cohort excludes employment in farm enterprises, but includes household plots with commercial orientation) work 21.8 hours per week, whereas "household workers ," i.e., individuals whose only occupation is farming the household plot, work as little as 16.3 hours per week. No comparable data are available for peasant farms as a separate group. However, official statistics indicate that the farm operator and other family members work on average 142.6 man-days per year, which is only 57% of the 250 work-days in a full year (or 64% if we allow family members to take a 4-week vacation, like all workers).

Table 7.20. Length of work week in agriculture, 2002

	Hours worked per week	Reference standard
Agriculture	34.5	39.4 (industry) 38.6 (average for the economy)
Corporate-farm sector: farm enterprises	39.4	
Individual sector		
Commercial household plots	21.8	
Subsistence household plots	16.3	
Peasant farms (operator and family members)	142.6 man-days/year (1999)	250 man-days/year (available)

Source: Agriculture, industry, and the economy from *Trud i zanyatost' v Rossii 2003*, Moscow: Goskomstat (2003), p. 99; farm enterprises from MinAg; commercial household plots from *Obsledovanie naseleniya po problemam zanyatosti, November 2002*, Moscow: Goskomstat (2003), p. 382; subsistence household plots from *Obsledovanie naseleniya po problemam zanyatosti, November 2003*, Moscow: Goskomstat (2004), p. 152; peasant farms from *Agricultural Activity of Private Farms in Russia 2000*, Moscow: Goskomstat (2000), p. 34.

Fig. 7.12. Relative work-time input per person: corporate and individual farms

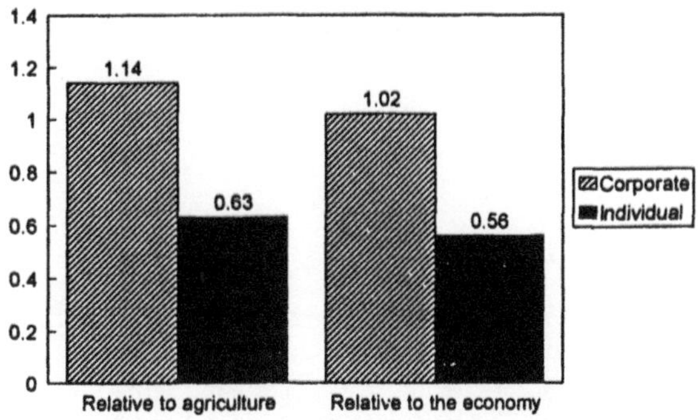

It is clear from these data that the work-time input per person in the individual sector is substantially less than in farm enterprises and in agriculture on average. While the work week in farm enterprises is 14% longer than the average for agriculture, the work week in the individual sector is nearly 40% shorter than the average for agriculture. Relative to the national average, the work week in farm enterprises is 1.02 and the work week in the individual sector 0.56 (Figure 7.12).

Wages

Agricultural wages have always lagged behind the average wage in the Russian economy, probably because rural households traditionally supplement their incomes with food products from the household plot. Moreover, this phenomenon is not unique to Russia: agricultural workers are typically regarded as low-skill labor and therefore earn less than the average wage everywhere in the world. Thus, in the United States, agricultural wages and salaries ran at the level of 60% of the average wage in domestic industries in the 1990s.[2] In Russia, however, the agricultural wage gap became much more pronounced during the transition.

All through the 1980s agricultural wages were actually increasing relative to the average wage in the economy: they gradually rose from 82% of the average wage in 1980 to a peak of 95% in 1990. After 1990 the trend reversed and agricultural wages began a long slide relative to the rest of the economy: from a peak of 95% in 1990 they slid to 40% of the average wage in the economy in 2000-2002 (this is shown by the vertical bars in Figure 7.13). The earning power of agricultural workers relative to the rest of the

---

[2] *Statistical Abstract of the United States 1998*, table 691: "Annual Total Compensation and Wages and Salary Accruals Per Full-Time Equivalent Employee, by Industry".

labor force in the economy was eroded by more than one-half between 1990 and 2002.

Fig. 7.13. Agricultural wages and average wage in the economy

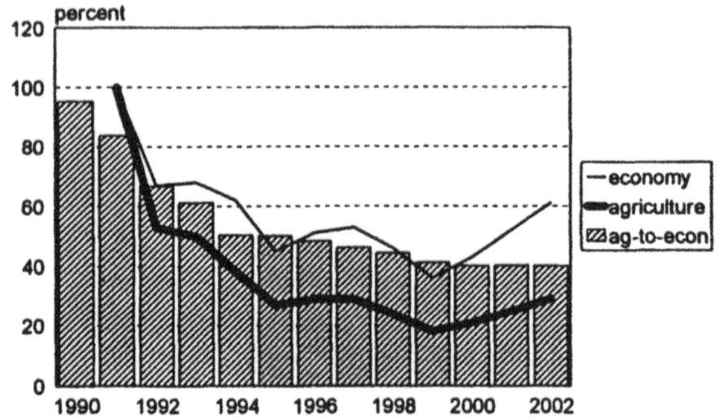

The erosion of earnings experienced by agricultural workers is also clearly seen from the changes in real wages (i.e., wages adjusted for inflation) over time. Figure 7.13 shows the index of inflation-adjusted wages in agriculture and in the economy as a whole. The real wage in agriculture in 2002 is 29% of its level in 1991. The real wage for the economy as a whole is 61% of the 1991 level and for industry 64%. Real agricultural wages dropped more than 70% between 1991 and 2002, while other sectors of the economy experienced a drop of "only" 40%. A different view of the inadequacy of agricultural earnings is obtained by comparing the wages to the minimum standard of living—a kind of poverty line measure (Table 7.21). In agriculture, wages are below the poverty line for 80% of the workers, compared to only 20%-30% in industry and 40%-60% in education and health services (between 1995 and 2002). Thus, on the basis of wages alone, agriculture is not competitive with other industries and cannot attract labor.

Table 7.21. Percentage of workers with wages at or below the minimum standard of living

	1995	1999	2000	2001	2002
Agriculture	80.2	82.0	84.6	81.3	75.0
Industry	32.5	26.5	25.7	23.9	18.9
Health care	63.7	67.2	65.7	61.0	38.8
Education	64.6	70.5	67.5	61.3	41.4

Source: *Trud i zanyatost' v Rossii 2003*, Moscow: Goskomstat (2003), pp. 406, 407.

The large and growing gap between agricultural wages and the rest of the economy could be sustained only because of the supplementary income opportunities that household plots provided for the rural population. Thus, according to a rural household survey carried out by the author in 2004 in Lukha District, Ivanovo Oblast, the value of farm products from the household plot adds fully 60% to the per capita cash income, raising it from 12,800 rubles per year (including earned income and social transfers) to 20,500 rubles. This, however, is still well below the minimum standard of living of 28,000 rubles per capita per year in Ivanovo Oblast at the end of 2003. As a result, only 26% of households in the survey had per capita incomes higher than the minimum standard of living.

The generally low level of agricultural wages hides two phenomena that are reflections of relatively large inequality in agriculture. First, the inequality between the high-earnings decile and the low-earnings decile is much sharper in agriculture than in other sectors of the economy (Table 7.22). Second, managers and administrative workers have enjoyed greater salary increases than all other workers in agriculture, while the salaries of agricultural production workers—the core cadre of agriculture—are shrinking relative to other cadres (Table 7.23). The relative increase in the wages of nonagricultural production workers is probably a reflection of competitive pressures in the labor market and the attempts of the reorganizing farm enterprises to diversify into nonagricultural activities as a survival strategy.

Table 7.22. Ratio of the mean wage of workers in the upper and lower decile (times)

	1995	1999	2000	2001	2002
Agriculture	23.8	31.6	35.6	48.6	29.6
Industry	20.6	21.5	20.8	25.4	21.2
Health care	13.5	15.1	14.9	16.9	13.2
Education	13.0	16.1	16.7	18.6	13.5

Source: *Trud i zanyatost' v Rossii 2003*, Moscow: Goskomstat (2003), pp. 406.

Table 7.23. Relative wages for different groups of workers in farm enterprises, 1990 and 2002 (in percent of the average wage)

	1990	2002
Average wage over all farm enterprises	100.0	100.0
Managers and specialists	122.9	142.6
Production workers:		
Agriculture	101.1	93.6
Nonagricultural activities	91.2	100.5

Source: Consolidated annual reports of farm enterprises, MinAg.

Additional evidence of competitive market pressures on wages is provided by the 2003 BASIS survey, which covered corporate farms operating at different distances from the urban center in their district (Table 7.24). Corporate farms close to district centers pay higher wages than farms at the periphery, presumably because they need to keep their labor force despite the greater accessibility of alternative jobs in nearby towns. Moreover, periurban farms pay most of the wages in cash, as is the common practice for hired labor in market economies. Remote farms, on the other hand, include a higher proportion of payments in kind in the compensation package, again because their workers have very limited alternatives.

Table 7.24. Average monthly wages by distance from urban center

	Average wage (including social benefits), rubles	Payment in kind, %
All farms surveyed	1,144	15.9
Farms near district center	1,285	10.2
Farms at medium distance from center	1,107	10.6
Remote farms	1,048	25.4

Source: 2003 BASIS survey in Ivanovo, Nizhnii Novgorod, and Rostov oblasts.

Is Agricultural Labor Underpaid? Evidence from Surveys

We have approached this question in two ways. First we looked at the return to labor from formal employment for rural families working in two different sectors—agriculture and rural social services. Then we compared agricultural wages to the value of marginal product generated by workers in farm enterprises. In both analyses, the results suggest that agricultural labor is indeed underpaid relative to other sectors and relative to its marginal product.

Table 7.25. Return to labor in agricultural and non-agricultural families

	Agricultural families	Non-agricultural families
Total family income* (thou. rubles)	55.0	54.4
Total labor input* (man-hours/year)	4,104	3,881
Wage employment (%):		
Wage income	51.6	77.9
Labor input	65.2	72.1
Return to labor (rubles/hour):		
Total	13.4	14.0
Wage employment	10.6	14.8
Household-plot farming	18.6	11.1

*Includes wage employment and household-plot farming.

Source: Rural household survey in Lukha District, Ivanovo Oblast, 2004. Agricultural families included 48 households with 73 respondents; non-agricultural families, 17 households, 27 respondents.

Table 7.25 presents comparative data on labor inputs and wage earnings of two average families: a family for which the main formal employer is a corporate farm (i.e., an agricultural family) and a family whose wage earners mainly work in rural social services (a non-agricultural rural family). Both families supplemented their wage income with earnings from the household plot. However, the data enabled us to disaggregate the earnings and the labor inputs into a component associated with formal wage employment and a component attributable to farming activities on the household plot. The total time input and the total earnings are very close for both families. The difference in total return to labor is not statistically significant. But disaggregated data show a different picture. The return to labor in household-plot farming is two-thirds higher for agricultural families, presumably because of the support they receive from the corporate farm with input supply and product marketing. On the other hand, the return to one hour of labor in formal employment is 40% less for agricultural families than for families employed outside agriculture—a reflection of low wages in agriculture.

Estimation of various production functions for corporate farms from the data of 2003 BASIS survey in three oblasts consistently produced values of marginal product of labor that were higher than the price of labor (agricultural wages). According to these estimates (see Chapter 10), an additional worker would produce 20,000-26,000 rubles worth of output (per year) while being paid about 12,000 rubles (also per year). The allocative efficiency ratio for labor is accordingly between 1.7 and 2.2 (this is the ratio of the value of marginal product to the marginal price). The traditional interpretation of these findings is that corporate farms can benefit by employing more labor, assuming that the price of labor stays at 12,000 rubles per year. In our context, however, we can interpret these results as evidence that agricultural labor is underpaid by a factor of 2: agricultural wages are very low relative to the value of marginal product that agricultural labor is observed to produce.

Table 7.26. Value of marginal product of labor estimated by production function analysis using different databases

Researcher	Database	Value of marginal product, thou. rubles	Average wage, thou. rubles	Allocative efficiency ratio
Uzun	Goskomstat data: all farm enterprises	22.6-27.1	18.7	1.21-1.45
Epshtein	Goskomstat data: all Leningrad Oblast farms	60.7	43.8	1.38
Lerman	BASIS survey	22.1-26.4	12	1.84-2.20
Bogdanovskii	BASIS survey	42.5 (per worker) 3.0 (per ruble of wages)	12 1	3.54 3.0

It can be argued that 12,000 rubles does not reflect the full earnings of a corporate-farm worker. Agricultural employees receive part of their wages as payment in kind, which is usually valued for accounting purposes at cost, and not at market prices. Valuation of the in-kind component at the higher market prices would increase the nominal wages by 3%-4% above 12,000 rubles. Employees of corporate farms additionally benefit from various subsidized and even free services, especially support of their faming activities on the household plot. These benefits also increase the effective wages above 12,000 rubles. The author's estimates for corporate farms in Mtsensk District (Orel Oblast) show that the combined impact of all these subsidies, services, and support, granted with the management's formal authorization or gained by means of informal shadow procedures, ranges around 20-25% of the nominal wage. If we accordingly assume that agricultural workers earn 15,000 rubles, or even 18,000 rubles (50% higher than the reported 12,000 rubles a year), this is still substantially below the value of marginal product of labor, which only strengthens the conclusion that agricultural labor is underpaid.

Table 7.26 presents some additional results for allocative efficiency of labor. Although the specific numbers vary, the value of marginal product is always substantially higher than the average wage, and the gap typically exceeds 40%. This essentially means that even when the wages are corrected for possible hidden subsidies enjoyed by corporate-farm workers, they remain below the value of marginal product, possibly signifying underpayment of agricultural labor.

Changing Motivation of Labor

The uncompetitive wages received in formal agricultural employment is one of the factors that have led to a shift in emphasis from the formerly dominant corporate farms to the individual sector. As a result, rural people devote a substantial part of their time to work on the household plot, and income from the household plot makes a significant contribution to the family budget.

This effect is particularly noticeable in unprofitable corporate farms, which face considerable difficulties meeting their payroll (however low the wages). Thus, in a series of surveys of unprofitable and low-profitability corporate farms conducted by the author in 1997-2001, agricultural employees were found to devote more than one-third of their working time to the household plot, which contributed 54% of total family income, while wage earnings accounted for only 36% (Table 7.27). For employees of profitable corporate farms, wage earnings still account for more than half the family income (58%), but income from the household plot has already reached 36%. It is notable that agricultural workers employed in unprofitable corporate farms devote much more time to farming the household plot than workers in profitable farms, whose wage earnings are apparently more secure.

Employees in unprofitable corporate farms are primarily motivated by non-market factors, which include payments in kind, income illegally siphoned from the corporate farm, and supplementary earnings from the household plot (Table 7.28). Profitable farms, on the other hand, seem to motivate their employees primarily by market mechanisms, such as payment of wages in cash and considerations relating to the organizational environment and economic conditions on the farm. These are typical workplace characteristics that are important for hired workers in any market economy. Unfortunately, such farms are a relatively small minority in Russia today.

Table 7.27. Time allocation and structure of family income for employees of corporate farms

	Unprofitable farms	Profitable farms
Number of farms surveyed	**11**	**4**
Number of respondents	**680**	**512**
Work time, weekly hours:		
wage employment	42.5	42.8
household plot	22.7	14.8
Structure of family income, %	100.0	100.0
wage earnings	35.7	58.5
household plot	54.1	35.9
non-agricultural business activity	2.3	0.5
pensions and social transfers	5.2	2.3
other income	2.7	2.8

Source: Author's surveys in corporate farms in Tambov, Saratov, Orel, and Moscow oblasts, 1997-2001.

Table 7.28. Ranking of motivations as perceived by employees of corporate farms*

	Unprofitable farms	Profitable farms
Payment in kind	1	3
Illicit income from corporate farm	2	4
Income from household earned with support from corporate farm	3	5
Organizational environment and economic conditions	4	2
Cash payments	5	1
Social benefits from corporate farm	6	6
Lease payments, dividends	7	7

*Ranking by average scores as assigned by the respondents to each motive.
Source: see Table 7.27.

Productivity of Labor

In principle, productivity of labor is simply calculated as the ratio of agricultural output to the number of people employed in the production of this output. In reality, the calculation for Russia is not simple at all. The difficulty is associated with the statistical discrepancy between the base of people producing the reported agricultural output, on the one hand, and the reported agricultural labor force, on the other.

While the reported agricultural output as shown in Table 7.29 for 1990-2002 reflects the total production of all types of units in Russian agriculture—corporate farms, peasant farms, and household plots, the agricultural employment statistics excludes all the "household workers" who produce goods (and services) for family consumption only. The entire contingent of people farming subsistence household plots and thus contributing to total agricultural output is omitted from the number of agriculturally employed as reported by statistical organs. The agricultural labor force is underestimated and the productivity calculated on the basis of this labor force is overestimated.

Table 7.29. Labor productivity in Russia's agriculture 1990-2002

Year	Agricultural labor, millions	Gross agricultural product (GAO), billion rubles*	Percent of 1990			Corporate farms	
			Labor	GAO	Labor productivity	Labor, millions	GAO, billion rubles*
1990	9.7	158.0	100.0	100.0	100.0	8.3	**116.0**
1991	9.7	150.9	100.0	95.5	95.5	7.9	105.6
1992	10.1	136.7	104.1	86.5	83.1	8.1	87.4
1993	10.1	130.7	104.1	82.7	79.4	8.1	79.3
1994	10.3	114.9	106.2	72.7	68.5	7.4	66.6
1995	9.7	105.7	100.0	66.9	66.9	6.7	56.4
1996	9.3	100.3	95.9	63.5	66.2	6.2	50.7
1997	8.6	101.9	88.7	64.5	72.7	5.7	51.6
1998	8.7	88.5	89.7	56.0	62.5	5.3	40.7
1999	8.5	92.1	87.6	58.3	66.5	5.1	42.9
2000	8.4	99.2	86.6	62.8	72.5	4.7	45.7
2001	7.9	106	81.4	67.1	82.4	4.2	50.3
2002	7.7	107.6	79.4	68.1	85.8	3.8	51.3

*In constant 1990 prices.
Source: Labor and GAO from Statistical Yearbook of Russia 2003; index numbers from author's calculations.

Bearing this caveat in mind, we calculated the productivity of agricultural labor on the basis of the time series in Table 7.29 and converted it to index numbers in percent of 1990. Despite the statistical problems, the calculated trend is a reasonably faithful representation of reality if we are willing to assume that the number of people farming subsistence household plots did not change substantially during the transition. This is a plausible assumption (despite the doubling of land in household plots), because the rural population—the main pool from which subsistence farmers are created—has virtually remained constant since 1990 (see Table 7.1). After the initial transition-induced decline, agricultural output bottomed out in 1998 and began to recover, rising from 56% of the 1990 level in 1998 to 68% in 2002. Agricultural labor, on the other hand, had a characteristically sticky phase at the beginning of transition, which then gave way to a steady decline after 1994. Because of the different trends of agricultural output and agricultural labor, the productivity of labor initially declined, dropping to 66% of the 1990 level in 1996, and then began to rise, reaching 86% of the 1990 level in 2002.

Fig. 7.14. Gross agricultural product: corporate and individual farms

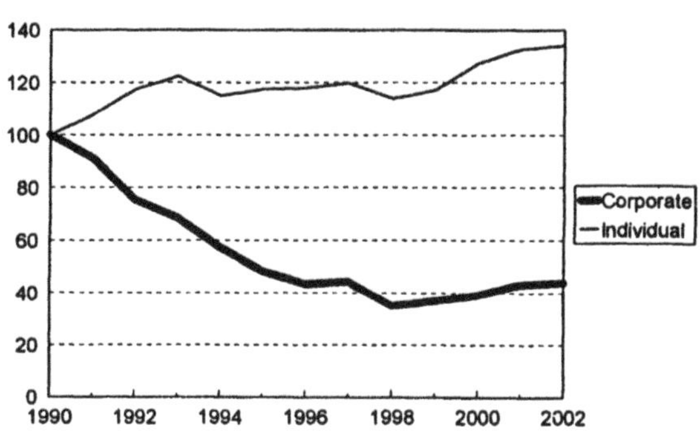

More precise data are available for calculating the labor productivity in corporate farms. Annual reports submitted to MinAg and Goskomstat contain both the value of agricultural output and the number of people (in annual-average full-time equivalents) employed in the production of this output. These data are presented in the last two columns of Table 7.29. The trends observed for corporate farms essentially replicate the trends for Russia's agriculture as a whole: initial decline of agricultural output changing to recovery (Figure 7.14, thick curve); continuing shrinkage of the labor force (Figure 7.15, thick curve). Agricultural labor productivity in corporate farms therefore declined up to 1998 and then began to improve. By 2002 it had essentially recovered to the same level as in 1990 (Figure 7.16, thick curve).

Fig. 7.15. Agricultural labor: corporate and individual farms

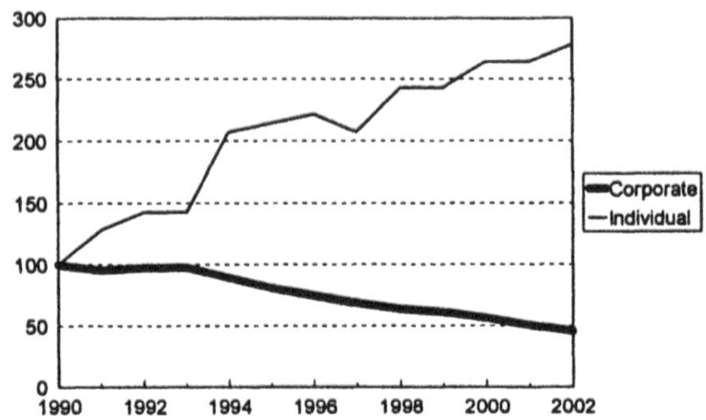

Fig. 7.16. Agricultural labor productivity: corporate and individual farms

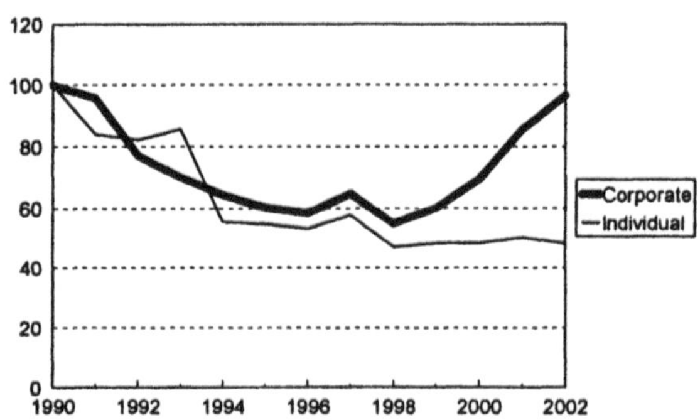

Given the production and labor time series for agriculture as a whole and for the corporate-farm sector, we can estimate the productivity of labor for the individual sector. Here, however, the estimates are very rough, because it is the individual sector that absorbs the full impact of the missing labor components (subsistence household plots are part of the individual sector). The agricultural production in the individual sector increased in the early 1990s, when additional land was allocated to household plots and peasant farms as part of the land reform, then it remained fairly steady until 1998, after which it resumed its growth in response to the general improvement in Russia's economy (Figure 7.14, thin curve). The individual sector did not contribute to the overall decline of agricultural production, which was entirely associated with the deteriorating situation in corporate farms.

Agricultural labor in the individual sector, on the other hand, sharply increased between 1990 and 2002 (Figure 7.15, thin curve). The exact magnitude of the increase is difficult to estimate because there is considerable uncertainty in the individual labor statistics in 1990-1992 (in addition to all other statistical problems with agricultural labor). It is clear, however, that the number of employed in individual agriculture rose sharply between 1990 and 2002 as people were ejected from corporate farms without alternative opportunities for non-agricultural employment. As a result of the increasing labor force in individual farms, the labor productivity of the individual sector declined over time, dropping by more than 50% between 1990 and 2002 (Figure 7.16, thin curve). Our results actually underestimate the decline in agricultural labor productivity, because the labor force used in these calculations is underestimated (possibly by as much as 40%-50%) due to the exclusion of pure subsistence farmers.

Corporate and individual farms thus exhibit different patterns of behavior by the three variables of agricultural output, agricultural labor, and agricultural labor productivity. In corporate farms, agricultural output initially dropped and then began to recover, whereas in the individual sector it generally increased over time. Agricultural labor shrank continually in corporate farms and increased sharply in the individual sector (even without the subsistence contingent). As a result, agricultural labor productivity has been improving in the corporate-farm sector (since 1998) and has deteriorated in the individual sector. These different patterns are shown in the three panels of Figure 7.14.[3]

Table 7.30. Agricultural labor productivity in farms of various types, 2002

	Agricultural output, mln. rubles[1]	Labor input		Labor productivity	
		mln. man-hours	mln. man-days	rubles per man-hour	rubles per man-day
Corporate farms	409,937	7,385[2]	1,019[2]	55.5	402.3
Peasant farms	38,164	–	129[3]	40.8[5]	295.8
Household plots	581,154	24,800[4]	–	23.4	–

Source: [1] Statistical Yearbook of Russia 2003, pp. 399, 400.
[2] Consolidated annual report of corporate farms for 2002, MinAg, Moscow.
[3] *Agricultural Activity of Private Farms in Russia 2000*, Moscow: Goskomstat (2000), p. 36; extrapolated to 2002 based on the number of farms.
[4] Calculated on the basis of updated data received by the author from Goskomtat (replacing the data previously published in *Obsledovanie naseleniya po problemam zanyatosti, Noyabr' 2002*, Moscow: Goskomstat (2003), p. 199).
[5] Calculated in the same proportion to labor productivity in man-days as for corporate farms.

---

[3] Because the individual labor data in the early years of transition are unreliable, we have recalculated the changes in agricultural labor productivity since 1994, omitting the sharp increase in individual labor between 1990 and 1993. The absolute magnitudes were naturally affected, but the general pattern of change remained the same as for the entire period 1990-2002 shown in Figure 7.14.

The recent changes in statistical data collection enable us to obtain alternative estimates of agricultural labor productivity in farms of different types, but for one year only (not for a time series, as above). Using the available data for 2002, we find that corporate farms achieve higher productivity of labor than the individual sector, and within the individual sector peasant farms have higher labor productivity than household plots (Table 7.30).

We have to caution against misinterpretation or misrepresentation of these findings. Agricultural labor productivity is just a partial measure of productivity, and the fact that individual farms have lower labor productivity than corporate farms does not necessarily signify that they produce less efficiently overall. Thus, corporate farms with their huge expanses of land produce less output per hectare than the much smaller individual farms. In other words, the partial productivity of land is lower in corporate farms than in the individual sector. What determines efficiency is not partial productivity of labor or land, but total factor productivity that examines output in relation to the entire basket of inputs—labor, land, livestock, machinery, fertilizer, fuel, power, etc. (this is the approach in the chapter on *Allocative Efficiency*). The decreasing productivity of labor in the individual sector is merely a manifestation of the "labor-sink effect" of individual farming: people who lose their jobs in corporate farms cannot readily find alternative employment because of inadequate skills, barriers to urban migration, and lack of non-agricultural job opportunities in rural areas. They are thus forced to shift to individual farming, which saves them from unemployment and provides a kind of a safety net, but at the cost of lower productivity.

Conclusion

Russia's rural population is faced with shortage of alternative employment opportunities, extremely limited mobility of labor, and total lack of political lobbying power. These factors have produced a specific Russian model of agricultural labor markets in which unemployment is held in check by reducing the effective working time and lowering the wages. As a result, there has been a massive shift of agricultural labor from wage employment in corporate farms to subsistence and semi-commercial farming on household plots. Russian agriculture is thus characterized by large-scale underutilization of labor, primarily in subsistence household farming, which only deepens rural poverty by creating an illusion of employment, but without sufficient income. Even these mechanisms, however, have been unable to reduce rural unemployment, which in 2003 reached 10% compared with 8% for the urban population.

A number of policy interventions appear to be appropriate in the present circumstances. First, rural non-agricultural activities should be encouraged and expanded by restoring the normal functioning of social services, as well as by promoting small entrepreneurship through credit facilities, tax breaks, and simplification of administrative requirements for small business creation. Second, education and training programs should be established to teach the rural population relevant marketable skills. Third, the large cohort of subsistence farmers working on their household plots as the main or sole occupation should be given the legal status of "economically employed" with all the associated rights for pension, medical insurance, unemployment benefits, and other forms of social protection. This will naturally involve eliminating the sweeping tax exemption that household plot operators continue to enjoy, but at the same correct one of the major distortions between different categories of individual farms. Last but not least, efforts should be made to raise the political conscience of the rural population and encourage it to start actively lobbying for better wages and better living standards. This may sound like a non-market mechanism, but active political awareness is one of the important tools that all constituencies employ in developed market economies to achieve their goals.

Chapter 8

Markets for Purchased Farm Inputs

Evgeniya Serova and Olga Shick

Russian policy makers usually argue that agriculture suffers from severe decapitalization due to financial constraints faced by producers. This view is the basis of the national agricultural policy, which emphasizes reimbursement of input costs to producers and substitutes government or quasigovernment organizations for market institutions. The present chapter evaluates the availability of purchased farm inputs, the efficiency of their use in agriculture, and the main problems in the development of input markets. It analyzes supply and demand, the emergence of market institutions, and the impact of government policies. Purchased farm inputs include a wide range of goods. To keep a proper balance between breadth and depth of analysis, we focus on five groups of purchased inputs: farm machinery (tractors and harvesters), fertilizers, fuel, seeds, and animal feed. Three sources of information are used in the analysis: official national statistics, data from the 2003 BASIS survey of agricultural producers (to evaluate the demand for farm inputs), and data from the survey of farm machinery and fertilizer manufacturers conducted in 2001 by the Analytical Center for Agri-Food Economics in Moscow (to evaluate supply; referred to as the 2001 AFE survey).

In the Soviet period, all sectors of the economy, including agriculture, were served by a state supply system, which delivered inputs in centrally planned quantities and provided financing through state banks. In this system, every enterprise was allotted a certain quantity of inputs sufficient to produce its planned output. A farm that was required by central planners to produce a certain quantity of grain was automatically entitled to receive an appropriate quantity of fuel, fertilizers, pesticides, new machinery, spare parts, and so on. At the same time, funds were earmarked in the state banking system to enable the farms to pay for these inputs. The farms paid lower prices for inputs than the manufacturers' wholesale prices, and the difference was covered from the state budget. The relative input/output prices were kept fixed, favoring agricultural producers at the expense of industrial manufacturers as another way of subsidizing agriculture.

Fig. 8.1. Terms of trade in agriculture 1990-2003

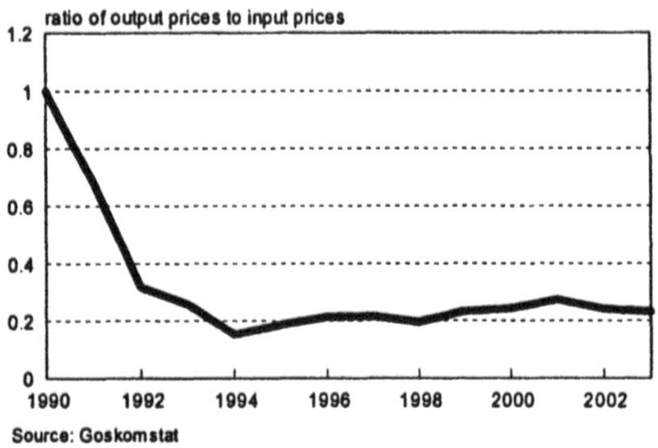

Fig. 8.2. Decline in production of farm machinery 1990-2002

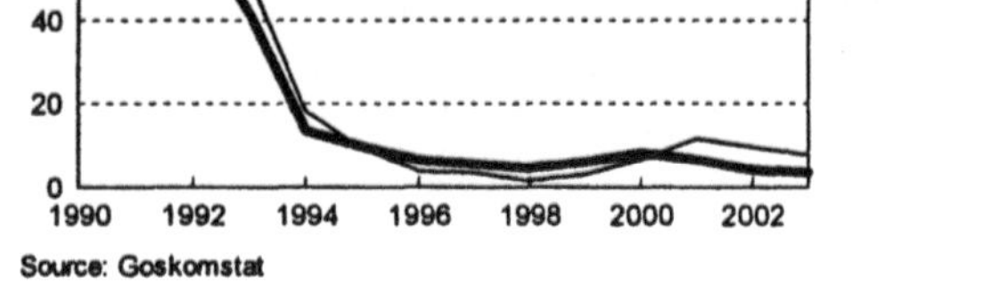

After 1991, liberalization of industrial prices resulted in an immediate increase of the cost of purchased inputs to farms, whereas food prices continued to be controlled by the government as a concession to consumers. The terms of trade in agriculture therefore deteriorated dramatically as input prices aligned themselves with the average world level while output prices did not even catch up with inflation. The main shock came between 1990 and 1994, when the terms of trade index for agriculture (the ratio of output prices to input prices) dropped to 0.2 of its 1990 level (Figure 8.1). After that, the terms of trade appear to have stabilized as the changes in food prices generally tracked the changes in input prices. Since 1998 we even observe a slight improvement in terms of trade in agriculture, as food price increases began to outstrip input prices in the wake of the 1998 devaluation.

The inevitable rise in the relative prices of farm inputs led to shrinking machinery stocks and extensification of agricultural production, significantly jeopardizing the potential for growth. The declining demand for inputs affected the supply of domestically manufactured farm machinery (Figure 8.2), while fertilizer manufacturers shifted to exports, reducing their dependence on the domestic market. It is only after the 1998 financial crisis that agricultural recovery has spurred a renewed demand for farm inputs leading to a certain growth in domestically manufactured machinery.

Supply of Farm Inputs

Farm machinery and fertilizer manufacturing in Russia are highly concentrated industries. Thus, five manufacturers produce nearly 90% of tractors, and just two manufacaturers produce 95% of grain harvesters. In the fertilizer industry, ten manufacturers produce 74% of nitrogen fertilizers, four manufacturers produce more than 70% of phosphorus fertilizers, and just two manufacturers produce the total output of potassium fertilizers. The concentration trend in the farm machinery industry is continuing, as the giant machinery manufacturers have recently begun creating vertical and horizontal holding structures, probably in anticipation of impending competition from importers.

Sales of farm machinery are also highly concentrated: in 1999-2000, fully 75% of the market was controlled by Rosagrosnab (a former state monopoly privatized in the mid-1990s) that operated jointly with the regional administrations; the remaining 25% were sales through agroholdings and other corporate farms. Since 2000 the market share of Rosagrosnab and regional governments has dropped to 55%, primarily due to the rapid growth of agroholdings and corporate farms as channels for machinery sales. Independent dealer and service networks also began to develop in 2000, after Rosagrosnab had lost its monopolistic position. These initiatives often followed reorganization and a management shakeup in the manufacturing plant, or takeover by a new investor (as in the Rostov and Krasnoyarsk farm machinery manufacturers). Up to 1999, the standard payment mechanisms of the large suppliers and manufacturers were mutual account offsets or barter deals. Since 1999, virtually all payments have shifted to bank transfers.

Table 8.1. Sales channels of medium and small machinery manufacturers (percent by volume)

	1998	1999	2000	2001
Mini-tractors				
Regional administration	—	—	6	5
Dealers	—	6	10	8
Commercial firms	5	13	23	19
Corporate farms	33	2		
Peasant farms	32	31	21	17
Household plots	30	38	40	51
Hitched implements, trailers				
Regional administration	—	—	4	5
Dealers	—	5	9	10
Commercial firms	26	20	20	22
Corporate farms	—	2	—	—
Peasant farms		25	17	20
Household plots	74	48	50	43
Spare parts, components, assemblies				
Rosagrosnab	90	95	28	40
Dealers	—	—	20	10
Commercial firms	10	5	52	50

Source: 2001 AFE survey.

Alongside the giant tractor and harvester manufacturers there is a group of medium and small manufacturers that produce a broad range of farm equipment (mini-tractors, hitched implements, spare parts and components) and accept repair and maintenance contracts. The sales channels of these medium and small manufacturers are much more diversified than those of the giant tractor and harvester manufacturers (Table 8.1). They began creating own dealer and service networks earlier than the giant manufacturers, probably because their products were never entitled to government support. The wider range and the generally lower cost of equipment sold by the medium and small manufacturers have also resulted in a greater diversity of payment arrangements (Table 8.2). These manufacturers accept cash payments and continue to rely on barter deals.

Table 8.2. Structure of payments for medium and small machinery manufacturers

	Percent of respondents	Trend
Bank transfers	60	Increasing
Cash payments	17	Increasing
Barter	16	Decreasing
Other	7	Decreasing

Source: 2001 AFE survey.

During the Soviet period, Belarus and Ukraine were Russia's main sources for farm machinery. After the dissolution of the Soviet Union, imports from countries outside CIS have increased substantially, and in the late 1990s the share of imported tractors was 67%. High price is the main obstacle to wider penetration of imported machinery in Russia. Thus an imported tractor costs 50% more than a tractor assembled in Russia or CIS. Imported machinery usually has a significant quality advantage, but the cost/benefit ratio still remains better for domestic machinery. Moreover, federal and regional subsidies are available only for domestic machinery. Three major international companies (John Deere, Case, and Claas) are vitally interested in expanding their market share in Russia. So far, however, their attempts to launch manufacturing or assembly plants have not been successful.

Russia is one of the leading fertilizer manufacturers and exporters in the world. It ranks first, second, and fifth in world exports of N, P, and K fertilizers, respectively. About 85% of Russia's fertilizer output is exported and only 10% is sold domestically (the remainder is used in further processing). Similarly to farm machinery, the fertilizer industry has experienced a significant growth since 1998. This growth, however, has not been in response to demand recovery: it is attributable to advantageous world prices and the entire additional output is exported.

Table 8.3. Sales channels of private fertilizer manufacturers (percent by volume)

	1998	1999	2000	2001
Regional administration	0.05	0.14	0.08	0.05
Corporate farms	5.5	6.7	4	3
Peasant farms	0.02	0.03	0.04	0.04
Exports	94.4	93.2	95.8	96.9

Source: 2001 AFE survey.

Table 8.4. Structure of payments for private fertilizer manufacturers

	Percent of respondents	Trend	Prices received
Bank transfers	94	Increases	Lowest
Cash payments	0.1	Decreases	Lowest
Barter	5.9	Decreases	Highest

Source: 2001 AFE survey.

All fertilizer manufacturers have been privatized, except those in Bashkiria and Tatarstan. The 2001 AFE survey has shown that in state-owned enterprises the domestic sales are mainly to the regional administration. Private fertilizer manufacturers, on the other hand, sell very little to regional administrations and most of their domestic sales are directly to agricultural producers, without any intermediaries (Table 8.3). Export payments are naturally settled by bank transfers (Table 8.4). Domestic sales, on the other hand, are predominantly in the form of barter transactions (e.g., "fertilizer for grain").

Domestic Demand

With the onset of market reforms in the early 1990s, agriculture was faced with a severe cash crunch, which bordered on a total financial collapse. Lack of financial resources constrained the purchase of farm inputs, and purchased inputs began to be replaced with land and labor. Land was treated as a free input because of the prohibition of land sales and the long-standing Marxist tradition that did not attach any costs to the use of land; labor was relatively cheap because of the rapid and uncontrolled slippage of agricultural wages (see Chapter 7). Russian agriculture was thus launched on the dangerous path of decapitalization and extensification. The stock of tractors and combines has decreased sharply since 1990; the consumption of fertilizers—the main factor sustaining intensive agricultural production—dropped by a staggering 85% between 1990 and 2002 (Figure 8.3).

Fig. 8.3. Fertilizer consumption and machinery availability 1990-2002

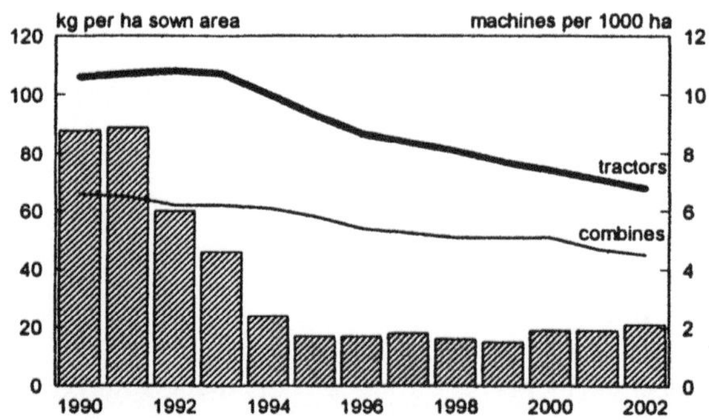

Source: Goskomstat

However, price increases have also encouraged a more efficient use of purchased inputs. In the Soviet era, fertilizer losses on farms reached 40% of deliveries, while farm machinery had a very short lifetime because of low quality, careless exploitation, and poor maintenance. Today, all the fertilizer delivered to the farm is actually applied; farm machinery remains in use for a much longer time, and old equipment is often cannibalized for maintenance and repairs. The reduction of fuel and power consumption in agriculture outstripped the decrease in production (Figure 8.4): while agricultural gross product decreased by 40% between 1991 and 2001, the use of gasoline dropped by 76%, diesel fuel by 63%, and electric power by 51%. These numbers also point to a more efficient use of purchased inputs in agriculture.

Fig. 8.4. Input use and agricultural production 1990-2002*

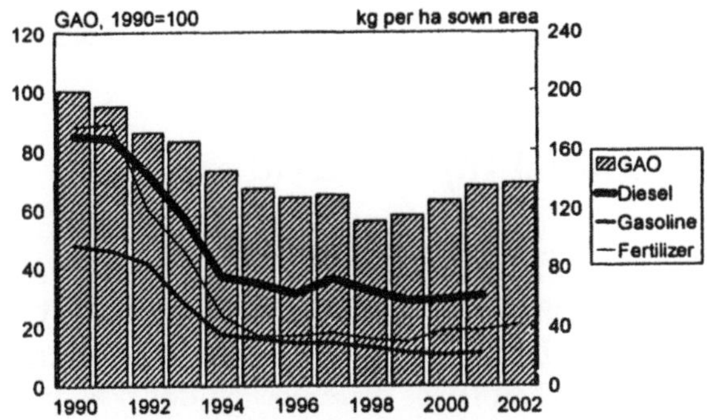

*Input use in kg per ha of sown area; GAO in percent of 1990

In the mid-1990s, growing liquidity constraints and high indebtedness of agricultural producers led to proliferation of various barter arrangements for provision of farm inputs. Federal and regional programs offer various commodity credits, while traders and processors provide inputs to farms that supply them with produce and raw materials. Even though the farms do not purchase these inputs directly for themselves, the credit commodity arrangements actively influence the market demand for inputs by channeling the actual purchases through market intermediaries. Another active channel for the purchase of farm machinery and other inputs began to develop in 1998 in the form of vertically integrated agroholdings, which purchase these factors of production for their affiliated farms (see Chapter 4).

An opposite effect on the demand for farm machinery can be traced to changing patterns of machine use in Russian agriculture. In the past, it was taken for granted that each farm had to have a full complement of machinery as prescribed by its technology, land endowment, and the cropping cycle. Even machinery that was needed for a very short time during the year had to be acquired and held by the farm in accordance with central planning. To this very day, MinAg continues to calculate sufficiency and shortage of farm machinery based on this "total coverage" approach. And yet this is no longer necessary, because various commercial entrepreneurs (both domestic and foreign) are offering machinery rentals and custom farming services, often in return for a share of the harvest. Waves of Turkish tractor and harvester teams move every year from the south to the north, following the harvest frontier with their own machinery and equipment. Corporate farms and especially individual farmers are also willing to rent out their machinery stock with or without an operator to any producer who happens to suffer from some shortage. This change of machine-use practices has sharply reduced the overall demand for new farm machinery, reinforcing the trend triggered by the decline of agricultural production and the increasingly parsimonious behavior due to rising relative prices. Moreover, machinery purchases are no longer limited to agricultural producers: upstream and downstream businesses have also begun to enter this market.

Use of Purchased Inputs: Evidence from the BASIS Survey

The use of purchased inputs by agricultural producers is one of the issues addressed by the 2003 BASIS survey of corporate and individual farms in three oblasts (the individual farms were broken down by peasant farms and household plots). Purchases of diesel fuel and to a lesser extent gasoline are reported by respondents with the highest frequency. These seem to be the essential inputs, while all other inputs are purchased much less frequently. Corporate farms generally show a higher frequency of input purchases than individual farms (Table 8.5; they also purchase larger quantities of inputs because of their larger size).The gap between corporate farms and peasant farms is particularly noticeable for purchases of fertilizers and machinery.

Table 8.5. Use of purchased inputs by corporate and individual farms (percent of farms reporting input purchases)

	Corporate farms (n=142)	Peasant farms (n=223)	Household plots (n=202)
Fertilizer	74	57	5
Gasoline	96	61	32
Diesel fuel	98	98	20
Seeds	49	49	16
Animal feed	32	4	90
Farm machinery	29	7	1
Mechanical field services	18	17	40
Farms without machinery*	4	15	80

*Tractors, harvesters, feed combines, and trucks.
Source: 2003 BASIS survey.

Among individual farms, there is a striking difference in the frequency of input purchases between the larger peasant farms and the smaller household plots. Peasant farms show a higher frequency of input purchases for most inputs (Table 8.5). Two notable exceptions are animal feed and mechanical field services. Virtually all household plots purchase animal feed because of their high reliance on livestock production and their small size (leading to insufficient capacity for feed production). Very few households have farm machinery of their own, and this in turn explains the high frequency of household plots that purchase mechanical field services (plowing, tilling, harvesting). The availability of machinery is highest among the corporate farms, which apparently continue to use the old machinery stocks accumulated during the Soviet period, while peasant farms rely on new machinery acquired during the last decade. Still, 85% of peasant farms have at least one piece of motorized machinery (a tractor, a harvester, a feed combine, or a truck). Overall, the data in Table 8.5 seem to suggest that the use of purchased inputs increases with the increase of farm size from household plots to corporate farms.

Table 8.6. Supply channels for corporate and individual farms (percent of inputs purchased from each channel)

	Gasoline		Diesel fuel		Fertilizer	
	Corporate	Individual	Corporate	Individual	Corporate	Individual
Regional government	1	4	2	3	4	1
Gas station	35	85	23	55	16	17
Trade	28	8	30	24	68	76
Buyer	20	2	28	4	6	2
Mother company	2	—	2	—	2	—
Other	14	1	15	14	4	3
Total quantity	100	100	100	100	100	100

	Machinery*	
	Corporate	Individual
Leasing	28	21
Manufacturer	14	0
Trade, dealers	24	35
Buyer	0	3
Mother company	15	—
Other farms (used)	19	41
Total	100	100

*Percent of all reported machinery purchase transactions in the sample (104 transactions for corporate farms, 34 transactions for individual farms).
Source: 2003 BASIS survey.

In sharp variance with the Soviet practice, the state no longer plays a major role as a supplier of farm inputs. The share of input purchases from regional authorities is very low (Table 8.6). The emphasis has shifted to commercial trade channels. These include wholesalers, fertilizer manufacturers, and gas stations. The reliance on commercial suppliers is greater for individual farms (the table combines peasant farms and household plots into one category). Corporate farms have access to two new supply channels that appeared during the 1990s: they receive inputs from buyers of agricultural commodities (dairy and meat processors, vegetable marketers, grain elevators) and also from "mother companies," i.e., commercial holding companies that acquire farms as part of their business strategy (see

Chapter 4). The reliance on the "mother company" is particularly noticeable for machinery purchases, where fully 15% of the reported transactions are organized in this novel way.

The share of inputs purchased through the mother company is substantially higher in Rostov Oblast, which is a "hotbed" of agroholding activity (see Chapter 4). Thus, holding structures supply 16% of gasoline, 7% of diesel fuel, and 9% of fertilizers to corporate farms in Rostov Oblast (compared with about 2% of these inputs in all the three oblasts surveyed). Together with inputs supplied by buyers of agricultural commodities, the share of vertical integration arrangements in Rostov Oblast approaches 40% of the total quantity of purchased inputs. These new channels may be regarded as part of a trend toward vertical integration of input supply and agricultural production, which has become possible only with the progress in market reforms.

The payment arrangements in the survey are primarily cash and bank transfers, with individual farms emphasizing cash transactions to a greater extent than corporate farms (Table 8.7). The prevalence of barter transactions, which characterized the early 1990s, is gone. Mutual offsets of payables and receivables—another payment method that emerged in the atmosphere of severe cash shortages at the beginning of the reform—is still practiced by corporate farms for 10%-15% of their purchases, but practically by none of the individual farms.

Table 8.7. Forms of payment for fertilizers and machinery in corporate and individual farms

	Fertilizer*			Machinery**	
	Corporate	Individual		Corporate	Individual
Commodity credit	9	7	Commodity credit	8	5
Barter	6	2	Barter	4	5
Cash	25	54	Cash	22	66
Bank transfers	51	34	Bank transfers	51	24
Mutual offsets	9	3	Mutual offsets	16	0
Total quantity	100	100	Total quantity	100	100

*Percent of quantity purchased (98 corporate farms and 138 individual farms).
**Percent of all reported machinery purchase transactions in the sample (104 transactions for corporate farms, 34 transactions for individual farms).
Source: 2003 BASIS survey.

Despite anecdotal claims, there is no evidence of price discrimination against individual farmers in input markets. On the contrary, individual farms surveyed generally appear to pay lower prices for inputs than corporate farms, but the differences in most cases are not statistically significant. Only the prices of diesel fuel and concentrated feed are statistically significantly lower for individual farms (but only by about 5%-10%). This may be due to the fact that individual farms purchase these inputs at market prices, whereas corporate farms often receive diesel fuel and concentrated feed as part of government commodity credit programs, which charge a higher markup.

Government Support Programs for Purchased Inputs

Reimbursement of input purchase costs is one of the main tools of government support to agricultural producers. One-third of the agricultural support funds in the federal budget is earmarked for input cost reimbursement. Oblast budgets supplement this allocation in varying degrees depending on regional policy priorities. Cost reimbursement programs include subsidies for fertilizers, fuel, electric power, elite seeds, and breeding livestock. Soil amelioration activities are also entitled to government support. An important segment of agricultural subsidies consists of programs that partially reimburse the interest expense of producers on commercial loans (this is the only form of credit subsidy in Russia today). Another category of support programs provide medium-term loans that allow machinery leasing. These loans are administered by regional leasing monopolies subordinated to the state leasing agency Rosagrolizing and do not go through the government budget (see Chapter 9).

The various support programs—both federal and regional—typically incorporate conditions that severely restrict the functioning of input markets. Thus, to be entitled to federal subsidies for fertilizer purchasing and machinery leasing, the producer must deal with suppliers and manufacturers from a limited list approved by the government. Most regional support programs incorporate similar restrictions, although some oblasts with relatively liberal policies (most notably, Vologda and Perm) allow producers to sign contracts with any supplier. Fuel subsidy programs often take the form of commodity credits, stipulating payment by delivery of farm products (e.g., grain) to federal or regional stocks.

The effectiveness of support programs can be assessed (at least in principle) by comparing the subsidized prices with general market prices. An attempt has been made to conduct such a comparison for the corporate farms in the 2003 BASIS survey. The results are summarized in Table 8.8 in the form of percentage differences between the subsidized prices and the market prices: positive differences indicate that subsidized prices are higher than market prices (ineffective subsidies), whereas negative differences indicate that subsidized prices are lower than market prices (effective subsidies).[1] We see from the table that input subsidies have a mixed effectiveness record across inputs and across regions. In Rostov Oblast, most subsidized prices appear to be higher than market prices; only the price of leased tractors is less than the market price (the price of subsidized gasoline is essentially equal to market price). In Nizhnii Novgorod Oblast, on the other hand, subsidized fertilizer and fuel are cheaper, while leased tractors are more expensive. In Ivanovo Oblast, subsidized diesel fuel is cheaper, leased harvesters are more expensive, while the price of subsidized gasoline and leased tractors is essentially equal to the market price (there are no general fertilizer subsidies in Ivanovo). As we have noted previously, this price information is highly unreliable: it is based on very small numbers of respondents and suffers from large data errors. Yet even this crude evidence is sufficient to raise serious doubts concerning the effectiveness of input subsidies programs. There is clearly a strong need for a careful analytical assessment and revision of the existing support mechanisms.

Table 8.8. Subsidized input prices compared with market prices for corporate farms in three oblasts*

	Rostov	Ivanovo	Nizhnii Novgorod
Fertilizer	5.9	n.a.	−11.0
Gasoline	−0.3	0.1	−3.3
Diesel fuel	5.2	−3.2	−3.1
Tractors	−5.1	0.0	46.4
Grain harvesters	47.8	91.0	n.a.

*Price differences in percent: +/− according as subsidized price higher/lower than market price.
Source: 2003 BASIS survey.

Since detailed examination of input subsidies produces such a mixed picture, we have combined several characteristics of regional support programs into a single index that can be used to rank the oblasts by the level of administrative intervention in agriculture (Table 8.9). Vologda, Nizhnii Novgorod, and Perm appear to be the most liberal oblasts, characterized by the lowest government intervention level. The agricultural policies of Chuvashia and Chelyabinsk, on the other hand, are highly interventionist. Ivanovo and Rostov fall somewhere in the middle on the administrative intervention scale. Table 8.9 also shows that, on the whole, wealthy regions adopt liberal agricultural policies, while poorer regions are more interventionist. This result is based on a very small non-representative sample of 7 out of 77 Russian regions and perhaps we should not be surprised that it is not entirely consistent with other published findings, which usually indicate

[1] The calculations for Table 8.8 are in principle equivalent to Nominal Protection Rates (NPR) for the respective resources (except for the sign).

that agriculturally rich regions are more conservative in their policies. Because of the different analytical definitions used in these studies, the relationship between agrarian policies and regional wealth requires further study.

Table 8.9. Scores for level of administrative interventions in agriculture and regional wealth in selected regions 2002 (0 = lowest, 3 = highest)*

	Intervention level	Regional wealth
Vologda	0.39	2.75
Nizhnii Novgorod	1.01	2.26
Perm	1.02	3.00
Ivanovo	1.59	0.00
Rostov	1.67	1.55
Chuvashia	2.20	0.81
Chelyabinsk	2.78	2.05

*Intervention level score based on number of support programs for general services; share of agricultural budget expenditure on food funds; number of restrictive government decisions affecting agriculture. Regional wealth score based on share of transfers from federal budget in regional budget revenues; gross regional product per capita; ratio of average per capita income to minimum standard of living.

Source: Calculated by the authors from data provided by regional and federal statistical organs and by the Russian Ministry of Finance.

Conclusion

In parallel with the development of markets for farm products, we are witnessing the emergence of new market channels for farm inputs. The state no longer has a role as a direct supplier of inputs to agricultural producers. This function has shifted to wholesalers, traders, and manufacturers, who sell mainly for cash and bank transfers, not barter. The strong imperfections that still prevail in input markets have encouraged vertical integration, with fertilizers, fuel, and machinery delivered in substantial quantities through internal channels of large holding structures.

Fuel is the one input that is purchased by most producers. Fertilizer purchases are reported less frequently, whereas seeds and animal feed are mostly used from own production (despite lower quality). It seems that cash shortages are forcing farms to substitute land and labor—the two cheapest factors of production—for some purchased inputs (fertilizers, seeds, feed), a process that inevitably leads to extensification of farm production and abandonment of productivity improving technologies.

Although the government no longer delivers farm inputs, it has a strong negative influence on input markets through a wide range of federal and regional support programs. Government-sponsored leasing programs with their restrictions of approved suppliers and models have created severe obstacles to the development of dealer networks, which will have a detrimental effect on the competitiveness of Russian manufacturers in the long run. The cost reimbursement policy for fertilizers only increased the demand for this input and encouraged the export-oriented manufacturers to raise prices in the domestic market. In regions characterized by lower levels of government intervention we are witnessing significant growth of competitive trading in both machinery and fertilizers.

Chapter 9

Farm Finances and Access to Credit

Olga Yastrebova, Aleksandr Subbotin, and David Epshtein

Russia's booming financial sector has bypassed agriculture. Agricultural financial markets are underdeveloped and thin. The institutional framework is mainly geared for credit delivery to the much more attractive manufacturing, trade, and service sectors. Agriculture is unattractive for credit providers because of its low profitability, which in turn leads to debt repayment difficulties and forces increasing reliance on government support programs instead of commercial banking.

Russia's farm sector today consists of three distinct organizational components—corporate farms, independent peasant farms, and household plots. Yet financial statistics are available only for corporate farms. The financial situation of the individual sector (peasant farms and household plots) can be studied only through surveys, and even these typically produce very meager results because of the respondents' reluctance to answer delicate financial questions. We accordingly start this chapter with a review of the financial performance of corporate farms as a mirror of Russian agriculture, discuss government support schemes and availability of banking services—also mainly for corporate farms, and then present an analysis of the financial issues reflected in the 2003 BASIS survey, making an effort to deal with both corporate and individual farms.

Corporate Farm Finances: Profitability and Debt

The sector of corporate farms—the new organizational forms that have evolved from former kolkhozes and sovkhozes—is characterized by low profitability and high frequency of farms reporting losses. While the economy as a whole registered profitability rates of 10%-15% of sales between 1995-2002, agriculture reported losses of up to 25% of sales between 1996-1998 and then recovered to positive profit margins of about 5%-10% of sales. Despite the recovery after 1998, the profit margins in agriculture have remained consistently below the average for the economy (Figure 9.1).

Fig. 9.1. Profit margins: agriculture and the economy

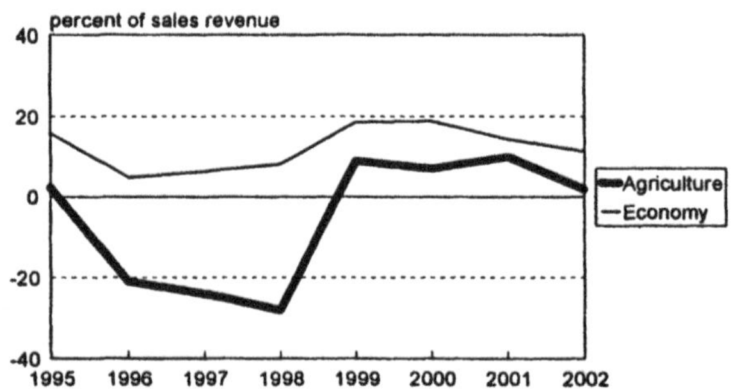

Fig. 9.2. Percentage of enterprises reporting losses: agriculture and the economy

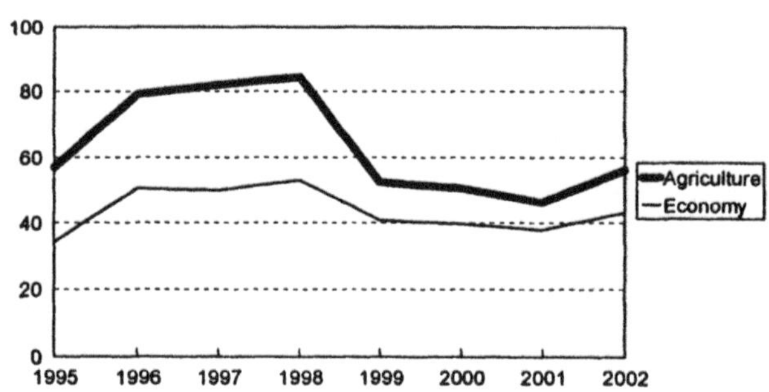

The low profitability of the sector is a reflection of the high frequency of unprofitable farms[1]: the percentage of corporate farms reporting losses increased from 60% in 1995 to a peak of more than 80% in 1998; after that, the general situation in agriculture seems to have improved, and the percentage of unprofitable corporate farms stabilized at around 50%. This essen-

[1] In Russia, as everywhere in the world, scholars often argue that the reported accounting profit is not equal to the "true" economic profit. Specifically, Karlova shows that the reported profitability of Russian farms is understated due to excessive depreciation charges and double counting of the cost of inputs produced on the farm (N. Karlova, "Correction of profitability rate in agriculture for depreciation and double accounting," *Byulleten' Tsentra APE* No. 3[5]: 25-32 (2000) [http://www.iet.ru/afe/bulls/Bull5r.pdf].

tially means that 12,500 corporate farms in Russia continue to report losses in their annual financial statements. Here again, despite the noticeable improvement after 1998, the percentage of unprofitable corporate farms is substantially higher than the percentage of loss-reporting enterprises in the economy (Figure 9.2).

The comparison between the profit performance of agriculture and the rest of the economy (Figures 9.1, 9.2) indicates that agriculture is apparently much less attractive than other sectors for financial capital. It is unable to compete for investment funds, which generally flow where the rates of return are highest.

The low profitability and the high incidence of losses among corporate farms have led to an increase in the level of indebtedness of the large farm sector. Farm liabilities are reported in inflated nominal values, which naturally do not provide a true picture of the changes in debt levels over time. One technique to obtain the growth of liabilities over time is by dividing the total nominal liabilities each year by the (nominal) sales revenue for that year. In this debt-to-sales ratio, both the numerator and the denominator are expressed in the same nominal rubles, and the behavior of the ratio over time is therefore free from inflationary effects. This ratio is expressed in units of time: it is sometimes called credit-years, as it gives the number of years that the farms would need to repay their total liabilities if all sales revenue were used for repayment.

Fig. 9.3. Farm indebtedness 1995-2002

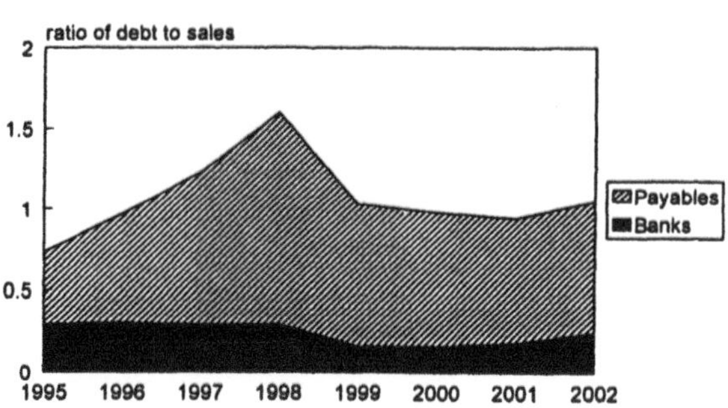

The top frontier in Figure 9.3 traces the debt-to-sales ratio of corporate farms between 1995-2002. Farm indebtedness increased dramatically from 0.8 in 1995 to a peak of 1.6 in 1998, and then declined stabilizing at a level of about 1 between 1999-2002. While the situation between 1995 and 1998 can be characterized as dramatic deterioration of indebtedness—it required 1.6 years of sales revenue to repay the outstanding debt in 1998 compared with less than a year of sales in 1995, the situation since 1999 constitutes significant improvement. Today, total liabilities of corporate farms run at about the level of annual sales, which is far from crushing indebtedness for the sector as a whole.

Yet this is an average result, and many corporate farms have debt-to-sales ratios much higher than the average of 1. Table 9.1 shows, for instance, that in 40% of the regions the average debt-to-sales ratio of corporate farms exceeds 1.5. Thus, many of Russia's 85 provinces suffer from high farm indebtedness.

Table 9.1. Distribution of the debt-to-sales ratio by Russia's regions, 2002

Debt-to-sales ratio	Number of regions (out of 85)	Examples
Up to 0.8	15	Krasnodar, Moscow, Leningrad, Novgorod, Yaroslavl'
0.8-1.0	16	Stavropol', Novosibirsk, Rostov, Tyumen
1.0-1.5	20	Belgorod, Ivanovo, Nizhnii Novgorod, Ryazan', Saratov
More than 1.5	34	Voronezh, Lipetsk, Kursk, Orel, Samara, Tambov

Source: MinAg.

Another measure of indebtedness is provided by the ratio of debt to assets. The normal practice is to calculate the ratio of total debt (both long-term and current liabilities) to total assets. However, the fixed-asset component of the denominator is very sensitive to inflation. As long as total assets are revalued periodically, it can be argued that total assets are reported approximately in current nominal values. If the fixed assets are not adjusted for inflation by revaluation, they continue to be reported at historical cost, which becomes progressively eroded in an inflationary environment. Asset revaluation was sporadic between 1990 and 1997 and it was discontinued in 1998. The ratio of total debt to total assets therefore produces a highly distorted picture because of the inflationary erosion of the denominator. In an inflationary environment, it is more appropriate to calculate the ratio of total debt to current assets (mainly accounts receivable and inventories), so that both the numerator and the denominator are expressed in nominal current values.

Both ratios are presented in Table 9.2. The debt-to-total assets ratio produces a very disappointing picture: it upsets the positive results obtained after 1998 for farm indebtedness relative to sales. The ratio of debt to total assets increased sharply from 20% in 1998 to 40% in 2002. The explanation for this phenomenon, however, is to be found in the last column, which indicates that the fixed assets were not revalued between 1998 and 2002, and the ratio is strongly biased upward by the erosion of the denominator. The ratio of debt to current assets, on the other hand, presents a picture which is basically consistent with that in Figure 9.3. According to this measure, farm indebtedness increased from 1995 to 1998 and then declined between 1998-2002, just as it increased and then declined by the ratio of debt to sales in Figure 9.3. Having said that, we should add a strong word of caution: farm indebtedness during 1998-2002 is much higher than it was during the earlier period 1990-1996. Farm debt since 1998 is greater than the value of current assets, which indicates serious liquidity problems.

Table 9.2. Ratio of debt to current and total assets 1990-2002 (percent)

	Debt to Current Assets	Debt to Total Assets	Revaluation policy
1990	79.9	21.7	
1991	44.2	14.2	
1992	51.0	10.6	Asset revaluation begun
1993	74.2	40.2	No asset revaluation in 1993
1994	93.0	18.9	Asset revaluation resumed
1995	82.1	11.9	
1996	85.5	8.9	
1997	98.7	13.2	
1998	122.1	20.5	Asset revaluation discontinued
1999	117.8	25.8	
2000	118.3	32.1	
2001	107.2	35.1	
2002	110.8	40.2	

Source: Calculated by Vasilii Uzun from the database of farm financial reports.

Figure 9.3 breaks the indebtedness of corporate farms into two components: debt associated with bank credit (the bottom layer) and debt associated with accounts payable—mainly suppliers and the government (the top layer). It is clear from the relative weights of the two components in Figure 9.3 that bank credit contributes a relative small portion of total indebtedness, and the main contribution (both before and after 1998) comes from accounts payable.

A more detailed view of the changes in the relative importance of bank loans and accounts payable is given in Figure 9.4, which shows the percentage structure of corporate farm liabilities. In recent years bank credit contributes a relative small portion of total indebtedness, and the main contribution comes from accounts payable. Banks provided a greater share of external financing until about 1995-1996, as long as the bank system was basically controlled by the state and agriculture enjoyed soft budget constraints. The shift to private commercial banking after 1995 resulted in tightening of financial discipline, which led to significant reduction of bank loans available to generally unprofitable agriculture. The share of bank credit shrank from 40% of liabilities to 20% between 1995-1998 and has remained at this reduced level since then. Improvement in farm profitability after 1998 resulted in certain expansion of the availability of bank credit to agriculture. Yet accounts payable contribute 75%-80% of total liabilities of corporate farms during the recent years.

Fig. 9.4. Structure of farm liabilities

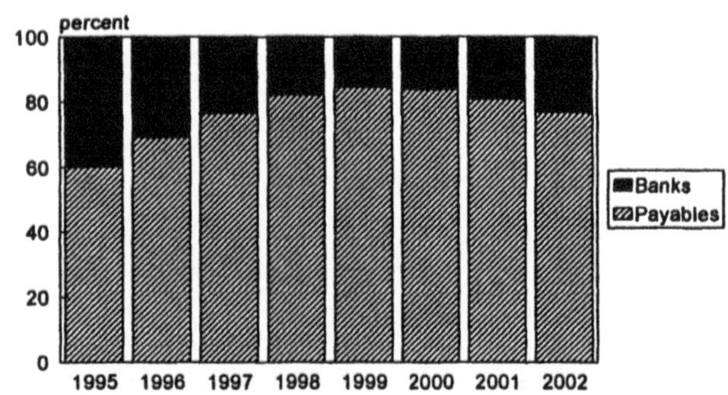

The structure of accounts payable is shown in Table 9.3 (the numbers are averages for two five-year periods 1993-1997 and 1998-2002). In recent years (1998-2002), 35% of accounts payable is supplier credit, i.e., amounts owed to suppliers for inputs and services purchased in the course of normal business operations, while more than 50% is debt to the government, partly in the form of taxes (13%) and more importantly in the form of accrued social deductions owed to the pension fund or to social and medical insurance (40%). The structure of payables in agriculture appears to be different from the structure of payables in industry, where the proportions are reversed: suppliers represent nearly 50% of payables and debt to the government is 30%. The structure of accounts payable in industry is thus much closer to the structure that prevailed in agriculture in previous years (1993-1997). Figure 9.5 visually demonstrates the dramatic change in farm accounts payable between 1993-1997 and 1998-2002: a sharp increase in the share of social deductions and tax arrears (from 32% to 44%), balanced by a decrease in supplier credit, payroll arrears, and other payables.

Table 9.3. Structure of accounts payable of corporate farms 1993-1997 and 1998-2002

	1993-1997	1998-2002
Supplier credit	44.3	35.1
Social deductions	22.0	40.2
Taxes	10.3	13.4
Payroll	14.0	5.7
Other	9.4	5.6

Source: Calculated by Vasilii Uzun from the database of farm financial reports.

Fig. 9.5. Structure of payables in corporate farms 1993-1997 and 1998-2002

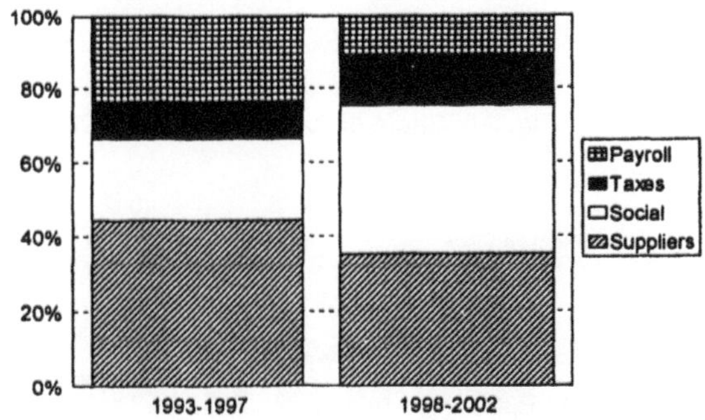

Corporate farms thus derive their non-equity financing mainly through supplier credit and even to a greater extent through accrual of unpaid obligations to the state (profit tax, payroll taxes, social deductions). Banks play a relatively minor role as suppliers of credit to corporate farms. This structure reinforces our conclusion that agriculture is not particularly attractive to financial capital (as represented by bank credit).

Overdue Debt

As noted previously, the corporate farms do not appear to suffer from a crushing debt burden, and yet nonpayments and overdue liabilities are perceived as a serious problem in Russian agriculture. Indeed, overdue liabilities increased from 40% of total debt to more than 70% in 1998-1999, and declined thereafter to about 60% (the black curve in Figure 9.6). Most of the overdue liabilities come from accounts payable, and the nonpayment rates for bank credit are substantially lower (peaking at 40% in 1998-1999 and then dropping back to 20% in 2002). Banks apparently manage to impose a much stricter payment discipline on corporate farms than other creditors, which are included in accounts payable.

Comparison of the structure of overdue payables and total payables (Table 9.4) also seems to suggest that, within the category of accounts payable, commercial structures command greater compliance than state creditors. Thus, the proportion of overdue supplier credit is less than the proportion of supplier credit in general, whereas the proportion of overdue debt to the state (overdue taxes and social deductions) is higher than the proportion of the state in total payables.

Nonpayments to the budget (mainly taxes and social security deductions) invoke severe penalties and fines, which are automatically added to liabilities. Accrued penalties and fines as of the end of 2002 were estimated by MinAg at 60 billion rubles, or 25% of all payables (this estimate was the basis for the July 2003 write-off of 57 billion rubles in fines and penalties—see below). MinAg further estimated the level of penalties and fines at 30% of the amounts due for social insurance and 50% of the amounts due for other deductions.

Fig. 9.6. Share of overdue debt 1995-2002

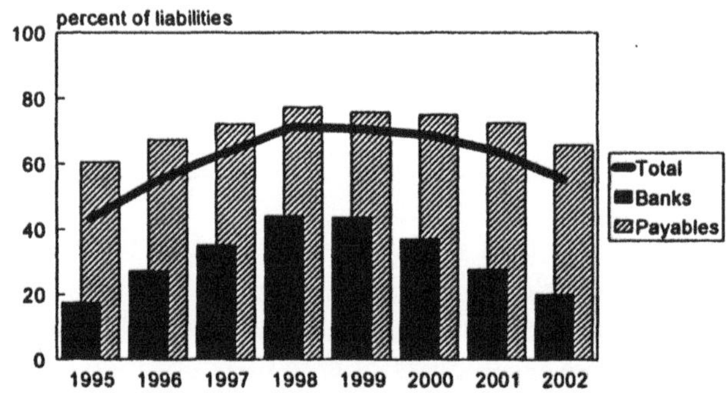

Table 9.4. Structure of payables and overdue payables 1998-2001

	Total payables	Overdue payables
Suppliers	38	34
Taxes	13	14
Social deductions	39	43
Other	10	9
Total	100	100

Source: Finansy Rossii 2002.

Since the early 1990s, the Russian government has imposed a draconian penalty mechanism for nonpayment of taxes and social deductions. Penalties are calculated at the fantastically high rate of 0.3% of overdue amounts per day, which is equivalent to 200% annually (until 1996 the rate was much higher: 1% per day or 3,600% annualized). The deadline for payroll taxes and deductions is calculated from the date when the payroll is accrued, and not from the date of actual payment to workers. Thus, farms that do not have cash to pay wages (including payroll taxes and deductions) on time find themselves in the paradoxical situation of accumulating fines on the deductions component of the payroll. These fines must be paid in cash, or else continue to grow at 0.3% per day. Finally, the authorities have the power to freeze the bank accounts of the delinquent farms and automatically confiscate all incoming deposits in partial payment of the amounts owing for taxes, deductions, and fines.

Needless to say, these draconian penalty procedures encourage a shift to payment systems with both suppliers and customers that bypass the banking system altogether. Many farms have adapted to the new reality. Thus, in a 1999 survey of 150 corporate farms in Chuvashia, Saratov, and Moscow Oblast conducted by TACIS,[2] 60% of respondents reported that they manage to survive without problems with blocked bank accounts, and 8% actually indicated that they had no need of bank accounts altogether. Cash-poor farms are thus being pushed into a "shadow" economy, probably with increasing reliance on barter transactions.

Bankruptcy

The next step after fines and penalties is the threat of bankruptcy. In the 2002 Bankruptcy Law, the debt threshold that allows a creditor to initiate bankruptcy proceedings is 100,000 rubles or about $3,500. According to judicial authorities in charge of bankruptcy proceedings (*arbitrazhnyi sud*—arbitrage court), 82% of all corporate farms were above this legal bankruptcy threshold in 2002 because of their arrears for taxes and social deductions. Given these dramatic figures, the bankruptcy law has been enforced fairly leniently. There were 5,500 farm bankruptcy cases pending in 2003, or 22% of all corporate farms in Russia (up from 2,100, or 8%, in 2001 and only 100 in the late 1990s). In Omsk Oblast, 90% of the 440 corporate farms were candidates for legal bankruptcy in July 2003; bankruptcy proceedings were actually initiated for 71 farms, or 18% of the candidates. Nationally, farm bankruptcy cases account for 20% of all bankruptcy cases in Russia (and that for a sector that produces 6% of GDP). Bankruptcy proceedings are usually initiated by tax authorities (for overdue taxes and deductions) and by major suppliers, such as fuel and power companies. In 2001, each of these two groups of creditors accounted for 50% of applications for bankruptcy (all data from arbitrage court).

Bankruptcy, which for many years had been regarded as a remote abstract possibility in agriculture, recently became a tangible threat for financially insolvent farms. The intensification of bankruptcy proceedings is a reflection of a change in government policy, or at least in the signals that regional authorities receive from the center. Previously, the central government was clearly indicating that agriculture should be left alone despite the growing farm debt. In recent years, however, the policy appears to have changed, probably due to improvements in farm profitability as well as the expressed interest of non-agricultural entities to enter agriculture (see Chapter 4). Farm bankruptcies are no longer taboo, especially in cases when liquidation of deeply indebted farms opens the road for large-scale takeovers

---

[2] EU Tacis RF 27 project "Assistance to the RF Ministry of Agriculture in the restructuring of farm debts" (February 2000).

by non-agricultural companies willing to enter agricultural production. Farm bankruptcy in Russia is thus beginning to assume its theoretical role of an asset redistribution mechanism in agriculture.

While bankruptcies certainly make some farm assets available for redistribution, specifically to non-agricultural interests, the problem of non-payments and overdue debt more generally appears to be one of the factors driving the creation of vertically integrated agroholdings—a new phenomenon in Russian agriculture that began to emerge in the second half of the 1990s. In principle, an agroholding is a corporate structure in which a non-agricultural entity assumes control of an existing agricultural producer and, among other things, guarantees its financing requirements. Farms can thus solve their solvency problems and avoid bankruptcy by joining an agroholding. Non-agricultural companies, such as input suppliers, power utilities, and even banks, in turn may be motivated to acquire their agricultural debtors as a strategy for salvaging at least some of their overdue receivables without going through lengthy and costly bankruptcy proceedings. Indeed, according to a recent survey in Rostov Oblast, 6 of the 14 agroholdings surveyed identified debt recovery from farms as the main reason for acquiring an agricultural producer (for more details see Chapter 4).

The legal bankruptcy mechanism includes several stages, which start with debt restructuring intended to restore the debtor's solvency and end with total liquidation. The entire process is allowed a maximum of five years. Experience so far shows that most farms entering bankruptcy proceedings are unable to restore their solvency and eventually go into liquidation. According to MinAg, only 16% of farms in bankruptcy proceedings in 2001 managed to restore their financial health, and the remaining 84% were liquidated. Farms are usually broken up into several pieces for liquidation, and only a small proportion are sold to new investors as a going concern.

Financial Solvency: Analysis of Corporate Farms in Leningrad Oblast

Two attempts have been made (by Uzun and Epshtein) to group Russian corporate farm by solvency or financial health measures. Uzun's solvency measure is a combination of profit and indebtedness; Epshtein's solvency measure reflects the ability to cover wages and depreciation from value added. The classification algorithms and the corresponding measures are presented in Table 9.5.

Uzun's algorithm has been applied to classify all corporate farms in Russia (24,000 farms with full data in the Goskomstat database of farm enterprises; see Chapter 2). Epshtein's algorithm has been applied to 200 corporate farms in Leningrad Oblast (from the same database; see Chapter 13). Table 9.6 presents the distribution of corporate farms by solvency groups for all of Russia and for Leningrad Oblast. Bearing in mind the caveat that the two classifications are based on different criteria, we note that the solvency pattern in Leningrad Oblast is characterized by regional features that set it apart from the all-Russia pattern. The distribution of Leningrad farms by solvency groups is fairly uniform (both in 1999 and 2001), whereas for all of Russia we observe a clear bunching of farms in the two extreme groups (best and worst). Moreover, the financial situation of Leningrad farms seems to have deteriorated between 1999 and 2001 (34% of farms in the two top groups in 2001 compared with 42% in 1999), whereas the overall situation of Russian farms generally remained unchanged.

Table 9.5. Algorithms for solvency classification of corporate farms

Groups	Uzun	Epshtein
	Reported profit (P) Indebtedness K = (payables − receivables)/revenue	K1 = (revenue − input costs)/(wages + depreciation) K2 = (revenue − input costs)/(minimum wages + depreciation)
1 (best)	$P \geq 0, K \leq 0.5$	$K1 \geq 1$
2	$P \geq 0, 0.5 < K \leq 1$ or $P < 0, K \leq 0.5$	$K1 < 1$ and $K2 \geq 1$
3	$P \geq 0, 1 < K \leq 2$ or $P < 0, 0.5 < K \leq 1$	$K2 < 1$ and $K2 \geq 0$
4	$P \geq 0, 2 < K \leq 4$ or $P < 0, 1 < K \leq 2$	$K1 < 0$ and $K1 \geq -0.3$
5 (worst)	$P \geq 0, K > 4$ or $P < 0, K > 0.5$	All others

Table 9.6. Distribution of corporate farms by solvency groups: all Russia and Leningrad Oblast

	All Russia (Uzun)		Leningrad Oblast (Epshtein)	
	2002	1999	2001	1999
1 (best)	24	22	14	21
2	18	17	20	21
3	15	16	30	25
4	16	18	14	14
5 (worst)	27	27	21	19

The availability of data for two years (1999 and 2001) has made it possible to calculate the transition probabilities between "good" and "bad" groups for Leningrad farms (Table 9.7). On the whole, in Leningrad Oblast, insolvent farms (groups 4-5) appear to be stuck in their insolvency: for farms in groups 4-5 the probability is over 95% that they will remain in the "bad" groups. "Good" farms also tend to stay "good," but much less resolutely: farms in group 1 (the best financial performers) have a nearly 30% probability of moving down to the "bad" groups, whereas farms on the next rung of the financial scale (group 2) have a nearly 50% probability of moving to the "bad" groups.

Table 9.7. Transition probabilities between "good" and "bad" solvency groups for Leningrad Oblast corporate farms (percent)*

Original group	Probability to be in "good" groups 1-2	Probability to be in "bad" groups 3-5 or drop out	Total
1	73.2	26.8	100
2	51.2	48.8	100
3	16.0	84.0	100
4	3.6	96.4	100
5	2.6	97.4	100

*Transition probabilities estimated as the actual changes in the composition of the five groups between 1999 and 2001.

Table 9.8. Distribution of corporate farms by solvency groups: all Russia and Leningrad Oblast

	All Russia 2002 (Uzun)			Leningrad Oblast 2001 (Epshtein)			
	Farms	Revenue	Overdue debt	Farms	Revenue	Profit	Overdue debt
1 (best)	24	55	7	14	48	59	7
2	18	19	11	20	27	29	17
3	15	12	17	30	16	10	24
4	16	8	22	14	5	2	15
5 (worst)	27	5	43	21	4	0	36
Total	100	100	100	100	100	100	100

The best farms account for most of the sales revenue and most of the profit (Table 9.8). Thus, about 20% of farms contribute 50% of total revenues and 60% of total profit. On the other hand, most of the overdue debt is concentrated in "bad" farms: 20%-25% of group 5 farms account for 40% of overdue debt while roughly the same percent of group 1 farms account for 7% of overdue debt.

Fig. 9.7. Profit margin and level of overdue debt Leningrad Oblast, 2001

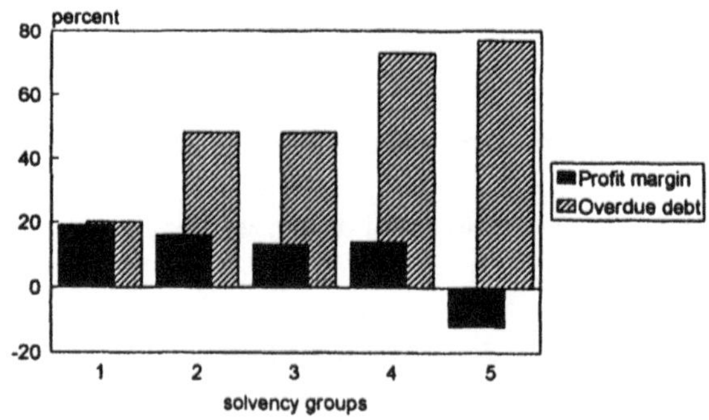

A self-explanatory pattern is observed for the distribution of profit margin and overdue debt across the five solvency groups in Leningrad Oblast (Figure 9.7). The ratio of gross profit to sales (the profit margin) decreases from "best" to "worst" farms, dropping to strongly negative values in group 5. The share of overdue debt (in percent of all farm debt) increases steeply from group 1 (20% overdue) to group 5, where more than three-quarters of debt is overdue.

Table 9.9. Investment in fixed assets by Leningrad Oblast corporate farms 2001

	All farms	Solvency groups				
		1	2	3	4	5
Total investment per farm, '000 rubles	4456	11703	4721	2819	1256	960
Percent financed from external sources	10	8	17	5	7	40
Investment in % of total fixed assets at beginning of year	7.1	13.3	8.7	4.6	3.3	3.4

Despite the dismal financial performance of more than one-third of corporate farms in the province, Leningrad farms continued to invest in 2001 (Table 9.9). Investment in fixed assets by corporate farms amounted to 7% of the total value of fixed assets at the beginning of the year. There are, of course, significant differences across the solvency groups: "good" farms invest much more than "bad" farms, but it is remarkable that net investment is observed for all farms in the oblast. Another remarkable feature is that all farms attract some external capital to finance their new investments. Most farms finance about 10% of their investments with external capital. However, insolvent farms (group 5) borrow 40% of their investment needs from outside sources. This is understandable in view of the low profitability of group 5 farms, but it is very difficult to understand how these practically bankrupt farms manage to borrow more.

The observed differences across the five solvency groups were tested for Leningrad Oblast by estimating a standard Cobb–Douglas production function with a dummy variable for the solvency group. The value of production was used as the dependent variable; the independent variables comprised a standard basket of inputs—labor, cost of purchased inputs, value of fixed assets, and agricultural land. The dummy-variable coefficients in this setting reflect differences in production efficiency (output produced by a given basket of inputs) for different levels of financial health as represented by the five solvency groups.

The coefficients of the solvency groups in the estimated production function were all positive and significantly different from zero (Table 9.10). This implies that group 5 (the base group in dummy-variable regression) is indeed the worst group among Leningrad corporate farms. Additional tests showed that the group coefficients are also significantly different from one another and that they are ranked in the expected order from group 1 (highest) to group 4 (lowest, but still significantly greater than zero) and then group 5 (zero). This is clear from an examination of the 95% confidence intervals of the estimated group coefficients in Table 9.9.

Table 9.10. Estimated Cobb-Douglas function for Leningrad Oblast corporate farms with solvency group dummies*

	Coefficients	Significance level	Left confidence limit (95%)	Right confidence limit (95%)
Labor	0.202	0.000		
Purchased inputs	0.851	0.000		
Fixed assets	0.005	0.799		
Agricultural land	0.005	0.703		
Group 1	1.029	0.000	0.88	1.15
Group 2	0.774	0.000	0.65	0.89
Group 3	0.588	0.000	0.48	0.69
Group 4	0.410	0.000	0.28	0.53
Group 5 (base)	0	—	—	—
Constant	−0.172	0.411		

*Dependent variable: value of production; all variables logged; $R^2 = 0.971$; 2001 data from Goskomstat database of corporate farms.

In a sense, the group dummies in Table 9.10 proxy for a missing management variable, as management quality is responsible (at least to some extent) for differences in financial health across solvency groups. The relationship between financial health and management quality was verified using a simple model in which financial health was expressed by the continuous solvency measure K1 (see Table 9.5) and the management quality variable was calculated as the ratio $Mng = Y_a/Y_e$. Here Y_a is the observed output (sales revenue) and Y_e is the estimated output from a Cobb-Douglas production function.[3] Farms with $Mng > 1$ outperform the norm predicted by the production function, presumably due to the superior quality of their management. Farms with $Mng < 1$ perform less well than the predicted norm, presumably due to the inferior quality of their management. Table 9.11 illustrates how the management quality decreases from the "best" to

[3] The Cobb-Douglas function for this analysis was estimated by regressing sales revenue in 2001 on standard input variables (labor, land, fixed assets, input costs) as well as a set of additional attributes characterizing natural conditions, distance from St. Petersburg and from the district center, farm specialization (mixed crop/livestock, feedlot, poultry, greenhouse, fur animals), level of regional and federal subsidies. In this regression with 195 observations $R^2 = 0.94$ and $F = 181.8$. This method of calculating management quality was borrowed from E. Heady and J. Dillon, *Agricultural Production Functions*, Ames, IA: Iowa State University (1972), p. 226.

the "worst" farms. The same table also shows the mean values of the continuous solvency variable K1. There is obviously a very strong positive correlation between the management quality and the solvency measure (the coefficient of correlation between Mng and K1 is 0.62).

Table 9.11. Management quality and solvency measure for the five solvency groups of Leningrad Oblast corporate farms in 2001

Solvency group	1	2	3	4	5	Oblast average
Management quality*	1.40	1.21	1.08	0.94	0.72	1.06
Solvency measure K1	1.48	0.72	0.24	−0.12	−0.71	0.27

*Defined as the ratio of actual sales to predicted sales from Cobb-Douglas production function.

To identify the determinants of financial health, we regressed the solvency measure K1 on the main variables from the Cobb-Douglas production function (labor, land, natural conditions, and farm specialization) adding the management quality as an explanatory variable. The conventional production model without Mng had a low explanatory power, $R^2 = 0.242$. The inclusion of Mng improved the explanatory power dramatically, raising it to $R^2 = 0.720$ in the full model with Mng. In the truncated model with Mng as the only explanatory variable we had $R^2 = 0.350$, which means that management quality alone accounts for almost 50% of the explained variability in the solvency measure K1 in the full model. Labor accounts for 28% of the explained variance in the full model and land for another 13%. All other variables combined account for less than 10% of the explained variability in solvency. Subsidies dropped out of the estimated model as statistically not significant

Management quality appears to be more important than physical endowments (labor and land) for success. Natural conditions and product choice play but a marginal role, while subsidies—perhaps the most popular policy mechanism in Russia—do not contribute at all to financial health. Government policies should therefore emphasize management quality through far-reaching training programs in finance, marketing, production management, and personnel management. A survey of corporate farms in Leningrad Oblast conducted in the spring of 2002 has shown widespread neglect of budgeting and cost control. The main managerial emphasis remains, as in the Soviet period, on production rather than on economic and financial performance.

Government Policies

In this section we examine some government policies that have a bearing on the main financial issues of the farm sector. These include policies relating to debt rescheduling, subsidized credit, and investment support. Debt rescheduling and write-off policies are "reactive" measures reflecting the government's response to the problem of mounting overdue debt and looming bankruptcies in agriculture. Subsidized credit and investment support programs, on the other hand, are "proactive" programs intended to improve the access of farms to the formal financial institutions and mitigate the observed shortage of agricultural investments.

Debt Rescheduling and Write-off

Implementing massive bankruptcies in agriculture is a very painful political decision, especially in a country such as Russia, where a high percentage of the population relies on agriculture for its living. Government policies over the years were accordingly aimed to allow negotiated restructuring of farm debt rather than enforce the bankruptcy law. Much of the debt overhang in the farm sector is attributable to arrears for taxes and social deductions, which are continuously inflated by accruing penalties and fines. Recognizing this situation, the government made repeated attempts over the years to write off the accrued penalties and reschedule the outstanding liabilities of the corporate farms to the state, thus allowing the frozen bank accounts to be unblocked. However, to be eligible for write-off of penalties and rescheduling of old debt, the farms had to start paying all their current liabilities to the state from a certain point in time (which was set anew in each succeeding resolution).

Table 9.12. Proportion of corporate farms that concluded debt restructuring agreements in 1999 (percent)

	Farms entitled to debt restructuring	Farms with debt restructuring agreements
Pension fund	89.2	3.4
Social insurance fund	84.7	26.6
Medical insurance funds	94.1	2.8
Unemployment benefits fund	83.8	4.8

Source: MinAg.

Most farms found this an impossible condition, or simply chose not to bother with the bureaucratic requirements, putting their trust in future write-off programs. Among the onerous bureaucratic requirements we can mention the need to sign separate debt restructuring agreements with each government agency (i.e., the pension fund, the medical insurance fund, etc.). As a result, only a very small proportion of farms among those entitled to debt restructuring according to the 1998 program actually joined the program (Table 9.12; note the significantly higher percentage of farms entering agreements with the social insurance fund, which from the start adopted a much more flexible and "debtor-friendly" policy than the other agencies).

One of the most recent attempts to restructure farm debt—the 2002 Law on Financial Recovery of Agricultural Producers—introduced uniform rescheduling requirements for both state and commercial debt. Yet the main intent of the law was to resolve the growing indebtedness of farms for taxes, social deductions, and obligations to state-owned energy suppliers (Gazprom, Electric Power System). For this reason, the law explicitly specified the write-off of fines and penalties as one of the instruments of debt rescheduling. The new law stipulated that agreement of creditors with claims to at least 75% of accounts payable was a precondition for debt rescheduling. Based on this condition alone, the large state creditors in most cases could negotiate debt rescheduling without the agreement of the commercial creditors. However, the law did not introduce any protection against initiation of bankruptcy proceedings by dissenting creditors, i.e., those who do not join the debt rescheduling agreement. Once the large state creditors had reached the required agreement and dropped out of the circle of outstanding claims, the remaining commercial creditors could start bankruptcy proceedings in the hope of recovering their (much smaller, but now "senior") debt through a distress sale of assets. This legal possibility constitutes a serious disincentive for the large creditors to reach a debt rescheduling agreement without the participation of virtually all the commercial creditors (with claims in excess of 300,000 rubles)—a consensus that is very hard to reach. In addition, the new law retained the previous controversial condition that farms would be eligible for debt restructuring only if they met all their current obligations to the state, i.e., taxes and social deductions.

These various requirements have virtually precluded the implementation of the debt rescheduling principles proposed by this law. As the next step toward possible resolution of the problem, a presidential decree of July 2003 promised 18,000 farms an unconditional write-off of 57 billion rubles in penalties and fines accumulated to January 1, 2002. To receive the write-off, the farms were "only" required to submit their fully substantiated applications before January 2004. According to MinAg, 30% of the eligible farms had submitted their applications by this deadline. The total outstanding debt of these farms to the state was 47 billion rubles, including 24.5 billion rubles in penalties and fines to be written off, i.e., just over 50% of the total write-off envisaged by the presidential decree.

Regional authorities also tried in some cases to help farms dig themselves out of the "debt trap." Thus, according to a 2002 program developed in Nizhnii Novgorod Oblast for 392 corporate farms, the oblast budget would repay the old debt to the state as long as the farms could meet all current payments in full.

An interesting remnant of the centralist bureaucratic system is observed in the treatment of collection of agricultural debt to the state: one state-owned bank—Rossel'khozbank—is responsible for collecting outstanding taxes and social deductions from all farms in Russia.

Subsidized Credit

Credit subsidies to Russian farms are based on partial reimbursement of the actual interest expense paid monthly to the lending banks. Two-thirds of bank interest charges are reimbursed, provided the bank rate does not exceed the Central Bank refinancing rate, which sets the ceiling on the subsidy.[4] This scheme has been in operation since 2000, when the government finally abandoned the traditional programs of allocating discretionary loans to "deserving" farms through regional authorities. All agricultural loans are handled by 291 commercial banks, of which just three control over 70% of loan agreements (56% in the privately owned Sberbank, 14% in the state-owned Rossel'khozbank, and 1% in Bank Vozrozhdenie; the share of the other banks is less than 1% each). This is in stark contrast to the situation during the Soviet period, when all lending to agriculture was handled by a single state-owned financial institution (Agroprombank).

Table 9.13. Amounts budgeted for interest-rate subsidies (million rubles)

	2001	2002	2003	2004 (projected)
Allocated in the state budget*	1,300	2,200	3,200	4,150
1-year credits	1,300	1400	2000	NA
3-year credits	—	800	1200	NA
Bank loans received**	18,400	25,600	38,800	50,000
Primary agricultural producers (%)	46.7	51.1	51.1	NA
Processors (%)	51.1	48.8	47.1	NA
Number of agricultural borrowers**	NA	10,100	12,200	13,500

*Federal Budget Law.
**MinAg estimates.

[4] The Central Bank refinancing rate has been steadily decreasing over time, following the decrease in annual inflation rates. It dropped from about 20% in late 2002, early 2003 to less than 15% in the first half of 2004.

Table 9.13 shows the amounts allocated in the state budget to interest-rate subsidies for agriculture (primary producers and processors). Each ruble of budget subsidies generated 10-15 rubles in loans from commercial banks. While most of the subsidies were directed to support short-term working-capital financing, about one-third was intended for medium-term investment loans (up to 3 years). Russian banks tend to lend short-term to agriculture and provide very little investment financing. Thus, in Bank Vozrozhdenie, 90% of agricultural lending is for working capital and only 10% are 3-year investment loans, all within the government's subsidized interest rate program.

Table 9.14. Government conditions for eligibility for interest-rate subsidies in agriculture

	Short-term (working-capital) credit	3-year credit	5-year credit
Program introduced in	2001	2002	2003 (May)
Lender	All agricultural producers, incl. private farmers	All commercial producers, incl. private farmers, provided no overdue debt for taxes and social deductions	All commercial producers, incl. private farmers, provided no overdue debt for taxes and social deductions. Selection by tender
Purpose	Purchase of farm inputs	Purchase of farm machinery and equipment according to MinAg approved lists	Construction of pig and poultry complexes.
Subsidy rate	Two-thirds of actual interest expense, but not more than two-thirds of the Central Bank refinancing rate		
Subsidy payment	Monthly against bank statements		
Required documentation	Loan agreement Interest calculation Proof of loan use for eligible purposes Proof of monthly interest payment Loan repayment schedule (excl. short-term loans)		
MinAg role	Allocation of total budget to regions Organization of subsidy disbursement through local organs Determination of list of eligible machinery and equipment (for 3-year loans)		
Role of regional authorities	Collection of applications from producers Disbursement of subsidies Reporting to MinAg		

Table 9.14 describes the institutional framework for subsidized credit to agriculture. First, on an encouraging note, there is no formal discrimination between producers of different organizational forms. Thus, peasant

farmers are eligible for all credit programs alongside corporate farms. However, only one bank—the state-owned agriculturally oriented Rossel'khozbank—services the financing needs of subsidiary household plots. Less encouraging, but perhaps perfectly understandable, is the emphasis on repayment capacity of eligible lenders, as reflected in absence of overdue liabilities to the state. This precondition for 3- and 5-year programs seriously restricts the scope of producers eligible for investment-credit subsidies (although it does not affect eligibility for working-capital subsidies). Finally, a striking negative feature of the whole institutional framework is the decisive role that MinAg and regional authorities continue to play in the allocation of subsidy budgets and the actual disbursement of monthly subsidy payments—a feature that creates multiple opportunities for corruption.

Table 9.15. Requirements to be satisfied by agricultural borrowers in commercial banks

Interest payment	Monthly
Collateral value	1.3-2.0 times the loan amount
Type of collateral	Livestock, machinery, stored commodities, real estate
General conditions	Good credit history
	No overdue debt or participation in debt rescheduling program
Additional conditions for medium-term loans	Business plan
	Guarantee of local administration in difficult cases

After passing through the fine sieve of government eligibility requirements, the potential borrowers have to win the approval of the commercial banks, which generally impose fairly strict conditionalities of their own before approving a loan in the framework of state support programs (Table 9.15). This is understandable, as the banks assume the actual risk of default, even though the funds come from government sources. Because of this double screening, the access to government credit programs is less than universal. Judging by the data in Table 9.15, we can roughly estimate that less than 7,000 farms received subsidized loans, which is about 25% of the total number of corporate farms in Russia.

Machinery Leasing Programs

We have seen in the previous section that, on the micro level, corporate farms continue to invest in fixed assets despite their financial difficulties (Table 9.9). Policy makers are convinced, however, that, on the macro level, low farm profitability, increasing burden of overdue debt, and lack of bankable collateral have had a damping effect on investment activity in agriculture. Indeed, the share of agriculture in total investment decreased from about 10% in the early 1990s to 3% in 1995-2002 (Figure 9.8). This is about one-half of agriculture's share in GDP, which runs at 6%-7% (also down from 10% in the early 1990s).

Fig. 9.8. Share of agriculture in investment and GDP

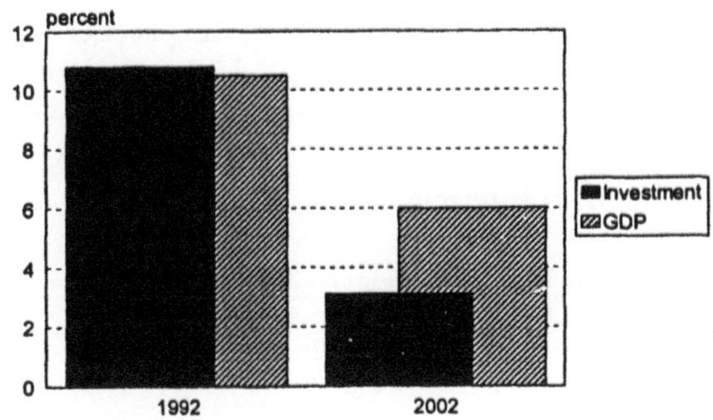

Source: Goskomstat 1999, 2003

These facts reflect a clear shortage of investment funds, which cannot but have a negative effect on long-term sectoral performance. Already now we are witnessing severe aging of the machinery park in agriculture (Table 9.16) and a persistent decrease in the availability of farm machinery per unit of arable land (Table 9.17). MinAg Agriculture is on record claiming that Russian farms have only 50% of the tractors and harvesters "they need" (whatever that means), that 8 million tons of grain (10% of total harvest) were lost in 2003 because of machinery shortages, and that agriculture must get 20,000 new harvesters each year to function properly.[5] The Russian government is therefore actively looking for ways to stimulate agricultural investment without returning to the old command methods of allocating machinery to farms according to plans and targets.

Financial leasing is regarded in Russia as the preferred way to inject new machinery into agriculture despite lack of ready investment funds. Financial leasing involves three parties: the equipment supplier or manufacturer, the leasing company, and the lessee. The leasing company is a financial intermediary: it purchases the equipment from the supplier and delivers it to the lessee—the end user (the farm). The equipment remains the property of the leasing company for as long as the farm uses it under a lease contract. Eventually the farm may buy the equipment from the leasing company or return it when the lease contract is terminated.

Financial leasing is regarded in Russia as the preferred way to inject new machinery into agriculture due to lack of ready investment funds. Fi-

[5] See www.rosagroleasing.ru/news/.

nancial leasing involves three parties: the equipment supplier or manufacturer ("the vendor"), the leasing company, and the lessee. The leasing company is a financial intermediary: it purchases the equipment from the supplier and delivers it to the lessee—the end user (the farm). The equipment remains the property of the leasing company for as long as the farm uses it under lease and its value in principle provides a collateral to the loan implicit in the lease contract (in Russia, however, the practice is to put the leased farm asset from the start on the lessee's balance sheet because of tax considerations: agricultural producers are exempt from property tax). Eventually the farm may buy the equipment from the leasing company or return it when the lease contract is terminated.

Table 9.16. Age structure of farm machinery, 2003 (percent)

	Up to 6 years	6-10 years	Older than 10 years
Tractors	8	21	71
Grain harvesters	11	25	64
Feed combines	17	15	68

Source: MinAg.

Table 9.17. Machinery park and machinery loads in agriculture 1992-2002

	1992	1995	1999	2002
Number of tractors, '000	1290.7	1052.1	786.8	646.4
Number of grain harvesters, '000	370.8	291.8	210.1	173.4
Hectares per tractor	92	108	130	NA
Hectares per grain harvester	160	173	195	NA

Source: Goskomstat 2003.

The government took upon itself to develop and introduce various leasing schemes that would be financed on the same principles as the subsidized interest-rate program and would provide an administrative channel allowing farms to access subsidized investment funds. Government intervention in investment was judged to be unavoidable, as many officials held the view that commercial companies would not be willing to finance leasing of machinery and equipment for low-profitability agriculture. In line with the prevailing privatization strategy, the Russian government opted for a strictly centralized monopolistic solution: in 1994 it established a leasing fund that was financed from the state budget and operated by Rosagrosnab—a parastatal monopolist which at that time controlled all farm input supply operations in Russia. In 2001, the entire leasing program was transferred to another specialized monopolist, Rosagrolizing. Between 2001 and 2004, the federal budget poured nearly 10 billion rubles into Rosagrolizing's equity—about 5% of the total volume of physical investments in agriculture during this period and comparable to the total amount allocated to the interest-rate subsidy program (see Table 9.13). Rosagrolizing's equity—the pool of capital from which lease contracts are made—rose from 5 billion rubles in 2001 to 14.3 billion rubles in mid-2004.

State-funded Rosagrolizing acts as the central banker of the agricultural leasing system. It distributes the available funds to regional leasing companies, which are responsible for arranging the lease contracts and collecting the lease payments from the farms on behalf of Rosagrolizing. These leasing agents are usually regional monopolists, although in some regions (e.g., Vologda, Kursk, Ivanovo) commercial competition is allowed. The tranche allocated by Rosagrolizing to a particular region is fully guaranteed by the regional government, and local authorities accordingly take a very active role in screening and approving the applications from their farms. Needless to say that this process is rife with favoritism and potential corruption. The list of machinery and equipment eligible for leasing is drawn up by MinAg in Moscow, and it is generally restricted to tractors and combines manufactured in Russia or Belarus. In principle, livestock also can be leased within this program.

The typical lease term is 5-7 years and lease payments are made quarterly. Rosagrolizing funds are loaned at 4% annually (down from 7% in 2002); in addition, the lessee pays an annual charge of 1.5% of residual asset value to Rosagrolizing and 1.5% to the regional leasing company. The farms are required to make a downpayment calculated at 7%-15% of equipment cost (which includes compulsory insurance charges). The regional leasing company adds a markup of about 4% to the base cost, but the machinery is usually purchased directly from the manufacturer at specially negotiated prices, which, according to Rosagrolizing, may be lower by as much as 10% than the MinAg reference prices. The total cost to the lessee farm over the entire term of the lease contract ranges from 18% of asset value for an ordinary tractor to 26% for advanced machinery (including combines) and 33% for breeding livestock.[6]

According to Rosagrolizing sources, the leasing scheme is fully operational in 60 of the 80 Russian regions. Between January 2002 and April 2004, nearly 1,300 lease contracts were signed (of them 120 livestock contracts) for a total value of 19.4 billion rubles; Rosagrolizing actually delivered 10,000 pieces of machinery worth 11 billion rubles, including 5,800 harvesters, 3,100 tractors, and 2 fishing boats.[7] MinAg estimates that the leasing program was responsible for one-third of the total volume of machinery and equipment acquired by farms in 2002 and the first half of 2003. This estimate is confirmed by the 2003 BASIS survey, in which machinery

[6] This and related information is available on the Rosagrolizing web site www.rosagroleasing.ru.

[7] www.rosagroleasing.ru/about/; *Krest'yanskie vedomosti*, Dec. 22, 2003; June 16, 2004.

leasing in 2002 was 33% of the value of all new machinery acquisitions for corporate farms and 36% for individual farms. In Leningrad Oblast, on the other hand, the leasing rate is somewhat lower: 16% of all new farm machines and equipment in 2001 was acquired by leasing.[8]

Rosagrolizing, like the rest of the agricultural establishment in Russia, is heavily biased toward corporate farms. Yet the federal leasing program is open to all agricultural producers, both corporate and individual alike, and Rosagrolizing tries to publicize its contracts with peasant farmers. It was recently announced that in 2003 Rosagrolizing had delivered 430 pieces of machinery to peasant farmers.[9] To put this announcement in context, we should note that 240,000 peasant farmers received 430 pieces of machinery while 24,000 corporate farms received 10,000 pieces of machinery. In the 2003 BASIS survey, the frequency of lease transactions is also much higher among corporate farms than among individual farms. Thus, leasing accounted for 29% of new machinery transactions among corporate farms and only 16% among individual farmers.

Farms apparently reserve the leasing option for relatively high-cost machinery and equipment, such as heavy-duty tractors, harvesters, and feed combines, purchasing less costly equipment (seeders, cultivators, plows) from other sources. In the 2003 BASIS survey, one piece of leased equipment on average cost more than one piece of equipment purchased through other channels (the difference is statistically significant for both corporate farms and individual farmers). This does not mean that leased combines cost more than purchased combines: it simply means that farms lease combines (expensive) and buy seeders (less expensive).

Table 9.18 compares the main institutional features of the leasing program with the subsidized interest rate program for investment credit. The main difference that may significantly increase the accessibility of the leasing program is that it does not require the applicants to prove a positive record history (the credit history and solvency requirements limit the scope of farms eligible for subsidized interest rate credit to 10%-15% of the total number of farms). Leasing approval relies on projection of future cash flows, not on collateral value as in bank loans. There is no clear-cut evidence to show which of the two investment financing options—subsidized credit or Rosagrolizing—is cheaper. It would appear that leasing is more expensive than subsidized credit, as in addition to paying a (subsidized) interest on state funds, the farm is required to pay the costs of at least two levels of intermediaries (central and regional) and provide a certain (albeit

[8] Estimated by David Epshtein from consolidated annual report of Leningrad Oblast corporate farms

[9] *Krest'yanskie vedomosti*, July 14, 2004.

minimal) downpayment from its own resources. Rosagrolizing publishes various calculations that attempt to prove the huge cost advantages of the leasing program (more than 25%), but a closer look at these calculations shows that the advantage relies on two highly contentious factors: an allegedly greater VAT refund and lower tax payments due to greater depreciation.[10]

Table 9.18. Access of farms to investment through subsidized interest rate credit and state leasing program

	Subsidized interest rate investment credit	State leasing program
Eligible farms	Financial solvency, positive credit history	Financial solvency, but less strict conditions (no credit history)
Contract term	3-5 years	5-7 years
Guarantees	Not required	Regional government or banks
Collateral	1.3-2 times the amount of the loan	Undertaking of proper use to avoid damage to value of leased asset
Business plan	Required	Required
Available machinery	MinAg approved lists, mainly Russian-made	Ditto, but a more restricted range of options
Role of MinAg/ Rosagrolizing	Main disbursing agent for regional bodies	Main disbursing agent for regional bodies
Role of regional/local authorities	Control the allocation of subsidies through application approval; accountable to Min Ag	Guarantee the regional allocation from Rosagrolizing
Repayment schedule	Negotiable	7% downpayment; quarterly or semiannual lease payments
Application procedure	Highly bureaucratic	Less bureaucratic

As noted above, the regional authorities are actively involved in the federal leasing program: they approve the regional leasing agencies, screen the applications, and most importantly guarantee the federal funds allotted to their region. The operational involvement of some regions in encouragement of farm investments through leasing started in 1997-1998, before the creation of Rosagrolizing. The regional support programs mostly supplement the federal leasing and subsidized interest rate programs, but some regions have introduced quite novel techniques of agricultural support. Some regions, instead of supplementing the federal leasing program, simply provide bank loans that allow farms to buy any equipment that they need from any supplier of their choice: under these "competitive" regional programs (operating mainly in Vologda, Kursk, and Ivanovo oblasts) the farms

[10] See www.rosagroleasing.ru/conditions/

are not limited to a centrally approved list or a monopolistic leasing agent. Other regions (e.g., Perm) have done away with state leasing, and instead subsidize equipment purchases by providing outright grants and partial reimbursement of lease payments. Rostov introduced a perverse twist into the system: it does not officially restrict equipment leasing to a list of approved suppliers and does not allocate funds to the federal leasing program, but only equipment purchased from the local machinery manufacturer (Rostsel'mash) is entitled to grants and loans from regional sources. Table 9.19 lists the main features of the leasing and investment credit programs—both federal and regional.

Table 9.19. Main features of leasing and investment credit programs in selected regions*

	Fed	Chel	Chuv	Vol	Perm	Ros	Ivan	NN
Single leasing company	+	+	+	–	–	+	–	–
Approved list of suppliers	–	+	NA	–	–	+	–	–
Approved list of machinery	+	+	NA	–	–	+	+	–
Reimbursement of federal program down-payment	–	–	–	–	+	–	–	+
Reimbursement of interest charges on investment loans	+	–	+	–	+	–	–	+
Reimbursement of machinery purchase cost	–	–	–	–	+	–	–	+
Linkage to delivery of farm products to regional stocks	–	–	–	–	–	+	–	–

*Legend: Fed—federal program; Chel—Chelyabinsk; Chuv—Chuvashia, Vol—Vologda, Perm—Perm, Ros—Rostov, Ivan—Ivanovo, NN—Nizhnii Novgorod.
Source: Based on information provided by regional agricultural departments.

Leasing programs look much larger than investment credit programs. Thus, in a sample of 12 regions,[11] the amounts allocated at the regional level

[11] Rostov, Samara, Nizhnii Novgorod, Chuvashiya, Tambov, Perm', Stavropol', Voronezh, Kursk, Novosibirsk, Moscow, Leningrad.

to equipment and livestock leasing in 2003 are nearly three times the expense of interest rate subsidies (Figure 9.9). This is a deceptive impression, however: the amounts in the leasing programs are virtually the full value of the leased assets, whereas the amounts shown for the investment credit programs are just the subsidized interest expense and they obviously support total loans that are an order of magnitude larger. A more relevant comparison is between the regional allocation of 1.8 billion rubles in 12 regions (Figure 9.9) and the total amount of 7-8 billion rubles that Rosagrolizing spends annually in 60 regions. In both cases, the average allocation per region is around 150 million rubles, so that regional support to leasing practically matches the national program. The total amount invested by state and regional leasing programs (around 10 billion rubles) is close to 25% of total investment in agriculture (42 billion rubles in 2002).

Fig. 9.9. Regional Support to Agricultural Investment 2003

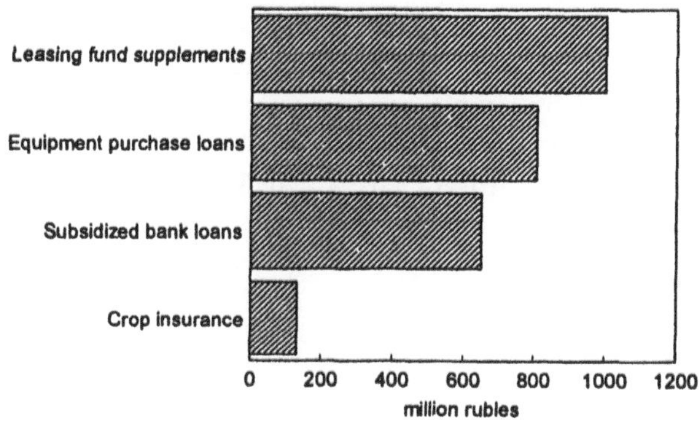

Although Rosagrolizing has an impressive record in promoting agricultural asset leasing, its opponents argue that the monopolistic nature of the contractual arrangements suppresses private initiative and competition. Thus, agricultural machinery represented a mere 2% of transaction volume among Russia's 30 largest leasing companies in 2003. When Rosagrolizing's monopoly was lifted in Perm' Oblast, 9 commercial leasing companies started supplying agriculture machinery and equipment during the first year. In Ivanovo Oblast, where competitive leasing is allowed (see Table 9.19), commercial transactions accounted for 10 of the 16 leasing contracts reported by corporate farms in the 2003 BASIS survey.

Access to Farm Credit: Findings of the 2003 BASIS Survey

The Russian establishment—politicians, agricultural officials, corporate farm managers, the media—firmly believe that inadequate access to credit is one of the major factors constraining the growth of the agricultural sector. In technical terms, they in effect claim that Russian agriculture faces credit rationing.[12] In this section we examine the issues of access and use of credit in Russian agriculture based on the data of the 2003 BASIS survey, which covered 142 corporate farms, 222 peasant farms, and 202 commercially oriented household plots in three oblasts. There were highly significant differences in size between the three types of farms, and we accordingly refer to the corporate farms as "large farms" and to the individual farms (peasant farms and household plots combined) as "small farms." The survey instrument included a special section devoted to farm finances, which specifically collected information on sources and uses of funds, access to credit from various sources, and the farmers' perception of credit constraints. The basic descriptive analysis of the financial findings for both corporate and individual farms is followed by simple econometric modeling of access to credit for corporate farms only.

Corporate farms and peasant farms usually keep accounts and produce financial reports (standard annual reports for corporate farms; a simpler format prescribed by statistical organs for peasant farms). It is therefore not surprising that virtually all respondents in these two categories provided fairly complete financial information: the response rate to financial questions for corporate farms and peasant farms exceeded 95% (Table 9.20). The situation was entirely different for household plots: despite their declared commercial orientation, the household plot operators do not keep accounts (at least not in the conventional format) and, being exempt from taxes, do not produce any tax reports. As a result, only one-third of household plots provided financial information on sources of funds; to the extent that household plots provided information on uses of funds, this was primarily cost of purchased inputs—a relatively straightforward expense item to monitor. The high proportion of missing data among household plots poses a difficulty with interpretation: we really cannot conclude that two-thirds of households have no access to any capital. The missing data are probably a result of genuine inability to provide financial information in a

---

[12] See J. Stiglitz and A. Weiss, "Credit rationing in markets with imperfect information, *American Economic Review* 73(3): 393-410 (1981); M. Carter, "Equilibrium credit rationing of small agriculture," *Journal of Development Economics* 28: 83-103 (1988).

non-business setting and only in part reflect lack of access to financial sources.

Table 9.20. Respondents providing financial information in the 2003 BASIS survey (percent)

	Corporate farms (*n*=142)	Peasant farms (*n*=222)	Household plots (*n*=202)
Sources	98	96	32
Own funds only	3	77	31
External funds	95	19	1
Of which: formal institutions	43	13	0
No information	2	4	68
Uses	97	99	89
Debt repayment	68	12	1
Investments	42	16	24
Purchased inputs	95	99	87

The most striking feature that distinguishes the three types of farms with regard to financial information is their different access to external funds.[13] Household plots do not borrow: they rely on own funds (i.e., internally generated earnings and household savings). Among corporate farms, on the other hand, virtually everyone uses external financing, and nearly half the respondents report access to formal financial institutions. Peasant farmers fall in the middle, with 19% reporting access to external capital (most of them from formal financial institutions). As a result, debt repayment is reported more frequently among corporate farms. Investments are also more frequent among corporate farms.

Financing Mix

The scope of the financial questions in the survey does not provide a full picture of sources and uses of funds. Thus, there is no information about suppliers' credit (accounts payable) among the sources of funds (except for overdue payables), whereas on the use side there is no information about the cost of utilities or the cost of hired labor. With these caveats in mind, we nevertheless calculated the structure of sources and uses as collected in the survey. The results are presented in Table 9.21 (these are weighted averages based on total amounts and total number of respondents in the survey). The structure of sources for corporate and peasant farms is also shown in Figures 9.10 and 9.11.

---

[13] In the 2003 BASIS survey external funds include bank loans, equipment leasing, government subsidies and commodity credits, and the overdue portion of accounts payable

Fig. 9.10. Structure of sources for corporate farms

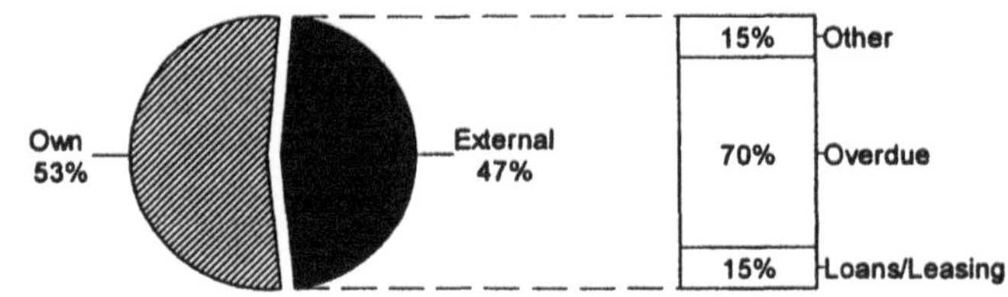

Fig. 9.11. Structure of sources for peasant farms

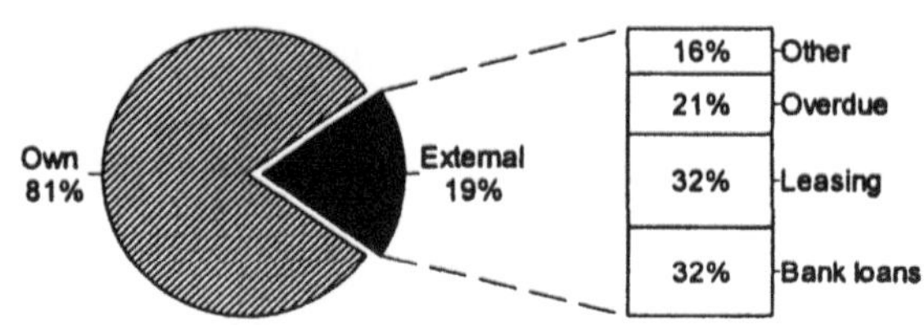

The sources of corporate farms are evenly balanced between own capital and external funds (with slight predominance of external funds; see Table 9.21), whereas individual farms use 80% own capital and 20% external funds in their sources and household plots rely almost entirely on own capital. The presence of bank loans in the financing mix provides evidence of access to the formal financial system.

The small household plots have no access to bank credit, and their minimal external funds are all derived from informal sources. On the other hand, both corporate and peasant farms borrow from banks, but the relative importance of bank credit is much higher for peasant farms, where it accounts for one-third of external financing sources. Equipment leasing is also reported as a very important source of funds for peasant farmers.[14]

[14] The share of equipment leasing in sources of funds of private farms surveyed (32%) is not necessarily representative: it is based only on 7 cases and one single case accounts for 50% of the total amount reported for equipment leasing. Without this one case the share of leasing in external sources would decline significantly from its present 32%, but still remain considerable.

Based on bank borrowing and equipment leasing data, we can definitely state that in the individual sector peasant farmers have access to formal financial markets, while household plot operators do not.

Table 9.21. Structure of sources and uses of funds (in percent of respective totals)*

	Corporate farms	Peasant farms	Household plots
Sources	100	100	100
Own funds	47	81	98
External funds	53	19	2
Bank loans	6	6	0
Equipment leasing	2	6	0
Overdue payables	37	4	0
Other external funds	8	3	2
Uses	100	100	100
Debt repayment	36	8	0
Investments	9	15	15
Machinery, buildings	6	13	4
Livestock	1	1	10
Other investments	1	1	1
Purchased inputs	52	77	84
Fuel	28	45	19
Fertilizers, chemicals	6	16	2
Seeds	6	10	2
Feed	12	1	61
Social sphere	3	0	0

*Sources and uses are calculated separately based on different numbers of respondents (see Table 9.20).

While corporate farms access formal financial markets through both bank credit and equipment leasing, it is overdue payables, primarily supplier credit, that constitute the main component of their external funds. This source of funds plays a less prominent role for peasant farms, but still it represents more than 20% of their external sources. These findings highlight the potential importance of current liabilities in the financing mix of both corporate and peasant farms, but unfortunately no data were collected in the survey on this particular component of sources of funds.

Surplus and Land

The financing mix in Table 9.21 is calculated separately for sources and uses because of differences in the number of respondents reporting the respective components (see Table 9.20). In the entire sample, the number of respondents reporting uses is greater than the number of those reporting sources, and yet the sum total of uses is less than the sum total of sources over a smaller number of respondents. This suggests the existence of potential surpluses that may go into accrued profits and household savings.

Table 9.22. Sources, uses, and surplus: per farm averages in matched subsamples (thousand rubles)

	Large farms (*n*=136)	Peasant farms (*n*=212)	Household plots (*n*=66)
Sources	10,462	305	45
Uses	6,191	189	20
Surplus (sources – uses)	4,271	116	25
Surplus in % of sources	41%	38%	56%
Non-positive surplus, % of farms	26%	17%	32%
Average farm size, ha	3,400	260	1.0
Surplus per hectare, rubles	1,234	451	26,600

To estimate the surplus as the difference between sources and uses, we restricted the analysis to matched subsamples of respondents reporting both sources and uses (Table 9.22). A regression analysis of uses as a function of sources gave a coefficient of 0.50 for small farms and 0.31 for large farms (both coefficients are statistically significant at 1%). This means that in matched samples of small and large farms reported uses are respectively 0.50 rubles and 0.31 rubles for each ruble of sources. Unless this can be explained by gross underreporting of uses among the respondents (which is not very likely, given the fairly comprehensive list of questions in the survey), the findings provide evidence of sizeable surpluses that go into savings.

Table 9.23. Impact of land and access to financial markets on sources, uses, and surplus

	Small farms (*n*=277)			Large farms (*n*=132)	
	Land	Borrowing**	R^2	Land	R^2
Sources	0.42 ($p = 0.000$)		0.53	0.61 ($p = 0.000$)	0.32
Uses	0.44 ($p = 0.000$)		0.55	0.72 ($p = 0.000$)	0.28
Surplus	0.33 ($p = 0.000$)	0.63 ($p = 0.019$)	0.30	0.39 ($p = 0.069$)	0.03

*Matches samples of sources and uses, log-log regressions.
**Upward shift relative to farms that rely only on own capital. Based on 200 observations with positive surplus.

The absolute values of sources and uses for corporate farms are an order of magnitude greater than those for peasant farms; the sources and uses for peasant farms in turn are an order of magnitude greater than those for household plots (Table 9.22). The same order-of-magnitude relationship holds also for the surplus (calculated as sources minus uses for the matching farms). The difference in financial variables is naturally explained by the orders-of-magnitude difference in farm sizes (3,400 hectares on average for large farms, 260 hectares for peasant farms, and 1.2 hectares for household plots), as sources, uses, and surplus all increase with the amount of land in agricultural use (see the regression results in Table 9.23, where the coefficient of land is positive and statistically significant for all three variables).

The surplus is around 40% of sources for both corporate farms and peasant farms (somewhat higher for household plots; Table 9.22). The partial productivity of corporate farms as measured by surplus per hectare is higher than that of peasant farms, but much lower than that of household plots. The very small household plots appear to be generating much greater surpluses per hectare of land than the larger peasant farms.[15] This is shown using averages in Table 9.22, but the same conclusion also emerges from regression analysis controlling for farm size. A log-log regression model with an organizational dummy variable (not reported in full) shows that, keeping the size constant, household plots generate 3.0 times the amount of sources and 3.5 times the amount of surplus generated by peasant farms.

Unfortunately, dummy-variable regression may not be entirely appropriate in this setting, because the sizes of the two organizational forms—household plots and peasant farms—constitute practically disjoint sets. There are hardly any household plots and peasant farms of the same size: 99% of household plots do not exceed 10 hectares while 98% of peasant farms are larger than 10 hectares. Our interesting results are therefore entirely in the nature of prediction outside the range of actually observed farm sizes, with all that this implies. It may be thus more relevant to explore the dependence of sources or surplus on farm size using a piecewise-linear regression model or a quadratic regression model with the pooled sample. This suggestion is consistent with the generally nonlinear appearance of the scattergram in Figure 9.12, which shows the surplus versus land in agricultural use (in log-log coordinates) for all farms reporting financial data (household plots, peasant farms, and large farms). Surplus clearly increases with farm size, and the increase is clearly nonlinear.

In quadratic log-log regressions with sources and surplus as the dependent variable, the coefficient of squared land was found to be positive and highly significant in the pooled sample (see Table 9.24 for the estimates obtained in some of these quadratic regressions). Larger farms (corporate and peasant) are thus catching up with household plots, but given the absolute differences in surplus per hectare (see Table 9.22) it is unrealistic to expect them ever to reach the same level of performance as household plots. The quadratic relationship is also consistent with the observation that

---

[15] It should be borne in mind, however, that the results for household plots are on the whole less reliable than for the other farms because of the high frequency of non-responses.

surplus per hectare is significantly greater for large corporate farms than for peasant farms (see also Table 9.22). This conclusion is further strengthened by the results of a regression analysis using an organizational-type dummy variable to distinguish between peasant farms and large corporate farms (Table 9.24; the small households plots were omitted from the pooled sample for this analysis). Controlling for farm size, the large farms achieve significantly higher surplus than peasant farms. Note that the farm-type dummy fully captures the nonlinear behavior of surplus with size: in the dummy-variable model the coefficient of the quadratic land term is not significant and this quadratic model produces the same R^2 as the linear model with the same dummy variable (which is of course substantially higher than the R^2 of the quadratic model without a farm-type dummy). In summary, the various lines of analysis seem to confirm that the large corporate farms in our survey use land more productively than the peasant farms.

Fig. 9.12. Surplus vs land for farms of different types

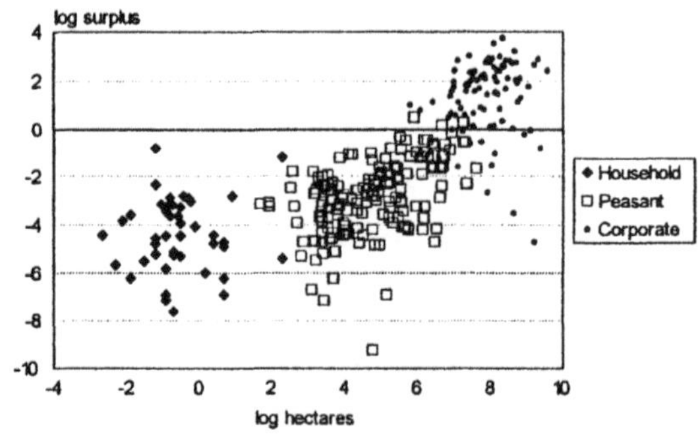

In addition to land, access to financial markets also appears to have a positive impact on surplus. This effect can be tested only for small farms, as practically all large farms in the sample use external financing (see Table 9.20). Among small farms, on the other hand, less than 10% of respondents report that some of their sources derive from external funds (mainly short-term bank loans). Individual farms that borrow achieve estimated surpluses nearly 90% higher than the surpluses of farms that do not borrow (keeping farm size constant), and this result is statistically significant at 5% (see Table 9.23). We have noted previously that surplus increases with farm size; moreover, individual farms that borrow are relatively large: 400 hectares on average compared with 115 hectares for farms operating with own capital only. Yet our regressions show that both factors are statistically significant for individual farms: surplus increases with farm size and, controlling for farm size, it is much greater for individual farms that have access to financial markets. There can be no technical objections to such dummy-variable regression in this case, because the sizes of individual farms with and without borrowing have a generous overlap.

Table 9.24. Impact of land on surplus in large farms and peasant farms (log-log regressions)

	Model 1 (quadratic without dummies)	Model 2 (quadratic with dummies)	Model 3 (linear with dummies)
Land	0.006 (p=0.986)	0.712 (p=0.032)	0.579 (p=0.000)
Land squared	0.089 (p=0.003)	−0.013 (p=0.679)	—
Type: large farms relative to peasant farms	—	2.375 (p=0.000)	2.312 (p=0.000)
R^2	0.614	0.678	0.677

For large farms, where practically everyone has access to external financing, we instead looked at surplus in relation to the share of external funds in total sources (a continuous variable with 53% weighted average according to Table 9.21). Controlling for land, surplus was not affected by the share of external funds in total sources for large farms. This seems to suggest that financial markets (including supplier credit) are not a constraint to the financial performance of large farms.

Perceived Ability to Access Credits

In addition to collecting information on the actual use of bank loans, the respondents were asked whether or not they were able to obtain credit (short- and long-term separately). In the individual sector (peasant farms and household plots), the respondents' perception of their ability to access the banking system is much more optimistic than the actual access rates (Table 9.25). Nearly two-thirds of the peasant farmers believed they would be able to borrow from banks (whereas only 11% reported actual loans). More than one-third of household plot operators believed they would be able to get loans (short- and long-term), whereas in reality none reported any borrowing.

The high rates of perceived accessibility of bank loans among the individual farmers contrast sharply with their relatively low rates of actual borrowing. This suggests that individual farmers do not feel any kind of credit rationing and the low incidence of borrowing reflects either adequacy of own capital or possibly the existence of various institutional constraints that prevent actual borrowing. The corporate farms, consistently with their broad use of external funds, have relatively free access to banks and other financial institutions: fully 33% of respondents report their bank loans. This picture of relatively wide access to borrowing is confirmed by the managers' perception of the availability of bank credit: 57% of respondents report that do not envisage difficulties borrowing (although borrowing short-term appears to be easier than long-term).

Table 9.25. Actual and perceived access to bank loans (percent of respondents)

	Corporate farms		Peasant farms		Household plots	
	Actual	Perceived	Actual	Perceived	Actual	Perceived
Access to loans	33	57	11	64	0	37
Short-term	25	56	9	63	0	37
Medium to long-term	9	29	2	33	0	21

Table 9.26. Perceived constraints on borrowing from banks: large farms and peasant farms

	Large farms (n=142)	Peasant farms (n=223)	HH plots (n=202)
Complex formalities	24	**54**	25
No collateral	27	**56**	11
Excessively short term	29	40	8
High default risk	34	18	23
No accessible agricultural banks	13	24	13
Interest rate higher than rate of return	19	9	16
Insufficient credit information	6	6	18
Poor credit history	**17**	5	2
Outstanding debt	**53**	5	1
Other	5	1	29

The perceived difficulties with access to credit are summarized in Table 9.26, which reveals clear differences in the perception of potential borrowing difficulties among the three types of farms. Managers of corporate farms appear to be more cognizant of the real institutional difficulties with banks. This is evident from the generally higher percentage of corporate-farm managers who address the questions of constraints on bank borrowing (Table 9.26). They are concerned with the potential effects of outstanding debt and credit history—the two factors that are virtually ignored by peasant farmers, who mainly worry about complex formalities and lack of collateral (the different pattern of perceived constraints as it emerges from the responses of corporate-farm managers and peasant farmers is highlighted by bold numbers in Table 9.26) and illustrated in Figure 9.13). Household plot operators are more deterred by risk, i.e., potential inability to repay the loan. High interest rates and lack of information on credit availability are also more troublesome for household plot operators than for the business-oriented peasant farmers. The high frequency of responses that household plot operators put in the "other" category suggests that the questions as formulated in the survey, while quite appropriate for peasant farms, may not have been entirely suitable for household plots.

Fig. 9.13. Perceived constraints on bank borrowing (percent of respondents)

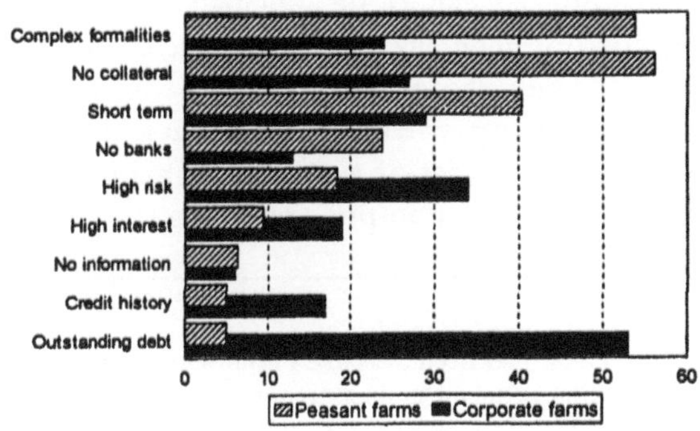

Determinants of Access to Credit: Econometric Analysis of Corporate Farms

We applied discrete regression analysis to study the determinants of access to credit for corporate farms. Corporate farms were chosen as the subject of study because they usually provided much more complete and systematic information that other farm types in the 2003 BASIS survey. Nevertheless, only a small percentage of corporate-farm managers provided information about the actual loan amounts from banks (see Table 9.20). Under such circumstances ordinary regression analysis using the amount of borrowing as the continuous dependent variable may produce highly biased results, as we in effect have a censored sample. It is preferable to shift to probit or logit regressions representing access to credit by a binary variable with the levels "borrowed" and "did not borrow." Farms that reported nonzero bank loans or borrowing through machinery leasing programs (see Table 9.21) were assigned to the category of farms with access to credit (41% of respondents); all other respondents were assigned to the category of farms without access to credit (59%). This binary variable could be used to model the probability of access to credit by logit regression according to well-developed standard procedures (see Box 9.1).

Based on the available literature, we chose five groups of factors that could potentially affect access to credit for corporate farms. Three groups focus on the activity of the farm: they include current repayment capacity (i.e., ability to generate profits); the "safety margin" that enables the borrower to smooth out negative fluctuations of wealth (i.e., asset endowments); and creditworthiness (as expressed by the credit history). Two additional groups of factors recognize the dependence of the access to credit on transaction costs (for the farm or the bank) and on regional policies. World experience suggests that both current repayment capacity and asset endowments should have a positive impact on access to credit. Transaction costs and poor credit history, on the other hand, reduce access to credit. The effect of regional policies cannot be anticipated in advance: it depends on local factors and is not generalizable. In the credit rationing paradigm, which is appropriate for describing situations with constraints on access to credit, the effect of current repayment capacity is associated with so-called price rationing, whereas the effect of asset endowments, creditworthiness, and transaction costs is linked to non-price rationing.[16]

Box 9.1. Logit Regression

If actual volumes of farm credit had been available in our survey database, we would have modeled access to credit using ordinary multiple regression with a continuous dependent variable (volume of credit in rubles). However, the available information on access of credit is limited to Yes/No answers: Yes, when the farm actually borrows (i.e., has access to credit); No, when the farm does not borrow (i.e., has no access to credit). Our dependent variable is discrete, taking nominal values, and it cannot be modeled by ordinary multiple regression. Logistic regression (or in short logit) is the standard modeling tool for situations when the dependent variable is nominal, i.e., takes the values Yes/No, Black/White, Employed/Unemployed. The nominal levels of the dependent variable are usually coded as 1 and 0 for analysis, but modern statistical packages work equally well with "value labels," such as "yes" and "no." While all these are examples of binary or dichotomous variables, so-called multinomial logit regression works with nominal variables that have more than two levels, such as "Red/Green/Blue" or "Elementary education/Secondary education/Technical education/Higher education."

Logistic regression models the probability of getting the different discrete values of the dependent variable. Let Y be the binary dependent variable characterizing access to credit: $Y = 1$ when the farm borrows and $Y = 0$ when the farm does not borrow.

As the single explanatory variable X in this example we take sales revenue in million rubles (a continuous variable). The probability of getting access to credit $P(Y = 1)$ is expressed as

$$P(Y=1)=\frac{\exp(\beta_0+\beta_1 X)}{1+\exp(\beta_o+\beta_1 X)}$$

Here the coefficients β_0 and β_1 are estimated by the regression engine.

The complementary event is not getting access to credit ($Y = 0$), and its probability is thus $P(Y = 0) = 1 - P(Y = 1)$. The ratio of the probabilities of these two events is called the odds of getting credit:

$$odds(Y=1)=\frac{P(Y=1)}{P(Y=0)}=\exp(\beta_0+\beta_1 X)$$

[16] J. Foltz, "Credit market access and profitability in Tunisian agriculture," *Agricultural Economics* 30: 229-240 (2004).

Odds (a term derived from gambling) is the probability of getting access to credit in relation to the probability of not getting access to credit (at given sales revenue X). Thus, if 25% of farms with sales revenue $X = 1$ million rubles get credit, while 75% of farms with the same sales revenue do not get credit, then the odds of getting credit given sales revenue of 1 million rubles are 1-to-3 (0.25/0.75=1/3).

We are naturally interested in comparing the odds of getting credit for different levels of sales revenue. For purposes of such comparison, calculate the odds of getting credit for two levels of sales revenue $X = X_1$ and $X = X_2 = X_1 + \Delta X$ (where ΔX is the incremental revenue between the two levels) and take the ratio of these odds. Simplifying and taking an increment of one unit ($\Delta X = 1$), we obtain the odds ratio

$$\frac{odds(Y=1\|X=X_0+1)}{odds(Y=1\|X=X_0)} = \frac{\exp(\beta_0+\beta_1 X_0+\beta_1)}{\exp(\beta_0+\beta_1 X_0)} = \exp(\beta_1)$$

The exponent of the regression coefficient $\exp(\beta_1)$ is thus equal to the change in the odds of getting the value $Y = 1$ as a result of a one unit change in the explanatory variable X. In other words, $\exp(\beta_1)$ is the marginal change in odds when the continuous explanatory variable changes by 1. The same result is obtained when X is a discrete variable taking the values $X = 0, 1$ or $X = 0, 1, 2, 3$. With such a discrete explanatory variable, $\exp(\beta_1)$ is equal to the change in the odds of getting the value $Y = 1$ when X changes its values by 1 jumping between two adjacent level (0 and 1 or 2 and 3).

The last column in Table 9.28 gives the values of $\exp(\beta_1)$ obtained in our logit regression. The exponentiated coefficient of revenue (a continuous variable) is 1.028, which means that an increase of revenue by one unit (i.e., by one million rubles) increases the odds of getting access to credit by 2.8% . The exponentiated coefficient of the discrete variable "distance from regional center" is 0.646. As the discrete variable changes its value from 0 (near the regional center) to 1 (far from the regional center) the adds of getting access to loans drops by 35%.

These five groups of factors affecting access to credit were operationalized by variables based on survey answers. Current repayment capacity was represented by sales revenue and profit margin (i.e., ratio of gross profit to sales revenue). Availability of collateral, or more generally the farm's ability to fall back on its stock of fixed assets for debt repayment in case of insufficient sales or profits, was represented by agricultural land in use, the pool of farm machinery (in horsepower units), and the livestock herd. To avoid potential multicollinearity problems due to the high correlation between the number of animals and sales revenue, the livestock factor was proxied by the share of livestock products in total farm sales as reported in the survey.

The farm's credit history was represented by its overdue debt (in absolute amounts), which is an important component of both government and bank conditionality for access to various credit programs. A location variable qualitatively characterizing the farm's distance from the regional center (far/not far) was used as a proxy for transaction costs on the assumption that farms farther away from the center face higher transaction costs than farms closer to the center. Finally, regional policies were characterized by availability of subsidized credit in recent years and by an explicit regional dummy for the three provinces surveyed (Ivanovo, Rostov, Nizhnii Novgorod).

Table 9.27. Explanatory variables used in regression analysis

Variable	Definition and units	Expected effect on access to credit	Sample mean: farms with access to credit (*n*=43)	Sample mean: farms without access to credit (*n*=62)
Group 1:	*Current repayment capacity*			
Revenue*	Sales revenue, million rubles	+	13,764	6,349
Profit margin*	Ratio of gross profit to sales revenue, percent	+	45.6	14.5
Group 2:	*Availability of collateral (stock of assets)*			
Land	'000 ha in use	+	3,582	4,026
Machinery	Horsepower	+	6,487	5,119
Livestock share	Percent of sales revenue derived from livestock	+	53.6	44.3
Group 3:	*Credit history*			
Overdue debt	'000 rubles, including suppliers and banks	-	1,194	1,386
Group 4:	*Transaction costs*			
Distance from regional center	Far from regional center	-	27.9%	38.7%
	Not far from regional center		72.1%	61.3%
Group 5:	*Regional policies*			
Subsidized credit*	Credit at subsidized interest rates available to respondent in the last two years	+?	53.5%	33.9%
	Not available		46.5%	66.1%
Region	Oblast dummies (Ivanovo and Rostov relative to Nizhnii Novgorod)	?		

*Differences between the two categories are statistically significant at $p = 0.10$.
Source: 2003 BASIS survey.

The survey variables used in the regression are summarized in Table 9.27 (the means are calculated for 105 of the 142 cases that did not have missing values). The expected effect of each variable on the probability of access to credit is characterized as positive (i.e., an increase of this factor is expected to increase the probability of access to credit) or negative (an increase of this factor is expected to reduce the probability of access to credit). The univariate differences between the two categories of farms by each variable separately are generally not statistically significant. Only the three asterisked variables in Table 9.27 are statistically significantly different for farms with and without access to credit (sales revenue, profit margin, and use of subsidized interest-rate credit in recent years). In this setting, it is essential to proceed with a multivariate regression analysis to identify the determinants of access to credit.

The logit regression results are presented in Table 9.28. Overall, the goodness of fit is quite satisfactory, with R^2=0.461 and total correct prediction rate of 81%. The signs of the estimated regression coefficients generally turned out to be consistent with our expectations. Thus, revenue and profit margin (the two factors characterizing current repayment capacity) have positive coefficients; machinery and livestock herd—two of the three factors characterizing availability of collateral—also have positive coefficients; overdue debt (a credit history factor) and the location variable (proxying transaction costs) both have negative signs, as expected. Not all the estimated coefficients are significant, however. Revenue, profit margin, and livestock share are highly significant ($p < 0.05$). Machinery, subsidized credit, and the location variable are not statistically significant (although they all have correct signs).

Table 9.28. Logit Regression Coefficients for Corporate Farms

Dependent variable: Access to credit (Yes/No)

Factor group	Explanatory variables	Estimated coefficient (*b*)	*p*-value	Exp(*b*)
Group 1: Current repayment capacity	Revenue	0.042	0.028	1.042
	Profit margin	0.039	0.000	1.040
Group 2: Availability of collateral	Land	−0.155	0.282	0.857
	Machinery	0.008	0.919	1.008
	Livestock share	0.058	0.010	1.060
Group 3: Credit history	Overdue debt	0.017	0.886	1.017
Group 4: Transaction costs	Distance from center	−0.269	0.638	0.764
Group 5: Regional policies	Subsidized credit (Yes/No)	0.837	0.133	2.309
	Ivanovo*	−0.587	0.476	0.556
	Rostov*	3.892	0.020	48.994
	Constant	−6.177	0.001	0.002
Goodness of fit measures				
R^2	0.461			
Correctly predicted	85.5% without access to credit	74.4% with access to credit		81.0% overall

*Relative to Nizhnii Novgorod as the base region.

Source: 2003 BASIS survey

Factors reflecting economic efficiency are the main determinants of access to credit. According to Table 9.28, farms with higher profitability have a higher probability of borrowing from financial institutions. This suggests that the Russian rural credit system, however limited and thin, behaves (to a certain extent) according to market principles.

On the other hand, asset endowments, such as land and capital stock, have a very weak effect on the ability to borrow.[17] In fact, land is the only coefficient in this regression model whose sign is inconsistent with our expectations (the negative coefficient of land is not statistically significant, however). A possible explanation of this curious behavior is that land is not really used for collateral in the present legal system (see Chapter 5) and that it cannot be easily sold to repay outstanding debt (although there is some evidence of non-agricultural firms taking over farmland as a means to recover moneys owed by failing agricultural producers; see Chapter 4). For these reasons, land in Russia does not play the theoretical role of a store of realizable value that can be liquidated in times of adversity, and perhaps we should not be surprised that the actual behavior of land in the regression model is not consistent with the theory.

Another deviation from the pattern of market economies is the lack of impact of credit history on farms' ability to borrow. This may be due to the fact that overdue debt is actually not an appropriate measure of credit history in an environment with pervasive soft budget constraints. It may also reflect the uncertainty surrounding the very notion of credit history in a transition economy, where owners and managers change very often and very rapidly. Under such circumstances, it may be better to use total indebtedness as a measure of solvency affecting access to credit.

What can be said about the effect of regional policies? Availability of subsidized credit (from federal and regional sources) improves the probability of access to credit, but its effect is not statistically significant (Table 9.28). The conclusion is that subsidized interest rate, one of the main tools of Russian agricultural policy today, has no impact on the access of corporate farms to credit. It is the commercial bank that decides whether or not to lend to an agricultural producer who is entitled to subsidized interest rate, and their decision is affected by general risk and creditworthiness considerations, as the decision of any market institution would be. Oblast effects are significantly different from zero and separate tests reveal significant differences in access to credit across the three oblasts. The large positive coefficient for Rostov implies that the probability of access to credit in this province is significantly higher than in Nizhnii Novgorod and Ivanovo. The large negative coefficient for Ivanovo implies that the probability of access to credit in this oblast is significantly lower than in Nizhnii Novgorod and Rostov. Thus, keeping all other variables constant and concentrat-

---

[17] We may tempted to argue that large corporate farms on average perform worse than smaller corporate farms, so that the expected effect of endowments is cancelled out by the stronger effect of profitability. This is not necessarily true however, as Epshtein shows (Chapter 13) that in Russia larger corporate farms are generally more successful than smaller corporate farms (see also Chapter 2).

ing only on the regional factor, we can rank Rostov as the region with the highest probability of access to credit, Nizhnii Novgorod as the next region in the ranking, and Ivanovo as the region with the lowest probability of access to credit.[18]

An alternative regression analysis was carried out for the same corporate farms using financial ratios instead of raw variables as the regressors. Separate probit regressions were run for access to short-term credit and access to all credit—short-term credit combined with medium- and long-term credit (probit regression is conceptually identical with logit regression, except that it assumes a different error distribution). In the final models (keeping only the variables with statistically significant coefficients), current repayment capacity was represented by profit per worker; availability of collateral was represented by the value of fixed assets per worker; creditworthiness was represented by the ratio of overdue debt to sales and also by the ratio of the value of leased assets to sales. In a sense, leasing was regarded as evidence of creditworthiness. The inclusion of this variable was suggested by the strong association observed in the sample between participation in leasing programs and bank borrowing: 44% of farms with leased assets had bank loans, compared to only 31% for farms without leased assets.

Table 9.29. Probit regression of access to credit for corporate farms

Variable	Short-term credit		All credit	
	Coefficient	Prob.	Coefficient	Prob.
Constant	−0.0075	0.984	−0.6226	0.009
Profit/worker	0.0162	0.028	0.0131	0.039
Overdue debt/sales	−1457.7	0.009	−956.4	0.002
Leased assets/sales	3.3742	0.036	0.3859	0.076
Fixed assets/worker	−0.0039	0.015	—	—

Source: 2003 BASIS survey

The regression results are presented in Table 9.29, which shows only the statistically significant coefficients. Both regressions are statistically significant in the sense that the hypothesis of all zero coefficients is rejected (at 0.001 level). As in Table 9.28, profitability has a positive impact on access to credit, while overdue debt has a negative effect. It is interesting to

[18] This regression-based ranking is not entirely consistent with the simple univariate analysis, in which the percentage of farms with access to credit is actually slightly higher in Nizhnii Novgorod than in Rostov (34% and 29%, respectively). Such inconsistencies are often observed in econometric work, because regression analysis allows for all the relevant factors simultaneously and estimates the effect of a particular variable while keeping all other variables constant, whereas univariate analysis ignores the effect of all other factors.

note that participation in leasing programs (as a creditworthiness variable) also has positive effect: the higher the value of leased assets, the higher is the probability that the farm borrows. The coefficients of both creditworthiness factors are statistically significant. The influence of collateral factors is weak and uncertain (again as in Table 9.28): fixed asset endowment is statistically significant only the short-term model, where it has the wrong sign; land is not statistically significant in either model. Regional factors (as represented by region dummies) turned out to be statistically not significant in the two financial-ratio models (unlike the model in Table 9.28).

The two sets of regression results caution against generalizing the conventional financial patterns of market economies to transition countries. Russia is apparently characterized by specific fundamental features that require further attention before the familiar principles of the developed countries can be extended to its financial markets.

Conclusion

The general improvement in the financial situation of Russian farms since 1999 is attributable to overall agricultural recovery, not to any government support measures. Continuing government interventions in farm finances, often replicating the old centralized models, do not appear to have had a significant impact on agriculture.

Although subsidized interest-rate programs reduce the cost of borrowing and thus increase the demand for credit, administrative barriers (the regional quota system, the precondition of zero overdue debt) severely restrict the actual volumes of credit used under these programs. The commercial banks are clearly showing an increased interest in agricultural lending in the light of the post-1999 recovery, yet they adopt similarly restrictive criteria for loan approval.

Government-sponsored machinery leasing programs appear to be unfolding on a substantial scale across the country, although their cost advantages are not clear because of non-transparent charges and calculations. The leasing programs are seriously hampered by administrative regulations that limit the choice of suppliers and models to a centrally approved list. They furthermore show a strong bias toward large producers, neglecting the huge mass of peasant farms that on the whole have much less capital stock than the established corporate farms.

The marginal impact of government support programs requires careful rethinking of the role of government in agricultural finance. To be successful, Russian farms probably need less barriers to the proper functioning of market channels, not necessarily more support. Management quality is observed to be an important factor for success and financial health of Russian farms—more important than physical endowments (labor and land). Analysis of farm solvency has shown that natural conditions and product choice play but a marginal role, while subsidies—perhaps the most popular policy mechanism in Russia—do not contribute at all to financial health. Government policies should therefore emphasize management quality through far-reaching training programs in finance, marketing, production management, and personnel management. The main managerial emphasis must shift from production to budgeting and cost control as the standard tools that enhance economic and financial performance.

Part III

Productivity and Efficiency of Russian Agriculture

Chapter 10

The Allocative Efficiency of Input Use in Russian Corporate Farms

William M. Liefert

This chapter examines the empirical analysis of how allocatively efficient Russian corporate farms—the successors of former kolkhozes and sovkhozes—have been in their use of inputs. The discussion covers all the allocative efficiency (AE) tests made by participants in the BASIS project. The AE test used in all studies is to compare the value of an input's marginal product to its price. The chapter starts with a brief description of the methodology and data used in the AE analyses. Section 2 presents the results and discusses how the calculations might be biased. Section 3 examines possible causes of allocative inefficiency in input use, Section 4 discusses policies the government could adopt to improve input allocative efficiency, while the last section presents conclusions.

Methodology and Data

The method used to assess the allocative efficiency of input use is to compare the price of input i (P^i) with the value of the input's marginal product in production of commodity j (VMP^i_j). VMP^i_j equals the marginal product of input i in the production of commodity j (MP^i_j) times the price of commodity j (P_j). Russian agricultural producers will be using inputs efficiently (and at the profit-maximizing level) when

(1) $$P^i = VMP^i_j = MP^i_j \times P_j$$

If P^i is not equal to VMP^i_j, we will refer to the situation as one of disequilibrium. Specifically, if $P^i > VMP^i_j$, the use of input i in the production of commodity j should be decreased in order to improve allocative efficiency (and farm profits); if $P^i < VMP^i_j$, the use of input i in the production of commodity j conversely should be increased in the interest of improving allocative efficiency (and raising farm profits).[1]

Figure 10.1 depicts the optimal level of input use. The VMP curve for the input to a specific user constitutes the user's demand curve for the input. The market price for the input equals p^2, such that the input supply curve faced by the input user equals the horizontal line S. The allocatively efficient, as well as profit-maximizing, level of the input to use is q^2. If q^1 of the input were being used, the producer should increase purchase and use. This is because the rise in the producer's revenue from using more input to produce more output (v^3) exceeds the input's price p^2. Using one more unit of the input in production would increase profit by v^3v^2. Alternatively, if q^3 of the input were being used, the producer should reduce use. The rise in producer's revenue from additional input use (v^1) is less than the input's price p^2. Using one less unit of the input in production would increase profit by v^1v^2.

A more general statement of the allocative efficiency condition is:

(2) $$MFC^i = VMP^i_j$$

where MFC^i is the marginal factor cost to the producer of obtaining one more unit of input i. Two situations can exist involving the relationship between the MFC of an input to a buyer and the input's price. The first is that each buyer of the input purchases relatively small amounts and its purchase volume cannot affect the input's market price. From the point of view of individual buyers, the market price is fixed. This is the situation depicted in Figure 10.1, where the input supply curve is horizontal at the fixed market price of p^2. Since the market price is constant, the MFC to buyers of each unit of the input purchased is also constant, and also equals p^2. The supply curve S equals each buyer's MFC curve.

The second situation is that, if the input buyer purchases such a large amount of the input that its purchase volume affects the input's market price, the buyer has market power (if the buyer is the sole purchaser of the input, it would be a *monopsonist*). The input supply curve faced by the input user would not be horizontal (as is S in Figure 10.1), but rather upward-

[1] This method for assessing the allocative efficiency of input use has been commonly used in empirical research. For example, Griliches (1964) and Headley (1968) use it to examine input use in U.S. agriculture in the 1950s and 1960s. Griliches finds major underuse of fertilizer (VMP high above price), while Headley finds underuse of fertilizer and pesticides, and overuse of labor. The results are consistent with changes in input use during the period covered, as use of fertilizer and pesticides increased substantially while the agricultural labor force contracted.

sloping. Thus, as the buyer purchased more of the input, its purchases would drive up the price it must pay for the input. The MFC of obtaining the input would therefore rise, which means the MFC curve of the input to the user would also be upward-sloping.

The inputs for which allocative efficiency tests have been done in this project are labor, material inputs (disaggregated to fertilizer, seeds, feed, fuel, electric power, spare parts, and services), and land. With the exception of labor, we can assume that individual farms throughout Russia are "small" purchasers of all these inputs, such that they face horizontal input supply curves at fixed prices (such as S in Figure 10.1). The allocative efficiency condition presented in equation (1) and depicted in Figure 10.1 thereby holds.

Fig. 10.1. Allocative efficiency of input use

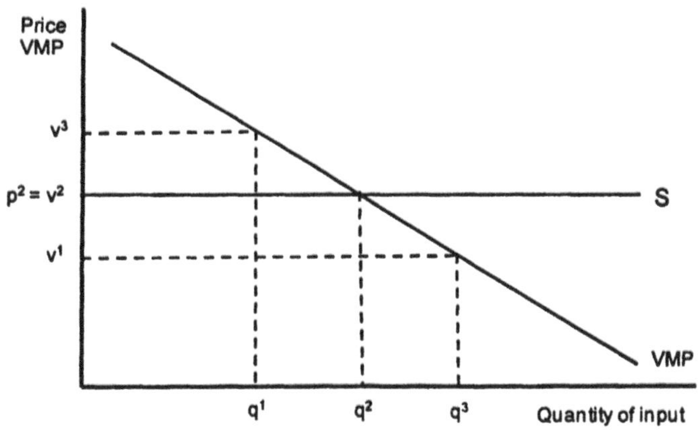

Labor is a possible exception, given that the former state and collective farms that still employ half of Russia's agricultural labor (see Chapter 7) are big, typically employing hundreds of workers, and labor mobility in the Russian countryside is poor. Farm managers can face constraints in both expanding and contracting their labor force, in the latter case because of their traditional social-welfare responsibilities and the concerns of regional government officials about rural unemployment. Farms' labor supply curve might in fact be more vertical (i.e., a fixed labor force regardless of price) than horizontal (i.e., a fixed price regardless of the number of employed). If so, farm managers might not pay all their workers a fixed "market-determined" wage. The consequences of such behavior for the AE analysis of labor will be examined later in the chapter.

AE calculations have been made within the project by five research teams, identified in Table 10.1. All the calculations reported in this chapter pertain to the behavior of the large corporate farms—former state and collective farms—in Russia.[2] This means that the volume of inputs used in estimating the production functions on which the AE calculations are based are the volumes used by the farms in their "collective" farm operations, while the output used in estimating the production functions are correspondingly the levels of output from farms' collective operations. Input use and output by the household plots associated with these farms are excluded in estimating the production functions.

Table 10.1. Allocative efficiency calculations

Researchers	Regions	Data source	Coverage	Time period
Uzun	All Russia	Goskomstat large-farm database	All corporate farms in Russia (24,000)	2001
Liefert, Gardner, Serova	All Russia	Statistical yearbooks	Aggregate data	2000
Lerman	Rostov, Ivanovo, Nizhnii Novgorod	BASIS survey	Sample of 142 corporate farms	2001
Grazhdaninova	Rostov, Ivanovo, Nizhnii Novgorod	BASIS survey	Sample of 142 corporate farms	2001
Epshtein	Leningrad	Goskomstat large-farm database	All corporate farms in Leningrad Oblast (195)	2001

The calculations vary by researcher with respect to (1) the regions covered; (2) the data used; (3) the specific output/input combinations examined. Uzun covers all of Russia, using a Goskomstat database of the annual reports of some 24,000 corporate farms. Liefert et al. also cover all of Russia, but they use aggregate national statistics and examine only the AE of fertilizer to produce grain. Lerman and Grazhdaninova cover the three oblasts where data were collected for 142 corporate farms as a part of the 2003 BASIS survey—Rostov, Ivanovo, and Nizhnii Novgorod. Epshtein's analysis is specific to Leningrad Oblast, using the annual reports of all 195 corporate farms in that region.

Some researchers examine the value of total farm output, including both crop and livestock production in essentially diversified Russian farms (Uzun, Epshtein, in part Lerman). In these analyses, the explanatory variables represent the total use of inputs, expressed in monetary values or in physical quantities. Other researchers analyze only crop or livestock production (Lerman), or focus even more narrowly on grain and sunflower,

[2] AE results for peasant farms in Tambov Oblast are reported in Chapter 3.

recognizing that these commodities are the main consumer of fertilizers (Liefert et al., Grazhdaninova). In these cases, the use of inputs is restricted to the enterprises or commodities analyzed, while the output is expressed in value terms (aggregate crops or livestock) or in physical quantities (grain).

The data for input and output prices used in the AE tests pertain to the specific regions covered in each analysis. The price data used by Grazhdaninova and Lerman are from the 2003 BASIS farm survey. Epshtein uses Leningrad Oblast prices, while Uzun and Liefert et al. use national prices published for the agricultural sector as a whole.

The main information needed for the AE analysis is inputs' VMP. The VMP used in all AE tests discussed in this chapter are from production functions econometrically estimated by the project researchers for the specific regions identified in Table 10.1. The only exception is Liefert et al., who use production functions estimated by Osborne and Trueblood (2002).

The key common feature of the estimated production functions is that they all take the Cobb-Douglas form, i.e.,

$$(3)\quad Y = aX_1^{b1}X_2^{b2} \ldots X_k^{bk},$$

where Y is the output produced using the bundle of inputs $X_1, X_2, \ldots, X_k$. The output may be given in physical units, e.g., tons of wheat, or in value terms (e.g., "total farm" output in aggregate). The marginal product of factor i is calculated by differentiating the product Y from (3) with respect to the input X_i at the point of the means:

$$(4)\quad MP_i = \Delta Y/\Delta X_i = b_i Y/X_i.$$

Here Y is the geometrical mean of the output (the mean of its natural logarithms); X_i is the geometrical mean of input i; b_i is the estimated coefficient of input i from the production function (the elasticity of input i). If the output Y is in physical units (tons), the marginal product is multiplied by the price of the output P_Y to obtain the value of marginal product of input i: $VMP_i = MP_i \times P_Y$ (see equation (1)). If Y is the value of the output, then the result of (4) is actually VMP_i.

To summarize, for each output/input combination, multiplying the estimated coefficient in the Cobb-Douglas function by the (geometrical) mean product of the input gives the input's estimated VMP.

Results

The AE calculations performed for various inputs by researchers participating in the BASIS project are presented in Table 10.2. The table identifies the specific type of agricultural output covered by the calculations. This means that the production functions used as the source for the VMP values in the AE calculations were estimated specifically for the commodity groups (or commodities) identified. The estimated production functions fall into two main groups: the group of models that use aggregated farm output expressed in rubles (the value of total farm product—GAO, or the value of output in crop and livestock enterprises separately); and the group of commodity models in which the output is the quantity of specific commodities (grain or sunflower) in physical units. The AE results are reported as ratios, which equal an input's VMP divided by its price or cost. Consequently, an AE value not equal to 1 indicates a disequilibrium and an inefficient volume of input use. A value greater than 1 means that the input is being underused from the point of view of both allocative efficiency and farms' profit maximization; a value less than 1 means that the input is being overused.

Labor

The first line in Table 10.2 presents the AE for labor. The results suggest a general, though not large, underuse of agricultural labor within Russia. For example, the AE calculation by Uzun covering total agricultural output in all Russia suggests that employing more agricultural workers would raise farms' revenue by about 20% more than the cost of employing the workers. Increasing employment would thereby raise farm profit (or if farms are unprofitable, reduce their losses).

Results showing that agricultural labor is being underused might seem surprising, given the commonly held belief that Russian agriculture suffers from a surplus of unskilled, aged, and demoralized labor. Yet, farm surveys done specifically for this project (in Rostov, Ivanovo, and Nizhnii Novgorod oblasts) indicate that although farms might have a surfeit of older and unskilled workers, they have a shortage of skilled laborers—such as those proficient in machinery use and repair, animal care (including knowledge of modern breeding and feeding practices and veterinary services), and low to middle level management activities (see Chapter 7). A disproportionate share of labor migration out of agriculture during the transition has been among younger and better educated workers. The high aggregate VMP of labor relative to wages could reflect increasing shortage of skilled workers.

Another explanation as to why agricultural workers' pay appears low relative to their productivity could be that the studies cited in Table 10.2 understate the "real wages" earned. The studies cover operations by former state and collective farms. These farms continue the obligation that existed during the Soviet period of providing social welfare services needed by their workers, such as health care, schools, housing, and recreation.

Table 10.2. Allocative efficiency ratios for main inputs*

	Aggregated output models (value units)							Commodity models (physical units)			
	Uzun Russia	Lerman Survey	Lerman[1] Survey	Lerman[2] Survey	Lerman[3] Survey	Lerman Survey	Epshtein Lenin-grad	Liefert et al. Russia	Grazhda-ninova Survey	Grazhda-ninova Rostov	Grazhda-ninova Survey
Output	GAO	GAO	crops	crops	crops	livestock	GAO	grain	grain	grain	sunflower
Inputs:											
1 Labor	1.19	1.28	1.12	0.97	1.32	1.28	1.38	–	–	–	–
2 Material inputs	0.80	0.55	0.70	–	–	0.44	0.92	–	–	–	–
3 Fertilizer	6.30	–	–	1.47	7.47	–	2.87	2.16	0.95	0.70	–
4 Seeds	−0.50	–	–	3.70	2.95	–	1.11	–	0.64	1.49	–
5 Feed	0.60	–	–	–	–	–	1.16	–	–	–	–
6 Fuel	1.00	–	–	–	–	–	1.49	–	–	–	–
7 Power	3.30	–	–	–	–	–	0.71	–	–	–	–
8 Parts	1.20	–	–	–	–	–	1.64	–	–	–	–
9 Services	1.60	–	–	–	–	–	0.84	–	–	–	–
10 Land	–	1.19	0.89	0.78	0.58	–	–	–	0.68	0.99	0.75

*The AE ratios are calculated as the ratio of input VMP to input price. All data for 2001 except Liefert et al. (2000). Survey regions are Rostov, Ivanovo, and Nizhnii Novgorod oblasts. GAO is gross agricultural output. A dash (–) denotes not computed.

[1]No disaggregation of material inputs, all measured in rubles.

[2]Fertilizer and seeds replace aggregated material inputs, both measured in rubles.

[3]Fertilizer and seeds replace aggregated material inputs, with seeds measured in rubles and fertilizer in physical units.

In the AE calculations, the value of these social welfare services should be included in workers' real wages. It appears that both the quality and quantity of such services provided by farms have been steadily declining, and the process of transferring these responsibilities from farms to local governments is beginning in some areas. Nonetheless, excluding from real wages the value of the remaining social services biases the AE calculations upward.

Another likely reason why real labor remuneration is understated is that workers on the large farms receive not only monetary wages and social services from their employing farms, but also in-kind payment. As discussed in previous chapters, all households on these farms continue to operate their own small plots, which account for about 10% of agricultural land in Russia, and yet in aggregate produce over half of the country's agricultural output (see Chapter 2). A reason for the plots' striking performance is that the plotholders receive inputs from their parent farms, either as legitimate in-kind payment or in some cases as stolen goods. Given that the plots specialize in livestock operations, inputs commonly transferred to plot use in this way are animal feed, use of vehicles by households to truck output to farmers' markets for sale, and fuel for the vehicles. Some fertilizer and seeds are probably also transferred from the corporate farm to the use of household plots, and grazing of plot animals on communal farmland is common.

The value of inputs transferred to use on household plots should be included in workers' real wages. Yet, given that much of the input transfer is unofficial, fully accounting for such reallocation is virtually impossible. The effect is that workers' real wages are understated, thereby again biasing the AE calculations upward. In work in the Mtsensk District of Orel Oblast, Bogdanovskii (Chapter 7) finds that farms' in-kind transfers and social welfare services to workers would increase base wages by 20-25%. Most of the AE calculations for labor show VMP values exceeding wages by 20-40%. If Bogdanovskii's markup in agricultural real wages applies to the entire country, it would appear that real wages are fairly close to workers' VMP.

Material Inputs

The second line in Table 10.2 gives AE calculations for material inputs in the aggregate, which can include fertilizer, seeds, animal feed, fuel, power, spare parts, and services (depending on which particular inputs are included in researchers' estimation of production functions). The lines that follow present the AE results for specific material inputs.

The results for material inputs in the aggregate strongly support a conclusion of overuse of inputs. All the AE ratios in the second line of Table 10.2 are less than 1. Based on these results, it would be efficient and profit-maximizing for farms to reduce in general their purchase and use of material inputs.

The calculations for material inputs, however, are all biased downward to some degree. In discussing AE analysis for labor in the preceding subsection, we argued that one of the reasons why labor's real wage is understated in the calculations is because it excludes material inputs that pass from farms to workers for use on their household plots. This input transfer distorts the AE calculations not only for labor, but also for the material inputs involved. The reason is that the bulk, if not all, of the transferred inputs are included in the estimation of farms' production functions, which exclude output on the household plots. If the inputs transferred to household plot use were also excluded in the estimation, the inputs' VMP would be higher.

The household plots specialize in production of livestock products to a much greater extent than corporate farms do. The downward bias in the estimated VMP for material inputs is therefore particularly strong for inputs used heavily in livestock production, such as animal feed, and for production functions whose output is livestock products. Correspondingly, the bias is weakest for production functions whose output is crops, given that the collective operations of the former state and collective farms account for most of Russia's output of bulk crops (for example, 88% of grain in 2001[1]).

The general downward bias in the AE calculations for all material inputs mitigates the conclusion that farms were overusing these inputs. The bias can help explain why Lerman's AE ratio of 0.44 for livestock products in Table 10.2 is particularly low. Epshtein's results should also suffer from a relatively large downward bias, given that they are for Leningrad Oblast, which specializes in livestock production. Taking this bias into account, it appears that farms in the Leningrad region were using material inputs fairly efficiently.

We now proceed to discuss the AE results for the disaggregated subcategories of material inputs presented in lines 3-9 in Table 10.2. Despite the previous conclusion that material inputs in the aggregate are overused, *fertilizer* specifically is underused (see line 3 in Table 10.2). Only Grazhdaninova's AE calculations for grain production are below 1. All the other calculations are decisively greater than 1. In fact, Uzun computes a ratio of 6.30 using the data for all Russian large farms, which suggests that fertilizer use at the margin was earning farms revenue more than six times the price farms had to pay for the fertilizer. Lerman computes an even higher ratio of 7.47 in one of his crop production models, when fertilizer and seeds are the only two material inputs specified in the production function, with seeds

[1] Statistical Yearbook 2002, p. 408.

measured in rubles and fertilizer in physical units. This high number sits oddly with Lerman's other calculation of 1.47, where the only difference is that both fertilizer and seeds are measured in value terms (rubles). Yet, even if Uzun's and Lerman's calculations are exaggerated outliers, the results support the general conclusion that fertilizer is underused.

The AE results for all the other specific subcategories of material inputs—seeds, animal feed, energy (fuel and electric power), spare parts, and agricultural services—are less conclusive than for fertilizer. This is particularly true for *seeds* (see line 4 in Table 10.2). Most of the results (and especially those by Lerman) indicate underuse, but Grazhdaninova suggests otherwise (for the three regions combined, though not for Rostov alone), while Uzun computes that the VMP for seeds for all Russia in 2001 was in fact negative. This is a strange result, implying that if Russian farms had used fewer seeds, they would have produced more output. Nonetheless, the calculation indicates that even if not negative, seeds' VMP was substantially below their price.

Uzun's AE result for animal *feed* (line 5 in Table 10.2), covering all Russian agriculture, indicates overuse, while Epshtein's calculations suggest underuse. However, Epshtein's work covers only Leningrad Oblast, where livestock production comprises a relatively high share of the value of total agricultural output, and which has a cold northern climate, far removed from the country's most fertile feed-producing regions. Perhaps the difficulty of obtaining feed in such a challenging area for livestock operations reduces feed use below what appears to be the optimal level. If so, Leningrad's AE results for feed are not very representative for the country at large. On the other hand, as discussed earlier, the downward bias in the estimation of VMP for all material inputs because of input transfer from farms to household plots is particularly strong for feed. This suggests that feed use on the large farms for all Russia based on Uzun's calculation might not deviate strongly from the optimal level, and that feed used in large farm operations in Leningrad Oblast is in shortage from an AE point of view.

The results for *energy*, both fuel and electric power, are also inconclusive (lines 6 and 7 in Table 10.2). Based on data for all Russian farms, Uzun obtains an AE ratio of 1.00 for fuel, which indicates that this input is being used at the optimal level. On the other hand, Epshtein's calculation specific to Leningrad Oblast shows underuse of fuel. The calculations of Uzun and Epshtein for power show even greater divergence. Uzun's calculation covering all Russian agriculture shows strong underuse, while Epshtein finds overuse of electric power specific to Leningrad Oblast.

The AE results for *spare parts* and *services* (lines 8 and 9 in Table 10.2) do not tell a consistent story either. The calculations of both Uzun and Epshtein indicate mild underuse of spare parts, while their calculations for services again show opposite results. As with the calculations for energy, Uzun's results covering all of Russia should perhaps dominate Epshtein's results, which are specific to the Leningrad region. This supports a tentative, though hardly conclusive, judgment of underuse of spare parts and services and optimal use of fuel.

Land

The AE ratios for land (line 10 in Table 10.2) are calculated by comparing the land's VMP to the rental price of land, as obtained from the 2003 BASIS survey. All AE calculations for land based on crop production models are below 1. The one exception is Lerman's estimate of 1.19 based on total agricultural output, which includes both crop and livestock production. This result is higher than the AE ratios for pure crop production models because livestock adds significantly to total output without using much land, thus disproportionately increasing the land's VMP. Yet, even in this calculation the land's VMP is only about 20% above its cost (as expressed by rental price in the survey).

All inputs are to some extent heterogeneous in qualitative characteristics, but in land heterogeneity is particularly important. Plots of land vary greatly in economic value because of differences in productivity (which determines its value in agriculture) and in location (which determines its value both in agriculture and in nonagricultural uses that agriculture must compete with). Indeed, in Russia arable land declined 8% between 1990 and 2002.[2] This suggests that during our survey period significant amounts of land in farms were idle, implying rental value of essentially zero.

In our regressions, the estimated marginal product of land is the difference in output across farms associated with differences in the area of land used in the farms, holding other input quantities constant. This does not generate a measure of the value of a marginal hectare of land at the extensive margin, where the next hectare is held idle, nor is it the value of marginal land in the sense of the lowest quality land in use. But the estimated marginal product might be correlated with decreases in the locational value of land if the larger farms had land with lower locational value, and might be correlated with productive quality of land if larger farms had lower quality land. If the farms sampled all have the same proportions of land with various locational and quality characteristics, then the estimated marginal product would measure the value of a hectare of land of average locational and productivity characteristics at the margin of farm size. But with heterogeneous land, the marginal land on the average farm will be of lower quality

---

[2] Statistical Yearbook 2002, p. 401.

than the average of all land in those farms. We can therefore interpret the shortfall of our estimated VMP below the average rental rate as an estimate of the difference between the value of land at the margin and the average rental value of all land in the sample farms.

This difference does not imply inefficiency in resource use. There could of course be mispricing of land, but this would not in itself imply inefficiency in the sense of resource misallocation. Inefficiency in resource allocation would occur, with too much land in agriculture, if the VMP of a plot of land in agriculture were less than its VMP in nonagricultural uses. Our regression results do not provide a test of this matter; but given the difficulties discussed above, it is surprising and encouraging that our estimated VMP is as close to the observed average rental rate as it is.

Biases in Allocative Efficiency Calculations

Previous subsections have examined biases in the AE calculations specific to particular inputs. This section examines two more general biases that could affect the calculations for all inputs. The first bias is that most of the studies are specific to the year 2001 (the only exception is Liefert et al. covering 2000). This was a good weather year for Russian grain production, with output at 82 million metric tons (mmt), compared to average annual production over 1996-2003 of 68 mmt.[3]

Using the results for 2001 to assess performance for a broader period of time around this year would bias VMP upward. This in turn would bias upward all the AE calculations given in Table 10.2. The VMP upward bias affects results based on production functions estimated for total agricultural output (such as Uzun and Epshtein), even more so the results specifically covering crops (such as Lerman for various inputs), and most strongly the results specific to grain (such as Grazhdaninova for various inputs). This upward bias in the AE calculations tempers the previous conclusion that labor is being underused in Russian agriculture, especially given that labor's VMP exceeds its price in most calculations by only 20-40%. The only other inputs for which a general finding of underuse is made are fertilizer and spare parts. The fairly mild degree of spare parts underuse may be mitigated or even eliminated altogether by the upward bias due to favorable weather. The empirical evidence of fertilizer underuse, however, is so strong that any upward bias does not challenge this conclusion. Yet, the upward bias can help explain the magnitude of underuse indicated, especially by the calculations of Uzun and Lerman.

---

[3] See *Production, Supply, and Distribution*, Washington, DC: U.S. Dept. of Agriculture, (www.fas.usda.gov/psd/).

Among all the specific output/input combinations for which AE calculations are made, this bias is probably strongest for grain and fertilizer. This particularly calls into the question the AE calculation by Liefert et al. (using domestic rather than trade prices), which at a value of 2.16 indicates fertilizer underuse. Yet, the calculation by Liefert et al. is for 2000, not 2001. Russian grain output in 2000 equaled 63 mmt, somewhat below average annual output for 1996-2003 of 68 mmt. If there is any bias in the calculation, it is downward. Given that Liefert et al. also finds underuse of fertilizer to produce grain, such a bias strengthens the finding.

Another possible measurement bias common to all the inputs stems from the classic "management bias" problem (see Mundlak, 1961). Farm managers vary in ability, with the more capable generating a higher VMP from inputs. This results in them using more inputs in production. A cross-sectional regression will therefore overstate the input's VMP on an average farm. The reason is that the estimated output elasticity measures the gain in output in better managed farms compared to poorly managed ones, and incorrectly attributes the higher output to greater use of inputs rather than correctly crediting it to superior management. As a consequence, the VMP estimated from the production functions will overstate the actual VMP. This will bias the AE calculations upward and thereby toward input underuse. The impact of management on farm performance is discussed in Chapter 9.

Summary of AE Results

The second column in Table 10.3 summarizes the results of the AE calculations as initially determined from the values in Table 10.2, but before discussion of measurement issues and conditions that could bias the results in some direction. The third column in Table 10.3 presents the direction of change in these results that can be expected once measurement errors and weather biases are allowed for in line with the foregoing discussion. Strong conclusions from Table 10.2 are possible for only the following inputs: labor appears underused, material inputs are overused in the aggregate, though fertilizer and spare parts are underused, while fuel is used at close to the optimal level. A qualification to the results for spare parts and fuel are that only two of the five research teams provide AE calculations for these specific inputs.

The underuse of labor is not strong, as most studies show a VMP exceeding wages by only 20-40%. Also, as preceding discussion indicated, the AE calculations for labor are biased upward, which results in an AE conclusion biased toward underuse (VMP > input price). The sources of bias are (1) the good grain year for 2001, which results in unrepresentatively high

VMP values for that year; (2) understated real wages for labor; and (3) the management bias problem (which exists for all the AE calculations). Taking these biases into account, it appears that labor is used close to its allocatively efficient level.

Table 10.3. Summary of allocative efficiency calculations

Input	Result based on estimates in Table 10.2	Results after conceptual adjustment for measurement errors and weather biases
Labor	Underuse	Closer to optimal use
Material inputs	Overuse	Overuse
Fertilizer	Underuse	Closer to optimal use
Seeds	Inconclusive	—
Feed	Inconclusive	—
Fuel	Close to optimal use	—
Power	Inconclusive	—
Spare parts	Underuse	Closer to optimal use
Services	Inconclusive	—
Land	Inconclusive	—

The AE calculations for material inputs in the aggregate are also biased in part toward a conclusion of underuse, because of 2001 being a good grain year and the management bias problem. Yet, a bias also exists toward input overuse, resulting from the largely unrecorded transfer of inputs by farms to workers for use on their household plots. The input transfer understates the VMP of using these types of inputs on farms' collective operations. This bias toward input overuse balances the biases toward underuse, such that the initial conclusion of overuse for aggregate material inputs appears still to hold.

The biases that affect the AE calculations for fertilizer and spare parts are mainly toward underuse. Two such biases again stem from the management bias problem and 2001 being a good grain year, which particularly affects AE analysis for fertilizer. The bias toward input overuse that results from input transfer to household plot use is not very strong for fertilizer, given that plots specialize in production of livestock products rather than bulk crops. These considerations in total mitigate the initial conclusion of fertilizer and spare parts underuse.

The above discussion shows that for the specific inputs for which AE calculations reveal inefficient use—labor, fertilizer, and spare parts, all toward underuse—the calculations are biased in that direction. When these biases are taken into account, a strong judgment of underuse cannot be made. Likewise, the examined results do not support an overall conclusion for Russian agriculture of either input overuse or underuse. Russia's performance with respect to the allocative efficiency of input use, though not wholly clear, appears fairly respectable.

Causes of Allocative Inefficiency

What could explain the allocative inefficiency that exists in Russian agriculture? We examine the following possible causes: (1) poor decision making by farm management; (2) farms have market power over their workers in setting wages; (3) farms lack the working capital to purchase inputs; and (4) farms are willing to pay higher prices to obtain more inputs, but suppliers ration the quantity available for domestic sales.

Poor decision making by farm management in using inputs has some plausibility, given that farm managers' goal during the planned period was not profit maximization, but rather achieving mandated output targets. Such a goal encouraged managers to lobby for low output targets and high input allocations, while being conservative with production techniques so as not to jeopardize meeting the short run output target. Managerial incentives did not encourage initiative and flexibility in changing output/input combinations and production methods to achieve maximum output from a given mix of inputs. Farm managers have probably not yet fully developed the skills necessary to achieve maximum allocative efficiency in input use.

A possible cause of the allocatively inefficient use of labor (as shown by low wages relative to productivity) is that farms have "market power" in setting pay levels. The earlier examination of farms' profit-maximizing level of labor use (equating labor's wage to its VMP) was based on the assumption that farms lack such market power vis-à-vis labor, such that wages are set by the "market." As explained earlier, though, if farms have market power, the profit-maximizing condition for labor use is equating labor's VMP not with the real wage paid, but rather with the marginal factor cost of employing labor. If an employer has market power, this cost increases as employment rises (reflecting an upward sloping supply curve for labor). Profit-maximization under such circumstances results in less labor hired at lower wages compared to competitive conditions.

Few employment opportunities exist in the Russian countryside outside of the large former state and collective farms. Farm workers also face difficulties in migrating to cities, such as obtaining permits to settle, finding low-cost housing, and most importantly finding employment (given their lack of marketable skills). Farm management might be helped in setting low wages in that the decision whether to uproot one's life by leaving the farm for uncertain employment in a distant city, especially in Russia's difficult economic conditions, might not be sensitive to marginal changes in the real wages paid by farms. With de factor market power, farms would maximize their profit by paying relatively low wages.

Another reason allocative inefficiency might exist, especially in the form of input underuse, is financial constraints, such that farms lack the working capital, financed from either their own revenue or credit, to purchase inputs. A well-operating commercial credit system does not yet exist for Russian agriculture and access to credit is limited (see Chapter 9). If production cycles are short, then a situation where an input's VMP exceeds its purchase price generates more revenue than cost and the input purchase becomes self-financing. In agriculture, on the contrary, inputs must be purchased months before output is ready for sale, and the unavailability of credit or working capital can preclude obtaining inputs whose acquisition might be profitable over the full production cycle.

Nonetheless, the above argument that input use can to some degree be self-financing suggests that in those cases where inputs' VMP highly exceeds their purchase price, farms would buy more inputs, if such purchases could be effected. This suggests that a general reason why material inputs are not being used at more efficient levels is that the markets for such inputs are undeveloped. Input suppliers are not responding fully to the localized demand by farms for more inputs.

A likely reason for domestic market disequilibrium for tradable inputs such as fertilizer and fuel is that farms are willing to pay higher prices for the inputs, but suppliers are averse to selling at even greater prices. The most likely cause of this behavior is that the Russian producers of these inputs receive higher prices if they export rather than sell domestically. In 2000, fertilizer producers exported nitrogen and potash fertilizer at average prices of 1,738 and 2,598 rubles per ton, respectively, compared to only 1,201 and 1,471 rubles per ton from domestic sales.[4] Another export incentive is that fertilizer producers can keep, and invest, their earnings abroad.

Responding to these higher export prices, the Russian fertilizer industry since the mid 1990s has been exporting about 80% of its output, mostly to West European countries. Consequently, use of mineral fertilizer per hectare of sown land in Russia has fallen dramatically, from 1990 to 2001 by 78%. The drop in total fertilizer use is even greater than that indicated by the above figure, given that during this period sown area itself declined by 28%.[5]

---

[4] Export prices computed from customs statistics in *Tamozhennaia statistika vneshnei torgovli Rossiiskoi Federatsii 2000*, Moscow: State Customs Committee (2001) by converting from U.S. dollars to rubles at the prevailing market exchange rates. Domestic prices from *Tseny v Rossii 2002*, Moscow: Goskomstat (2002), p. 132.

[5] Fertilizer use data from Statistical Yearbook 2002, pp. 411-412.

The severe drop in fertilizer use can explain why the VMP of fertilizer is now relatively high, while the disparity between domestic and world trade prices for fertilizer can explain why Russian farms have difficulty obtaining more fertilizer to close the gap between its VMP and domestic purchase price. To get fertilizer, Russian farms usually need the help of regional government, which either pays higher prices for it or commands delivery at low prices. Regional governments often "sell" fertilizer to farms at attractively low prices, in return for the farms' commitment to sell back their output, or at least sell within the region.

Russian domestic fertilizer markets therefore reflect not only a disequilibrium between input prices and input VMP, but also the fact that domestic prices differ so strongly from world trade prices. Liefert et al. calculates that in 2000, the AE ratio of Russian farms' VMP of fertilizer to produce grain to fertilizer's price was 0.99, when the price of both fertilizer and grain used in the calculation is the products' trade price. This result compares to the ratio of 2.16 (Table 10.2) when the fertilizer and grain prices are domestic prices. This indicates that from the point of view of trade, rather than domestic, prices, Russia was using fertilizer in the production of grain at the optimal level of allocative efficiency. In other words, if Russian farms paid world prices for fertilizer rather than low domestic prices, in 2000 they would have purchased almost exactly the amount of fertilizer they were actually using (in their grain-producing activities). The low domestic prices Russian farms pay for the fertilizer they receive explains the "disequilibrium" that exists between fertilizer's VMP and domestic purchase price. From the point of view of world prices, a disequilibrium of underuse does not necessarily exist.[6]

Like fertilizer, fuel and energy in general are products Russia exports in large volume. During the transition period, about half of Russia's hard currency export earnings have come from energy products (especially oil, oil products, and natural gas). Also as with fertilizer, Russia has sold its energy products on the world market at higher prices than on the domestic market. In 2000, Russia exported diesel fuel and electricity at average (unit value) prices of 6,825 rubles a metric ton and 507 rubles per thousand kilowatt

---

[6] Price differences might also explain a notable discrepancy in Table 10.2, where Grazhdaninova's AE ratio for the three survey oblasts points to (slight) overuse of fertilizer, contrary to the other results. All other researchers using the national average price of less than 1,500 rubles per ton obtained AE ratios greater than 1, indicating underuse of fertilizer. Grazhdaninova calculated the AE ratio using the survey mean price of 2,600 rubles per ton (adjusted for nutrient content), which was very close to the world trade price. This gave an AE ratio of 0.95 for fertilizer use in grain production in the three survey oblasts, which is very close to the point of allocative efficiency. The low AE ratio for Rostov oblast in Table 10.2 probably points to actual fertilizer overuse in this agriculturally productive region.

hours, respectively, while Russian agricultural producers purchased these inputs at only 5,444 and 337 rubles a ton/1,000 kilowatt hours.[7]

The gap between domestic and world prices for fuel and energy also motivates "non-market" distribution of such inputs to farms through the help of regional governments. Despite such assistance, the higher prices Russian energy producers can receive from exports than from domestic sales have resulted in a severe drop in deliveries to Russian farms. From 1990 to 2001, supplies of gasoline and diesel fuel to farms fell by 83% and 74%, respectively, while farm consumption of electricity decreased 62%.[8] Yet, unlike with fertilizer, the large decline in farm use of fuel and electricity does not coincide with AE calculations for energy that conclusively show underuse (high VMP relative to purchase price).

Policy Options for Improving the Allocative Efficiency of Input Use

The policy options for improving input allocative efficiency follow closely from the preceding examination of why current input use can be inefficient. One option is to try to improve farm management, especially in the area of economic decision making. Raising the allocative efficiency of input use requires that managers flexibly alter output levels and the mix of inputs used in production in response to changes in output and input prices. Such decision making was unnecessary, and therefore unnurtured, under the planned economy of the Soviet period.

Another option is to improve training for farm workers, in two ways. One would be to educate more workers in the specialized skills currently in shortage on farms. The second would be to train unskilled workers for employment outside of agriculture. The government could promote agricultural outmigration by reducing restrictions that cities place on relocation to their jurisdictions, and by increasing urban low-cost housing. These steps would reduce the "market power" that the large farms hold over their labor force.

The government could also help develop a more effective system of farm credit that would provide farms with working capital. Two features of Russian agriculture during the transition period have strongly discouraged commercial creditors from lending to farms. The first is that during the 1990s most Russian former state and collective farms were unprofitable,

---

[7] See note 4 for sources.

[8] Energy use data from Statistical Yearbook 2002, pp. 410-412.

and in 2001 the figure was still 46%.[9] Consequently, farms have gotten most of their working capital from state or state-affiliated creditors. Yet, most of the credits are never repaid, but rather are periodically written off, which means the credits become de facto subsidies. The second feature of Russian agriculture that discourages commercial lending is that farms have not been able to pledge land or other physical assets as collateral. However, some policy steps have been taken that strengthen property rights in land and could ultimately facilitate the collateralization of farm assets. These steps include the new land legislation passed in 2002 and the recent amendment to the mortgage law adopted in 2004 (see Chapter 5).

A related policy option is to strengthen input markets institutionally, such that suppliers could respond more effectively to farm demand. A major problem in the development of well-operating markets of any kind in the Russian economy has been payment arrears, whereby bills for delivered goods and services go unpaid for long periods (although the problem has been diminishing over time; see Chapter 9). The entire economy would benefit from a stronger system of commercial law that protected property, strengthened debt collection, and enforced contracts.

Another policy option is to decrease, or even wholly eliminate, the gap that exists between world market prices for exportable agricultural inputs, such as fertilizer and fuel, and the inputs' domestic purchase prices. By reducing the incentive of input producers to export rather than sell to domestic users, such a policy change would encourage the commercial and institutional development of domestic input markets. Narrowing the price gap would also improve the allocative efficiency of input use, with prices for tradable inputs now being determined by their more meaningful opportunity cost values—world trade prices. On a broader level, it would increase Russia's economic gains from trade according to comparative advantage and greater integration into world markets.

Though benefiting the economy at large, such a change would not be in the narrow interests of the agricultural sector. The jump in domestic purchase prices for such inputs would reduce input use, farm output, and farm profit.[10] Since economic reform began in the early 1990s, agricultural producers' terms of trade—the prices received for output compared to prices paid for inputs—have deteriorated, from 1990 to 1997 by 77% (Liefert and Swinnen (2002), p. 10). The reason is that the controlled price system of the Soviet period kept input prices low relative to output prices. Such price-

[9] Statistical Yearbook 2002, pp. 408.

[10] The calculation by Liefert et al. for fertilizer to produce grain for 2000 indicates that if domestic purchase prices moved to world trade levels, the amount of fertilizer used by farms would not fall. The farms, however, would lose their current de facto subsidy from paying fertilizer prices below world prices.

setting behavior means that in the pre-reform period, producers were subsidized not only through direct budget subsidies, but also indirectly through the price system. When price and trade liberalization began in the early 1990s, prices for inputs rose to levels closer to their real cost of production, and closer to world market levels. The increase in inputs' real (or relative) prices is a major reason why purchase and use of inputs by farms have declined so substantially.

Yet, domestic prices for tradable inputs such as fertilizer and fuel still remain below world market levels, which suggests that the transition-driven rise in their real prices has not yet ended. In addition to domestic economic and political pressure for these prices to increase, Russia is also facing such pressure in its negotiations for accession to the World Trade Organization (WTO). As a condition for accession, WTO member countries are demanding that Russia raise domestic energy prices closer to world prices. Member countries wish to reduce the subsidy received by Russian domestic producers, in both agriculture and industry, from paying energy prices below world market levels.

Conclusion

The firmest conclusions from the allocative efficiency analyses done by participants within this BASIS project are that labor, fertilizer, and spare parts appear to be underused (VMP exceeds the price of the corresponding input), fuel is used close to the optimal level, but material inputs in the aggregate are overused (VMP is less than marginal cost). However, the measurement errors for labor, fertilizer, spare parts, and fuel bias the results toward underuse, and without these measurement errors the AE calculations would show less underuse. Labor real wages are understated because they do not include the value of farm social welfare services and in-kind benefits. This biases the labor AE calculations toward underuse. The year for which most studies were done (2001) was a good weather year for Russian grain production. This overstates inputs' estimated VMP, and consequently biases the AE calculations toward underuse for all inputs, but especially for fertilizer and other inputs used mainly in grain production. On the other hand, a measurement bias toward overuse of material inputs results from corporate farms' transfer of inputs to workers for use on their household plots. This largely unrecorded reallocation understates inputs' estimated VMP on farms' collective operations.

After accounting for these biases, the results do not indicate that any particular input is being strongly overused or underused, nor do they show a pattern of either input overuse or underuse for agricultural inputs in general. Russian agriculture appears to be using agricultural inputs at levels not substantially far from the allocatively efficient volumes. The results therefore do not provide economic justification for the appeal by Russian farms that the government take action to correct the severe drop in their use of inputs during the transition period. In fact, in coming years Russian domestic prices for exportable inputs such as fertilizer and fuel are more likely to rise rather than fall in real terms, as domestic prices move closer to world market levels. Farms would respond to the price hikes by further reducing their purchase and use of these inputs.

Narrowing or eliminating the gap between domestic and world market prices for tradable inputs is therefore a policy the government could adopt to improve the economy's overall allocative efficiency of input use, as it would result in input prices being determined by their most meaningful opportunity cost values—world market prices. Yet, the government would have to face the consequences that the price increases would hurt farms, decreasing both their output and profit (reducing input use would minimize the financial harm suffered by farms from the price hikes, but farm profit nonetheless would fall). Other policy options for increasing input allocative efficiency include improving farm management; investing in training for farm workers, both to generate more workers with the specialized skills farms need, and to expand employment opportunities for unskilled workers outside of agriculture; developing a system of farm finance that would provide farms with working capital to buy inputs; and strengthening input markets institutionally such that suppliers could respond more effectively to farm demand.

References

Griliches, Z. (1964). "Research Expenditures, Education, and the Aggregate Agricultural Production Function." *American Economic Review* 54(6):961-974 (December).

Headley, J. (1968). "Estimating the Productivity of Agricultural Pesticides." *American Journal of Agricultural Economics* 50(1):13-23 (February).

Liefert, W. and Swinnen, J. (2002). *Changes in Agricultural Markets in Transition Economies.* Agricultural Economic Report 806, Economic Research Service, U.S. Dept. of Agriculture, Washington, DC, February; downloadable electronic version available at www.ers.usda.gov/publications/aer806/

Liefert, W., Gardner, B., and Serova, Eu. (2003). "Allocative Efficiency in Russian Agriculture: The Case of Fertilizer and Grain." *American Journal of Agricultural Economics* 85(5):1228-1233 (November).

Mundlak, Y. (1961). "Empirical Production Functions Free of Management Bias." *Journal of Farm Economics* 43(1):44-56 (February).

Osborne, S. and Trueblood, M. (2006). "An Examination of Economic Efficiency of Russian Crop Production in the Reform Period." *Agricultural Economics* 34(1):25-38 (January).

Lerman, Z., Kislev, Y., Kriss, A. and Biton, D. (2003). "Agricultural Output and Productivity in the Former Soviet Republics." *Economic Development and Cultural Change* 51(4): 999-1018 (July).

Osborne, S. and Trueblood, M. (2002). *Agricultural Productivity and Efficiency in Russia and Ukraine: Building on a Decade of Reform.* Agricultural Economic Report 813, Economic Research Service, U.S. Dept. of Agriculture, Washington, DC, July; downloadable electronic version available at www.ers.usda.gov/publications/aer813/

Sedik, D., Trueblood, M., and Arnade, C. (1999). "Corporate Farm Performance in Russia, 1991-1995: an Efficiency Analysis." *Journal of Comparative Economics* 27(3):514-533 (September).

Voigt, P. and Uvarovsky, V. (2001). "Developments in Productivity and Efficiency in Russia's Agriculture: the Transition Period." *Quarterly Journal of International Agriculture* 40(1): 45-66.

Chapter 11

Allocative Efficiency of Corporate Farms: Evidence from the 2003 BASIS Survey

Margarita Grazhdaninova and Zvi Lerman

The analysis reported in the present chapter applies the methodology described in Chapter 10 to the data of 144 corporate farms (generally former kolkhozes and sovkhozes) that participated in the 2003 BASIS survey in three oblasts (Rostov, Ivanovo, and Nizhnii Novgorod).

The allocative efficiency was estimated for "total farm" production as well as for crop and livestock production separately. The "total farm" production function was estimated using the value of output in money units ("gross output" model). Crop and livestock production functions in turn were estimated on two levels of aggregation: using the value of output in money units to estimate "whole crop" and "whole livestock" models; and using the data on the physical output of each commodity (in physical units) to estimate specific commodity models (grain, sunflower, beef, milk, and pork). The commodity models should be robust to intentional and unintentional distortions of financial performance measures that affect the value of output—a relevant factor for transition countries with a large shadow economy.

Allocative Efficiency Results

The estimated coefficients of production functions for specific crops and for crop production as whole are presented in Table 11.1. The different models produce similar results. All the coefficients are positive, which means that the output increases with the increasing use of each input (keeping the use of all other inputs constant). Most models described 70%-80% of the observed variability.

Table 11.1. Crop production functions

	Models in physical units			Models in value units**	
	Grain	Grain Rostov*	Sun-flower	Model 1	Model 2
Constant	1.848	0.920	-	1.062	1.510
Land	0.415	0.861	1.161	0.207	0.178
Labor	-	-	-	0.314	0.257
Purchased inputs (total value)	X	X	X	0.516	X
Fertilizer	0.101	0.060	-	X	0.076
Seeds	0.328	-	-	X	0.566
Weather	0.370	0.482	-	X	X
Pre-reform status					
Marginally profitable	0.319	0.601	0.587	X	X
Profitable	0.283	0.473	0.505	X	X
Unprofitable	0	0	0	X	X
Rostov regional factor	-	X	-	0.483	0.688
R^2_{adj}	0.73	0.64	0.80	0.70	0.70
Number of observations	103	46	45	114	107

Only factors that are statistically significant at 10% level are shown.
X — not included in the model; - — not statistically significant.
* Estimated only for Rostov Oblast.
** In Model 1, all purchased inputs are aggregated by value; in Model 2, fertilizers and seeds are included separately (in value units). Labor defined as the average annual number of workers engaged in crop production. Output defined as the value of crop production.

Land is estimated to be the most significant factor of crop production. Purchased inputs, and in particular fertilizer and seeds, also have a significant positive impact on output. Labor is statistically significant in the aggregated crop production models and not significant in the commodity models. This is probably due to the high correlation between sown land and labor, so that in some models the combined effect of both these factors is reflected in the estimated coefficient of one of them (land).

Crop output is seen to depend on weather conditions. Pre-reform performance is also significant: corporate farms that were profitable (and marginally profitable) in the past are estimated to produce more than the formerly unprofitable farms.

Farm machinery (tractors, grain harvesters, feed combines, trucks) is not statistically significant in any of the crop production models. This result shows that farm machinery (acquired mainly in the pre-reform period) is not a limiting factor in the sample. Machinery, however, can prove to be a limiting factor for the most efficient farms. This indirectly follows from the observation that a substantial number of corporate farms in the sample (though definitely not all the farms) actually purchase machinery.

The estimated coefficients of production functions for specific animal products and for livestock production as a whole are presented in Table

11.2. As in crop production, the livestock models have a satisfactory explanatory power. Labor and cattle numbers are the most significant factors in livestock production functions, which confirms the extensive character of the Russian livestock sector. The high estimated coefficients of these factors probably incorporate also the effect of fixed assets on output. Fixed assets were not directly included in the models because of low reliability of the available data.

Table 11.2. Livestock production functions*

	Beef		Milk	
	Model 1	Model 2	Model 1	Model 2
Constant	−1.731	−1.840	1.575	-
Labor	-	-	0.366	0.267
Number of cattle	0.763	0.680	0.705	0.801
Purchased inputs	X	0.054#	X	-
Weather	0.397	-	0.401	-
Ivanovo regional factor	0.454	-	-	-
Feed quality##	−0.673	−0.731	-	-
R^2_{adj}	0.851	0.815	0.909	0.938
Number of observations	77	75	72	50

	Pork		Livestock **
	Model 1	Model 2	
Constant	-	−4.867	1.319
Labor	0.791	0.507	0.280
Number of cattle	0.613	0.844	0.303
Purchased inputs	X	-	0.481
Weather	-	-	X
Ivanovo regional factor		-	X
Feed quality##	-	-	X
R^2_{adj}	0.811	0.752	0.760
Number of observations	45	47	98

Only factors that are statistically significant at 10% level are shown.
X — not included in the model; - — not statistically significant.
*In Model 1, feed is expressed in feed units from the survey; in Model 2, feed is expressed in terms of inputs used for its production (labor, fertilizers, and agricultural machinery). Feed is not statistically significant in any of the models. Output defined as production of beef, milk, or pork in physical units
**Output defined as the value of livestock production.
#Fertilizer only. ##Share of concentrated feed in ration.

The distance of the farm from the oblast capital does not have a significant effect on livestock output (the same result is observed in crop production). This is probably the outcome of two oppositely directed effects. On the one hand, the farm's distance from the oblast center limits its access to the markets for inputs and products, and to some extent reduces the availability of government support. On the other hand, remote farms do not have to compete for skilled labor. The pre-reform status of the farm is not statistically significant in livestock production (contrary to crop production, see Table 11.1).

The value of marginal product (VMP) for commodity models and for crop and livestock production as a whole is presented in Table 11.3. The estimated coefficients of the inputs are taken from Tables 11.1 and 11.2. For comparison, Table 11.3 also shows the value of marginal product for the "gross output" production function, which aggregates the values of both crop and livestock products.

Table 11.3. Comparison of values of marginal product with factor prices

Factor	Units	Model	Coefficient	VMP, rubles	Factor price, rubles
Land	hectares	Grain	0.414	682	500-1500[1]
		Grain (Rostov)	0.860	1,738	1,500-2,000[1]
		Sunflower	0.410	1,319	1,500-2,000[1]
		Crop production M1	0.207	332	500-1,500[1]
		Crop production M2	0.178	292	500-1,500[1]
		Gross output	0.401	446	500-1,500[1]
Fertilizer	tons (active substance)	Grain	0.100	2,496[2]	2,619[3]
		Grain (Rostov)	0.060	1,835[2]	2,619[3]
	rubles	Crop production M2	0.076	1,469	2619[3]
Seeds	tons	Grain	0.330	2,239	1,500-3,518[4]
	rubles	Crop production M2	0.566	3,702	n/a
Labor	man-years	Crop production M1	0.314	22,120	12,000-18,000[1]
		Crop production M2	0.257	19,270	12,000-18,000[1]
		Livestock production	0.280	25,628	12,000-18,000[1]
		Gross output	0.410	25700	12000-18000[1]
Livestock	head	Milk	0.705	4,297	n.a.
		Beef and veal	0.760	23,691	n.a.
		Gross output	0.303	2,037	n.a.
Purchased inputs	million rubles	Crop production M1	0.516	0.704	1
		Livestock production	0.481	0.439	1
		Gross output	0.163	0.546	1

[1] expert estimate; [2] calculated per ton of fertilizer based on the average active substance content of one ton of purchased fertilizers; [3] weighted average price per ton for fertilizer purchases in the survey; [4] lower limit is the cost of seeds from own production; upper limit is the average market price.

In the grain and sunflower models, the value of marginal product of land varies from 682 rubles/ha for all grain to 1,738 rubles/ha for grain in Rostov Oblast. These estimates are higher than the value of marginal product of land in crop production as a whole and in the gross output model, where VMP varies from 292 rubles/ha to 446 rubles/ha. The estimates are close to the market price of land, which suggests that land prices adjust to the actual economic productivity of land. This means that on the whole land is used efficiently by corporate farms.

The value of marginal product for fertilizer is 1,469 rubles/ton for crop production as a whole and 2,496 rubles/ton for grain. The return to fertilizer in grain production is much higher than the average because grain is one of the most profitable commodities. In the Rostov grain model, however, the value of fertilizer marginal product is lower (1,835 rubles per ton), which probably reflects the effect of decreasing returns to scale (the use of fertilizer is much higher in Rostov Oblast than in the other regions).

The weighted average price of fertilizer in Table 11.3 is 2,619 rubles/ton, higher than the value of marginal product. This indicates that fertilizer is overused given the market prices of inputs and farm products. This is a paradoxical result, because the use of fertilizers has decreased by more than by 50% since the beginning of the 1990s. Perhaps our result does not represent an the equilibrium in the domestic fertilizer market, but instead points to overpricing of fertilizer due to the emphasis of Russian fertilizer manufacturers on exports and the price-boosting impact of input subsidy programs (see Chapters 8, 10).

The value of marginal product of seeds is 2,239 rubles/ton in the grain model and 3,702 rubles/ton in the total crop production model. This does not mean, however, that the return on seeds for other crop products is higher than for grain. In the grain model the explanatory variable is the quantity of seeds used in tons, whereas in the total crop production model the explanatory variable is the value of purchased seeds. The cost of purchased seeds is only 18% of the total cost of seeds used in the sample. The mean price of purchased seeds is 3,518 rubles/ton, while the mean cost of seeds from own production is 1,500 rubles/ton. Given the estimates for the value of marginal product, it is inefficient to use purchased seeds at the current market price. The most probable explanation of this phenomenon is the almost total absence of a seed market in Russia. Faced with severe working-capital constraints, farms have sharply curtailed their seed purchases and have shifted to seeds from own production. These are lower-quality seeds, but at least their use does not require "ready money." The reduced demand for seeds has led to a sharp contraction of commercial seed production and virtual disappearance of the seed market, which is currently in an inefficient equilibrium characterized by low quantities and high prices.

Estimation of the value of marginal product of labor presents some difficulties. In most commodity models the coefficient of labor is statistically not significant and the value of marginal product of labor cannot be calculated. In total production models (two crop production models, livestock model, and "gross output" model), the coefficient of labor is statistically significant and the value of marginal product varies from 19,000 to 26,000 rubles per year. This is close to the actual level of agricultural wages, and we conclude that the use of labor in Russian agriculture today is allocatively efficient (despite the fact that agricultural wages are less than the minimum standard of living and many farm managers estimate that three-quarters of the employed are redundant).

The value of marginal product of purchased inputs estimated from the aggregated models in money units (total crop production, livestock production, and gross output) is between 0.439 and 0.704 million rubles per each million spent on purchased inputs. This again suggests that purchased inputs are overpriced relative to their marginal product, and the use of purchased inputs is inefficient at the current market prices.

Conclusion

Land is unquestionably the cornerstone of agricultural production. Its impact on output is positive and highly significant. The share of idle land is substantial and the price of land is close to the value of marginal product. In order to intensify land use and prevent degradation of land quality (which is typical for unprofitable farms), it is essential to encourage redistribution of land from least efficient to most efficient users. The government should actively move to simplify the procedures for buying, selling, and leasing of land so as to reduce the high transaction costs in the land market.

Efficient production requires purchased inputs. Their positive impact on output is proved by production functions. Yet the actual use of purchased inputs in Russia is substantially lower than in developed market economies. The analysis of allocative efficiency for purchased inputs produces a mixed picture. However, this analysis is based on actual market prices for products and inputs. Underpricing of farm products combined with overpriced inputs may lead to allocatively efficient use of purchased inputs, but at very small quantities. In medium- and long-term perspective, this may cause degradation of land and other negative phenomena.

Price distortions in input markets may be caused by various reasons. Thus, in fertilizer and fuel markets, pervasive monopolization (especially the existence of local monopolies) combined with high export demand strengthens the bargaining power of input manufacturers and leads to overpricing. To correct this distortion, it is necessary to encourage deliveries to the domestic market, while carefully avoiding direct bans that have already demonstrated their absolute inefficiency. Input subsidies do not correct price distortions, and sometimes even aggravate them. The solution of the problem probably requires development of appropriate market institutions that will increase the bargaining power of agricultural producers and at the same time lower the suppliers' risk in dealing with agricultural producers. The most obvious approach is through the creation of farmers' input-supply cooperatives capable of pooling the flows, reducing the risks, and possibly developing short-term credit schemes for input purchases.

Chapter 12

Allocative Efficiency of Corporate Farms: Evidence from National Data

Vasilii Uzun

In this chapter, we estimate production functions and allocative efficiency ratios for the entire cohort of corporate farms in Russia. The analysis uses the Goskomstat database with detailed information for 24,995 corporate farms. The actual estimations are carried out using the 2001 data for a somewhat smaller number of farms without missing values in all the relevant variables. The production functions are estimated on two levels of aggregation: "total farm" production functions, including the entire agricultural activity of the corporate farms; and separate production functions for crop and livestock enterprises of the same farms. A total of seven Cobb-Douglas models are estimated using different combinations of inputs: three "total farm" production functions, two crop production functions, and two livestock production functions. In all functions, the output variable is the appropriate sales revenue as reported in the Goskomstat database. The analysis is extended to examine performance differences across farms of different organizational forms and in different financial-health categories.

"Total Farm" Production Functions

The purpose of the analysis was to estimate the effect of main farm inputs on farm output as measured by the revenue from farm sales (both crop and livestock products). Three "total farm" production functions (models M1, M2, M3) were estimated, differing in the specific combination of inputs used. The variables used in each model are shown in Table 12.1.

Arable land and labor were included as main inputs in all three models. Beyond this commonality, the models primarily differed in their treatment of production costs. Model M1 emphasized endowments, so that in addition to arable land and labor it included the value of fixed assets (net of accumulated depreciation) and the value of current assets as a measure of working capital proxying for production costs. Models M2 and M3 replaced working capital with the actual cost of material inputs—aggregated into a single figure in model M2 and disaggregated into costs of component inputs in model M3. The depreciated value of fixed assets in turn was replaced models M2 and M3 with two variables: the original value of production assets (at cost, before depreciation) and the size of the livestock herd as another endowment fundamental to farm production. Models M2 and M3 also included an independent variable representing the amount of subsidies and grants received through government support programs.

Table 12.1. Variables included in "total farm" production functions (n = 22,438 corporate farms)

	Mean value#	Model M1	Model M2	Model M3
Output: Sales revenue	12,052	+	+	+
Arable land, ha	3,819	+	+	+
Labor, full-time equivalents	179	+	+	+
Production assets at cost	31,606		+	+
Fixed assets net of depreciation	22,073	+		
Current assets	11,048	+		
Livestock herd, standard head*	628		+	+
Subsidies and grants	620		+	+
Cost of material inputs (aggregated)**	9,621		+	
Feed	3,717			+
Seeds	937			+
Fertilizer	485			+
Fuel	1,723			+
Electric power	290			+
Spare parts	1,190			+
Services	1,108			+

#All variables in '000 rubles except land (ha), labor (full-time equivalents), and livestock herd (standard head).

* Standard head calculated using the following conventional weights: 1.0 for cow or horse; 0.6 for feedlot cattle; 0.3 for pigs; 0.1 for sheep and goats; 0.02 for poultry.

** The means for specific material inputs are calculated on the basis of the maximum available number of observations in each case and therefore do not sum to the aggregated cost of material inputs.

Although the Goskomstat database contained 24,995 farms, only 22,438 farms were used in the analysis. Farms with zero values for any of the main variables (sales revenue, land, labor, production assets at cost, fixed assets net of depreciation, and material input costs) were excluded; for all other variables listed in Table 12.1, zero values were replaced with a small positive value (0.1). This was necessary to allow us to proceed with estimation of Cobb-Douglas production functions without losing too many observations.

The estimated coefficients of the three production functions are given in Table 12.2. All coefficients are statistically significant at 10% (except land in model M1 and subsidies in model M2). All three regressions had a fairly high explanatory power (with R^2 greater than 0.8).

Table 12.2. Estimated coefficients for "total farm" production functions (n = 22,438 corporate farms)*

	Model M1	Model M2	Model M3
Constant	−0.275	−0.067	1.801
Arable land, ha	0.000#	−0.022	0.010
Labor, full-time equivalents	0.593	0.443	0.710
Production assets at cost	X	0.025	0.067
Fixed assets net of depreciation	0.013	X	X
Current assets	0.667	X	X
Livestock herd, standard head	X	−0.062	−0.153
Subsidies and grants	X	−0.003#	0.014
Cost of material inputs (aggregated)	X	0.793	X
Feed	X	X	0.136
Seeds	X	X	−0.044
Fertilizer	X	X	0.056
Fuel	X	X	0.190
Electric power	X	X	0.046
Spare parts	X	X	0.105
Services	X	X	0.056
R^2	0.82	0.88	0.83

*OLS estimation of Cobb-Douglas model in logged variables.
#Not significantly different from zero at 10%.

The estimation results for land are inconclusive. Although land is the main factor of production in agriculture, its coefficient is zero (more precisely, not statistically significantly different from zero) in model M1 and negative in model M2. A positive coefficient for land is estimated only in model M3, but even here it is very small: a 1% change in the area of arable land produces a 0.01% changes in farm sales. These results seem to suggest that land is not a scarce resource: there appears to be an excess of arable land in Russia's corporate farms. This is not surprising if we recall that corporate farms control 80% of agricultural land, while producing less than 50% of gross agricultural output (Chapter 2). Our findings are consistent with those of Epshtein (Chapter 13), who also estimated negative coefficients for land.

However, this conclusion regarding "unnecessary" land may be attributable to the effect of livestock in "total farm" production functions. The separate estimation of crop and livestock production functions in the next section in fact shows that the coefficient of land is positive and statistically significant for crop production (Table 12.5). Livestock production does not require land (or requires relatively little land), and the inclusion of livestock sales in the output variable may make land appear statistically not significant.

Contrary to land, working capital and material input costs have the highest coefficients in the estimated models. In model M1, an increase of 1% in working capital increases sales revenue by 0.7%; in model M2, an increase of 1% in expenditure on material inputs increases sales revenue by 0.8%. Shortage of working capital and financial resources for the purchase of inputs thus imposes a serious constraint on farm performance (as measures by sales revenue). Because of insufficient funds, seeds are not sown or sown into land that has been inadequately prepared and fertilized; the harvest remains in the field or is brought in with large losses; animals are left without feed; people are left without wages. Irregular or low wages encourage skilled workers to leave; insufficient or low-quality ration reduces livestock productivity and animal numbers. The available land, farm buildings, and other resources cannot be used efficiently because of financial constraints, and yet there is generally no mechanism to redistribute them to other users, which further exacerbates the financial difficulties of the corporate farms.

Labor is estimated to be the second most important factor of production. A 1% increase in the number of workers raises sales revenue by anywhere between 0.4% and 0.7% (depending on the specific model). The high coefficient of labor in "total farm" production functions shows that labor, like financial resources, is also a scarce input.

The effect of subsidies is uncertain: the estimated coefficient is statistically not significant in model M2 and very small in model M3. Subsidies apparently do not play a very important role in farm performance. This is consistent with the conclusions in Chapter 9.

The negative coefficient of livestock is understandable. In most corporate farms livestock is an unprofitable enterprise, and therefore a larger herd has a negative impact on sales.

Seeds have a negative coefficient in model M3, where disaggregated farm inputs are introduced. As with the case of land, this curious result may be attributable to the inclusion of livestock in "total farm" production: the coefficient of seeds is positive and statistically significant in the crop production model in Table 12.5.

An attempt to detect differences in farm performance as a function of financial health was made by introducing dummy variables for the five solvency groups discussed in Chapters 2 and 9. Farms in the "best" solvency groups (financially healthy farms in groups 1 and 2) were observed to achieve on average significantly higher levels of sales than financially less robust farms (groups 3-5). The estimated difference in sales between group 1 and group 5 farms was as high as 40%. The results are presented in Table

12.3 in the first column (Model M1 with dummies for financial groups). These results are consistent with the findings for Leningrad Oblast farms in Chapter 13 (see Table 13.1).

Table 12.3. Production functions with dummy variables for financial group (column 1) and organizational form (column 2)

	Model M1 with financial groups (1)	Model M1 with organizational forms (2)	Model M1 without dummy variables (Table 12.2) (3)
Constant	0.188	−0.167	−0.275
Arable land, ha	0.029	0.001#	0.000#
Labor, full-time equivalents	0.516	0.597	0.593
Fixed assets net of depreciation	0.081	0.014	0.013
Current assets	0.555	0.655	0.667
Group 1	0.339		
Group 2	0.129		
Group 3	0.043		
Group 4	−0.082		
Group 5 (base)	0.000		
State enterprises		−0.095	
Agricultural cooperatives		−0.056	
Limited-liability partnerships		+0.043	
Joint-stock companies (base)		0.000	
R^2	0.85	0.83	0.82

#Not significantly different from zero at 10%.

Fig. 12.1. Average level of sales by solvency groups relative to financially weakest group 5

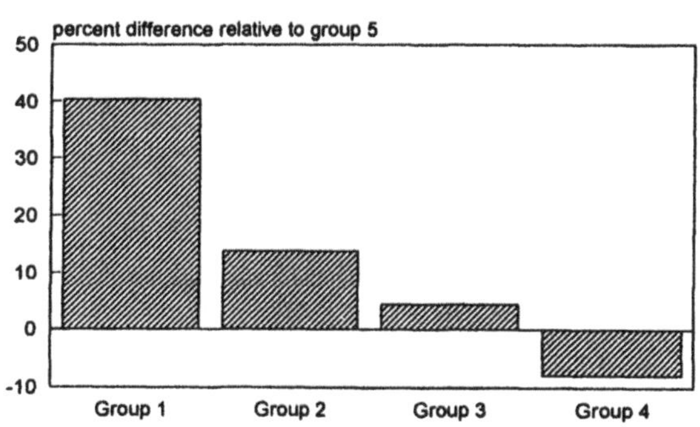

Source: Goskomstat database of corporate farms

Table 12.3 also shows the outcome of a similar exercise with dummy variables representing various organizational forms (the second column). Here state farms and agricultural production cooperatives are estimated to lag by 10%-15% relative to the average sales level of the more successful joint-stock companies and limited-liability partnerships. The level differences in percent relative to the base category (the group of financially weakest farms and the group of joint-stock companies, respectively) are shown in Figures 12.1 and 12.2.

Fig. 12.2. Average level of sales by organizational form relative to joint-stock companies

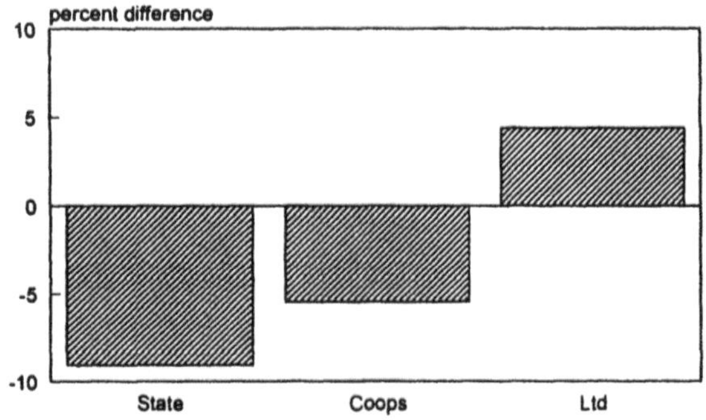

Source: Goskomstat database of corporate farms

Production Functions for Crop and Livestock Enterprises

The Goskomstat database of large corporate farms separates total sales revenue into sales of crops and livestock products. We thus have an output variable for estimation of separate production functions for crop and livestock enterprises. There is also sufficient information to separate most of the inputs into specific uses in crop and livestock production.

The variables used to estimate separate production functions for crops and livestock are listed in Table 12.4. Arable land, seeds, and fertilizer were naturally assigned in total to crop production, while animals and feed were assigned to livestock production only. All other variables were apportioned according to their reported use for crop and livestock production respectively. Labor (expressed in full-time equivalents in the "total farm" model) was replaced with wages, as this variable was reported separately for crop and livestock enterprises. Since fixed assets were not apportioned to crops and livestock in the Goskomstat database, we used instead the depreciation expense, which was reported separately for the respective enterprises. Current assets were omitted for the same reason, and only material input costs were used.

The number of observations in each regression was determined by the incidence of zeros in the database: farms with zero values for sales revenue, wages, depreciation, material input costs, land (in crop models), and animals (in livestock models) were excluded; for all other variables, zero values were replaced with a small positive value (0.1).

Table 12.4. Variables included in production functions for crop and livestock enterprises in corporate farms

	Crops (*n*=20,915)			Livestock (*n*=20,705)		
	Mean value#	Model M4	Model M5	Mean value#	Model M6	Model M7
Output: Sales revenue	5,165	+	+	6,601	+	+
Arable land, ha	3,963	+	+	–	–	–
Wages	1,298	+	+	1,455	+	+
Depreciation expense	452	+	+	307	+	+
Livestock herd, st. head	–	–	–	670	+	+
Subsidies and grants	299	+	+	279	+	+
Cost of material inputs	4,647	+	–	5,563	+	–
Feed	–	–	–	3,956	–	+
Seeds	990	–	+	–	–	–
Fertilizer	513	–	+	–	–	–
Fuel	1,382	–	+	445	–	+
Electric power	111	–	+	197	–	+
Spare parts	868	–	+	398	–	+
Services	730	–	+	440	–	+

#All variables in '000 rubles except land (ha) and livestock herd (standard head).

* Standard head calculated using the following conventional weights: 1.0 for cow or horse; 0.6 for feedlot cattle; 0.3 for pigs; 0.1 for sheep and goats; 0.02 for poultry.

Table 12.5. Estimated coefficients for production functions for crop and livestock enterprises in corporate farms*

	Crops (*n*=20,915)		Livestock (*n*=20,705)	
	Model M4	Model M5	Model M6	Model M7
Constant	−2.556	−1.077	−0.132	0.807
Arable land	0.097	0.133	X	X
Wages	0.512	0.623	0.335	0.469
Depreciation expense	−0.008#	0.021	−0.006#	0.023
Livestock herd	X	X	0.065	0.068
Subsidies and grants	−0.023	−0.015	0.027	0.031
Cost of material inputs	0.765	X	0.670	X
Feed	X	X	X	0.352
Seeds	X	0.089	X	X
Fertilizer	X	0.018	X	X
Fuel	X	0.287	X	0.036
Electric power	X	0.041	X	0.022
Spare parts	X	0.062	X	0.048
Services	X	0.068	X	0.033
R^2	0.71	0.70	0.88	0.86

*OLS estimation of Cobb-Douglas model in logged variables.

#Not significantly different from zero at 10%.

The estimated coefficients for the four crop and livestock models are presented in Table 12.5. The explanatory power of the crop production functions is somewhat lower that for livestock and "total farm" production, but still quite satisfactory ($R^2 = 0.7$). All the estimated coefficients are significantly different from zero at 10% (except the coefficient of depreciation expense in crop production model M4 and livestock production model M6.

As in "total farm" models, labor and material input costs play the most important role in determining the level of farm sales. Contrary to "total farm" models, arable land has positive coefficients that are significantly different from zero in the crop production models, while the number of animals has positive coefficients that are significantly different from zero in the livestock production models. The effect of these two production factors emerges very clearly when they are assigned specifically to enterprises for which they are relevant: land for crop enterprises and animals for livestock enterprises. In "total farm" models, on the other hand, their effect is partially masked (as for land) or distorted (as for animals) by the interference from other enterprises.

Value of Marginal Product and Allocative Efficiency of Selected Inputs

The estimated coefficients in Cobb-Douglas production functions enable us to calculate the value of marginal product (VMP) of each input and to assess the allocative efficiency of its use by comparing its VMP to its price or cost (see Chapter 10). For each input i with estimated coefficient b_i in the Cobb-Douglas production function, the value of marginal product VMP_i is calculated as

$$VMP_i = dY/dX_i = b_i(Y/X_i),$$

where Y/X_i is the average output per unit input use (this is the ratio of geometric means, i.e., the ratio of the exponentiated mean logs of sales revenue Y and input use X_i). The calculation results for VMP of the relevant inputs are presented in Table 12.6.

Labor has large significant coefficients in all "total farm" production functions. Based on these coefficients, the VMP of labor ranges between 17,000 rubles and 27,000 rubles per worker per year. The average VMP over all three models (22,200 rubles) is greater than the average annual wage (18,700 rubles), which indicates that on the whole labor is underused in corporate farms: they could improve their profit by employing more labor, assuming that the cost remained fixed at 18,700 rubles per worker per year.

The marginal product of land cannot be evaluated on the basis of "total farm" production functions, because one of the estimated coefficients is negative and another one is not statistically different from zero. However, in crop production models (M4 and M5), the estimated coefficient of land is statistically significant and positive. Multiplying the partial productivity of land (587 rubles per hectare) by the corresponding coefficients of land from these models, we obtain marginal product of 57 rubles per hectare and 78 rubles per hectare respectively.

To assess the allocative efficiency of land use in crop production, these values should be compared to the annual cost of land. An absolute lower bound on this cost is provided by the land tax, which runs at 30 rubles per hectare per year. At this price, land is grossly underused, but unfortunately 30 rubles per hectare is not a relevant yardstick. Most land in agricultural use is leased, not owned, by the users. Annual lease payments range from a high of 300-400 rubles per hectare per year for land leased from private owners (in the form of plots of land shares) to a low of about 150 rubles per hectare per year for land leased from the state through regional authorities. Since corporate farms rely heavily on land leased from the state, their annual cost of land is therefore around 200-250 rubles per hectare. Given these costs, land is observed to be grossly overused. In other words, land is not a scarce resource and there is a huge excess of land in Russian agriculture.

Table 12.6. VMP of main inputs from estimated production functions

	Output per unit input use, rubles	Model	Estimated coefficient	VMP, rubles per unit input use	Cost of unit input use, rubles
Labor	46,400	M1	0.593	22,600	18,700
		M2	0.443	16,900	18,700
		M3	0.710	27,100	18,700
		average		22,200	18,700
Land	587	M4	0.097	57	30-300
		M5	0.133	78	30-300
		average		68	30-300
Material inputs					
Aggregated	1.1	M2	0.793	0.8	1
Feed	4.7	M3	0.136	0.6	1
Seeds	10.9	M3	−0.044	−0.5	1
Fertilizer	112.8	M3	0.056	6.3	1
Fuel	5.5	M3	0.190	1.0	1
Electric power	71.4	M3	0.046	3.3	1
Spare parts	11.6	M3	0.105	1.2	1
Services	28.3	M3	0.056	1.6	1

Material input costs (in the aggregate) are not covered by incremental sales revenue: each additional ruble spent on material inputs generates only 0.8 rubles in sales. However, material inputs is a mixed bundle of specific inputs and the efficiency of use varies from one to another. Only feed and seeds appear to be overused, with incremental sales not covering their cost. Table 12.6 shows that most material inputs are actually underused in the sense that their value of marginal product is greater than their cost.

The gap is particularly large for fertilizer, where an additional ruble of costs generates 6.3 rubles in sales. And yet, 30% of corporate farms did not buy any fertilizer in 2001. All these farms had cropland, but they could not afford to buy fertilizer because of financial difficulties. The main fertilizer users were just 12% of farms that accounted for 70% of expenditures on fertilizer purchases in 2001. The inability to buy fertilizer leads to highly detrimental results, as farms that used fertilizer in 2001 generated 159,000 rubles per hectare in sales more than farms that did not use fertilizer. Farms that did not use fertilizer were smaller (121 workers and 8.2 million rubles in sales compared to the average of 198 workers and 13.7 million rubles in sales for all corporate farms) and generally reported losses.

Table 12.7. VMP of labor and land for corporate farms by solvency groups (2001)

	Labor			Land		
Solvency groups	Sales per worker, rubles	VMP of labor, rub/worker*	Average wages, rubles	Sales per ha, rubles	VMP of land, rub/ha **	Cost of land, rub/ha#
1 (best)	64,200	37,400	23,100	1,087	125	150
2	44,400	25,800	17,400	754	87	150
3	39,800	23,100	16,100	661	76	150
4	31,300	18,200	15,500	442	51	150
5 (worst)	19,800	11,400	13,100	232	27	150
All farms	46,400	22,200	18,700	587	68	150

*Average of "total farm" models M1-M3.
**Average of crop production models M4-M5.
#Average cost of leasing land from the state.

The allocative efficiency estimates discussed so far are averages for the entire cohort of Russian corporate farms. Just as negative average profitability of Russian agriculture hides a wide distribution ranging from highly profitable farms to highly unprofitable ones, we though it possible that the observed averages were hiding a pattern of differences among farms of different types. Given the classification of all corporate farms into five solvency groups (see Chapters 2, 9), which represent different levels of financial health and probably proxy for different management quality, we recalculated the value of marginal product of labor and land for each solvency group. The results presented in Table 12.7 show quite predictably that the value of marginal product of both inputs is the highest for the "best" farms and the lowest for the financially weak farms. However, the conclusions regarding allocative efficiency do not change: and the marginal product of labor remains above the average wage for four out of five groups (i.e., labor is underused in groups 1-4 and it is slightly overused only in the financially weakest group 5) and the marginal product of land remains below the cost of land for all solvency groups (i.e., land is overused in each group, as in the entire cohort).

The gap between VMP of land and the cost of land increases sharply from financially healthy to financially weak groups. This essentially implies that the financially weak groups are much more inefficient in their use of land than the financially healthy groups. There is much more excess land in low-solvency farms.

Chapter 13

Allocative Efficiency of Corporate Farms in Leningrad Oblast

David Epshtein

In this chapter we estimate the production functions for the corporate farms in Leningrad Oblast in Northwest Russia, where agriculture has to adapt to relatively unfertile non-chernozem soils. The estimation relies on a regional subset of the 2001 Goskomstat database used by Uzun in his analysis of all corporate farms in Russia (Chapter 12). There are 195 corporate farms in Leningrad that regularly file annual reports and are included in the Goskomstat database. For purposes of this study, we omitted all narrowly specialized producers (poultry factories, greenhouses, fur-animal breeders, feedlots operating on feed imported from outside the oblast) and kept the relatively homogeneous subpopulation of typical diversified farms that produce milk, potatoes, vegetables, and other crops. This screening reduced the number of farms in the analysis from 195 to 157. In specific regressions the actual number of farms may be less than 158 because of missing values in some variables. Three Cobb–Douglas production functions are estimated using different combinations of inputs. In all models, the output variable is the reported sales revenue. Only "total farm" production functions are estimated, and, contrary to Chapter 12, we do not differentiate between crop and livestock production.

A special feature of this chapter is the estimation of separate production functions for five solvency groups characterizing different levels of financial health. The solvency measure used to classify the farms into five groups reflects the ability to cover fixed costs (i.e., wages and depreciation) from value added. The classification algorithm is described in Chapter 9. In a sense, different solvency groups represent different levels of farm management quality, and the efficiency results by solvency groups shed light on the impact of the missing management variable in production functions.

Table 13.1 presents the mean characteristics of Leningrad farms by solvency groups (and for the oblast as a whole). All per-farm indicators, except land, decrease from group 1 to group 5. Thus, financially healthy farms are generally larger than financially ailing farms by sales revenue, number of workers, machinery stock, and production costs. They also pay much higher wages to their workers. Land paradoxically remains fairly constant across farms of all solvency groups, which immediately points to lower productivity in financially less healthy farms.

Table 13.1. Main characteristics of corporate farms in Leningrad Oblast (per farm averages, 2001)

	All farms	Solvency groups (1 = best, 5 = worst)				
		1	2	3	4	5
Number of corporate farms	157	11	35	53	24	34
Sales revenue, '000 rubles	25,860	83,236	39,641	18,077	18,157	10,684
Number of workers	234	515	315	186	212	148
Agricultural land, ha	2,861	3,397	2,997	2,578	2,624	3,158
Material costs (purchased and intermediate inputs), '000 rubles	21,273	50,307	28,424	16,273	19,105	13,844
Value of machines, equipment, '000 rubles	11,367	33,053	15,106	8,535	8,982	6,601
Annual wages per worker, '000 rubles*	31,700	44,336	40,181	30,323	29,624	22,474
Depreciation in percent of sales revenue	12	6	7	10	10	23

*Including social deductions.

While the overall objective was to estimate the effect of main farm inputs on farm output, we especially focused on the allocative efficiency of financial resources. The set of inputs in our analysis therefore included material costs expressed in rubles (and not in physical units), as well as a decomposition of material costs into cash and non-cash components (purchased inputs and intermediate inputs, respectively).

In addition to material costs, which are essentially an annual flow variable, the models also included endowments. Land and labor were expressed in physical units. The stock of fixed assets used by the farm was represented by the value of machinery, equipment, and vehicles (taken at cost, without depreciation). The capital component thus included mainly short-lived fixed assets, whose value is reported in farm financial statements with much higher accuracy than the value of longer lived buildings, for which the practice of periodic revaluations introduces considerable distortions.

Two production functions were estimated (Table 13.2). Model 1 is the basic production function with inputs including agricultural land, labor, fixed assets, and cost of material inputs (both purchased and intermediate). In Model 2, material costs were decomposed into two components: cash costs (purchased inputs) and non-cash costs (intermediate inputs).

Table 13.2. Estimated coefficients of Cobb-Douglas production functions*

	Model 1	Model 2
Dependent variable: sales revenue, '000 rubles		
Constant	0.264	1.468
Agricultural land, ha	−0.267	−0.202
Labor, full-time equivalents	0.378	0.358
Machinery and equipment, '000 rubles	0.132	0.116
Cost of material inputs, '000 rubles	0.796	
Cash costs (purchased inputs), '000 rubles		0.423
Non-cash costs (intermediate inputs), 000 rubles		0.360
Credits and loans, '000 rubles		
Material costs excluding credit, '000 rubles		
Number of farms	157	126
R^2	0.956	0.966

*OLS estimation of Cobb-Douglas models in logged variables. All coefficients significantly different from zero ($p < 0.01$).

All inputs except land have positive coefficients. This implies normal production function behavior, when increased use of each input (except land!) leads to a certain increase in output (sales revenue). Material inputs have the strongest impact on output: a 1% increase in material costs raises sales revenue by almost 0.8%.The next most important coefficient is labor producing 0.4% increase in sales for each 1% increase in the number of workers. The estimated coefficient of land is consistently negative (and statistically significant with $p < 0.01$). This implies that more land leads to less sales, or conversely, reducing the land area will increase sales. This type of behavior is often observed in countries with huge excess of land—a situation similar to that in Russia. Similar results were obtained by Uzun using the full Goskomstat database of all corporate farms in Russia (see Chapter 12, although the negative land effect there is less pronounced than in Leningrad Oblast).

Given the estimated production function coefficients in Table 13.2, we can calculate the value of marginal product (VMP) of each input and assess the allocative efficiency of its use by comparing its VMP to its price or cost (see Chapter 10). For each input i with estimated coefficient b_i in the Cobb-Douglas production function, the value of marginal product is calculated as $VMP_i = b_i(Y/X_i)$, where Y/X_i is the average output per unit input use (see formula (4) in Chapter 10). The calculations for the inputs entering Model 1 are laid out in Table 13.3.

The calculation of VMP is meaningful only for inputs that have positive coefficients in the estimated production function. The VMP of labor is 38.0, which means that each additional worker produces additional sales of 38,000 rubles. On the other hand, the average wages in Leningrad farms are 31,700 rubles per worker (see Table 13.1), so that farms can increase their profits by increasing the workforce. In terms of allocative efficiency, labor is underused by corporate farms in Leningrad Oblast at the present level of wages.

Table 13.3. VMP of main inputs from estimated production function (Model 1)

	Labor	Material inputs	Machinery	Land
Estimated coefficient b_i	0.378	0.796	0.132	−0.267
Mean use of input X_i	234	21,273	11,367	2,861
Mean sales revenue estimated from production function Y	23,522	23,522	23,522	23,522
$VMP_i = b_i(Y/X_i)$	38.0	0.88	0.27	n.a.
Price of unit input i	31.7	1	0.1	—
Allocative efficiency ratio of input i	1.2	0.9	2.7	—
Allocative efficiency of input i	Underused	Overused	Underused	Grossly overused

A different situation is observed with material inputs: an additional 1,000 rubles spent on materials inputs will increase sales by 880 rubles only. Increasing the use of material inputs farms only reduce their profits. From profit maximization considerations it actually pays to reduce the use of material inputs. In terms of allocative efficiency, material inputs in aggregate are overused by corporate farms in Leningrad Oblast.

An additional 1,000 rubles invested in machinery and equipment stock generates incremental sales of 270 rubles. As the price of machinery and equipment (a very mixed bundle of assets) we can approximately use the annual depreciation expense. Assuming that the investment in machinery is depreciated over 10 years, the annual depreciation expense associated with an investment of 1,000 rubles in new assets is 100 rubles. The new investment thus generates incremental sales of 270 rubles and involves incremental costs of 100. To maximize profits, the farm should increase its investment in machinery and equipment. In terms of allocative efficiency, farm machinery, equipment, and vehicles in aggregate are strongly underused, with a very high allocative efficiency ratio of 2.7. This is probably a reflection of severe financial constraints that restrict the ability of Leningrad farms to purchase new machinery and equipment as prescribed by production considerations.

The calculation of VMP was repeated for each of the five solvency groups using the coefficients of Model 1 from Table 13.2. In effect, the same production function was assumed for all five groups, but the calculations were carried out using the particular averages for each group. The results are presented in Table 13.4, which shows the allocative efficiency (AE) ratio of each input, calculated as the ratio of that input's VMP to its unit cost.

Table 13.4 also shows the AE status of each input in terms of its being underused or overused. Labor and machinery were previously found to be underused for all farms (see Table 13.3), and they are now seen to be consistently underused across all solvency groups. Farms of all financial profiles experience both labor and machinery constraints. The situation is different regarding material costs. Here we previously observed overuse for all farms on average. It turns out, however, that farms in the first two solvency groups, i.e., farms with the highest financial health, use material inputs efficiently: each additional 1,000 rubles spent on material inputs in these farms generates 1,100-1,300 rubles in incremental revenue. On the other hand, the financially troubled farms in solvency groups 3-5 generate only about 800 rubles of incremental revenue from each additional 1,000 rubles spent on material inputs.

Table 13.4. Allocative efficiency of main inputs for different solvency groups

Solvency group	Labor		Machinery		Material costs	
	AE ratio	AE status	AE ratio	AE status	AE ratio	AE status
1	1.38	Underused	3.3	Underused	1.32	Underused
2	1.18	Underused	3.5	Underused	1.11	Underused
3	1.21	Underused	2.8	Underused	*0.88*	*Overused*
4	1.09	Underused	2.7	Underused	*0.75*	*Overused*
5	1.21	Underused	2.1	Underused	*0.61*	*Overused*
All farms	1.20	Underused	2.7	Underused	0.88	Overused

Table 13.5. Allocative efficiency of cash and non-cash material costs for different solvency groups

Solvency group	Cash material costs (purchased inputs)		Non-cash material costs (intermediate and bartered inputs)	
	AE ratio	AE status	AE ratio	AE status
1	1.10	Nearly optimal	1.55	Underused
2	1.08	Nearly optimal	1.03	Nearly optimal
3	0.97	Nearly optimal	*0.78*	*Overused*
4	0.94	Nearly optimal	*0.60*	*Overused*
5	1.20	Nearly optimal	*0.38*	*Overused*
All farms	1.09	Nearly optimal	0.87	Overused

This is a counterintuitive result. Farms with a poor solvency rating experience severe financial constraints, and we would naturally expect them to avoid unnecessary spending and thus use their material inputs with maximum efficiency.

This is a counterintuitive result. Farms with a poor solvency rating experience severe financial constraints, and we would naturally expect them to avoid unnecessary spending and thus use their material inputs with maximum efficiency. To resolve the apparent paradox, we calculated the VMP and the allocative efficiency ratio separately for the two components of material costs: cash material costs (purchased inputs) and non-cash material costs (intermediate inputs and inputs acquired through barter arrangements). The results presented in Table 13.5 show that all farms use their cash material costs nearly optimally (with AE ratio close to 1). Differences are observed only for non-cash material costs, which are grossly overused by the low-solvency farms.

The analysis of cash material costs thus shows that all farms experience binding financial constraints. To improve their profits, low-solvency farm should substantially reduce their ratio of costs to sales. This is the only way to counteract the negative effect that deteriorating terms of trade in agriculture have on farm profitability.

Chapter 14

Technical Efficiency in Russian Agriculture

Gregory Brock, Margarita Grazhdaninova, Zvi Lerman, and Vasilii Uzun

For decades, Russian agriculture had had little technological progress and virtually no foreign investment, which resulted in a stable production possibilities frontier and made the sector ideally suited to production function analysis. The production function estimations reported in Chapters 10-13 add to a series of previous studies of the input/output relationship in Russian agriculture (e.g., Clayton, 1980, 1984; Gray, 1981; Johnson and Brooks, 1983), which generally followed the same methodology. In the late 1970s and the 1980s, however, the average response production functions gave way in the economics literature to more sophisticated production analysis techniques that measured not only productivity but technical efficiency as well (Aigner et al., 1977; Bauer, 1990). Some of the major methodological advances in applying technical efficiency analysis to individual firms were made by a joint Russian-American team in Moscow in the early 1980s (Jondrow et al., 1982; Danilin et al., 1985), but lack of data for many sectors of the Russian economy precluded the application of this technique until the end of the decade. When the Soviet Union collapsed, the initial optimistic expectation was that many sectors of the new Russian economy could rapidly achieve both higher productivity and higher technical efficiency once market forces prevailed. Our research attempts to understand why this has not happened in Russian agriculture in terms of technical efficiency.

For this purpose we apply standard technical efficiency algorithms to compute the production possibilities frontier (as distinct from the production function estimated in previous chapters) from the data on corporate and individual farms collected in the BASIS 2003 survey. National-level data on all corporate farms in Russia supplemented the survey analysis.

Technical Efficiency Framework

When measuring technical efficiency we recognize the one-sidedness of the production possibilities frontier. Farms that fail to use their inputs as efficiently as other farms fall short of the "best practice" frontier and are labeled technically inefficient. Inefficiency is measured within a sample, and so even farms that are observed to be 100% efficient among their "peers" may not be efficient compared with farms in other regions or other countries.

Several methods exist for measuring technical efficiency with the choice of method often depending on the data and the researcher's philosophical view of the importance of measurement error (Forsund et al., 1980). The various methods calculate a technical efficiency index (TE score), which measures the distance of the observed firm from a point on the production frontier. Firms lying on the production frontier are 100% technically efficient (with TE = 1), and the "inefficiency" of the remaining firms increases with the distance from the production frontier (for a concise description of the two approaches to technical efficiency see Coelli et al., 1998).

Two methods that particularly appealed to researchers in the 1980s were stochastic frontier analysis (SFA) and data envelopment analysis (DEA). SFA is a regression-like econometric method that often assumes a Cobb-Douglas production function with constant returns to scale (CRS) and constructs a linear production frontier in the input/output space. DEA is a linear-programming technique that constructs the production frontier as a convex envelope of the observed points in the input/output space without assuming a specific functional form for the production function and thus allows variable returns to scale (VRS).

DEA and SFA also differ in their treatment of errors. DEA is a strictly deterministic technique: it ignores the error term and treats the total deviation from the production frontier as inefficiency. SFA, on the other hand, assumes that the deviations from the frontier can be split into two components: a symmetrical ("two-sided") random error with mean zero (classical white noise) and a "one-sided" inefficiency component that takes only positive values from a truncated normal distribution with a positive mean (e.g., the half-normal distribution). Such an assumption decomposes the total error into a classical error term and an added one-sided error term. As a result of this different treatment of the error term, none of the observed points can by definition fall outside the DEA production frontier, whereas in the SFA model some points may definitely fall outside ("above") the production frontier if their classical error term is large enough. These stray points may still have a non-zero inefficiency score, which is determined by their one-sided error component.

Figure 14.1 illustrated the DEA production frontier constructed as a convex envelope of the sample of observation points in the input–output plane. Points C, F, A, B lie on the production frontier and are by definition efficient. Point D lies inside ("below") the production frontier and is inefficient. Its input-oriented inefficiency is represented by the distance DF from the frontier (this is the amount by which input use can be reduced at the given level of output); its output-oriented inefficiency is the distance BD (this is the amount by which output can be increased at the given level of input use). In this chapter all DEA results represent input-oriented inefficiency.

Fig. 14.1. DEA production frontier (schematic diagram).

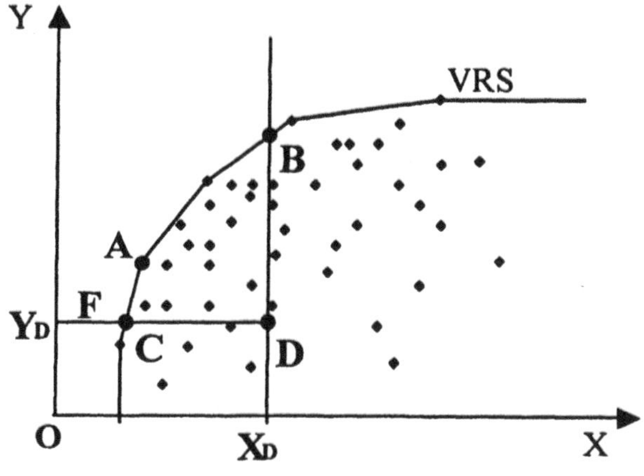

Fig. 14.2. SFA production frontier
for a sample of Ukrainian farms

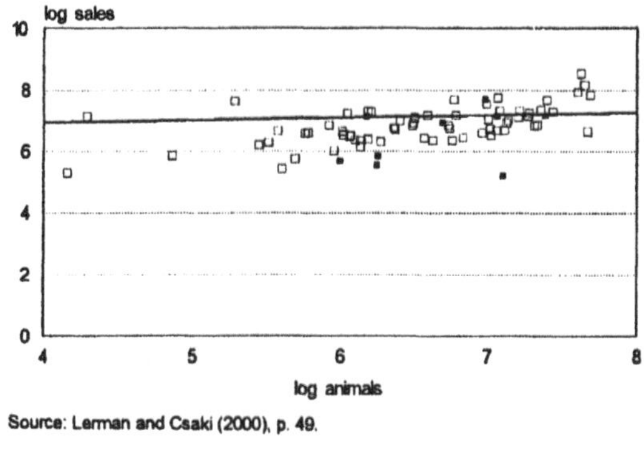

Source: Lerman and Csaki (2000), p. 49.

An SFA production frontier constructed using a sample of real data from a survey of Ukrainian farms is shown in Figure 14.2. The SFA inefficiency of a particular farm is not directly equal to its distance from the frontier: the inefficiency index is calculated by subtracting the random error component from the total distance from the frontier (the total error). Therefore, not all the points fall below the SFA frontier: for some points the random error ("white noise") component exceeds the inefficiency component and these points lie above the frontier.

As DEA requires no functional form on the input/output relationship, it is used by those who believe that imposing any functional relationship on the input/output mix is too restrictive. While DEA's non-parametric approach may appear more flexible, SFA has the advantage of explicitly accounting for measurement error in the classical error term, which if not included (as in DEA) means that any measurement error is incorrectly assumed to be technical inefficiency. So there is no clear, better choice between these two methods which often results in both methods being applied to the same sample.

The TE index produced by both methods actually measures "gross" technical efficiency, which in fact consists of three components: pure, scale, and congestion TE (Fare et al., 1985). Ignoring the distinction between gross and pure TE may be a problem when, as we do here, a "whole farm" versus "commodity by commodity" approach to farms are both considered. The different components of TE are affected by different combinations of regional and organizational factors. Therefore, different levels of aggregation in sample data and the varying institutional environment in which a farm is embedded could lead to TE indexes that appear to contradict each other due to differences in the three components.

In the 1990s, the technical efficiency literature expanded again with the growing use of Z-variables in the application of SFA. Prior to the 1990s, researchers would take estimates of TE and run auxiliary or "second step" regressions on a wide range of policy variables (so-called Z-variables) that might explain the measured technical inefficiency. The newer SFA method allows for the effect of these Z-variables simultaneously with the calculation of the TE scores in a one-step procedure (e.g., Audibert, 1997; Wang and Schmidt, 2002). The method enables analysts to better link technical inefficiency to policy by explicitly including in the estimation both economic variables and other variables (e.g., institutional or sociological factors) that fall outside standard production function analysis. This one-step extension is available for SFA only, while DEA must still use two steps if Z-variables are being considered.

Technical Efficiency of Russian Farms

Prior evidence on production efficiency in Russian agriculture suggested that, coming out of the Soviet era, large corporate farms exhibited high technical inefficiency within large samples from a single region (see, e.g., Brock, 1996/1997). In other words, there was much room for improvement in farm utilization of existing inputs given a constant state of technology.

Table 14.1. Comparison of the two datasets used for technical efficiency analysis

	2003 BASIS survey	Goskomstat national dataset
Number of farms	500	Over 20,000
Type of farms	Corporate Individual	Corporate only
Time frame	One year (2002)	Panel data 1995-2002
Disaggregation options	"Total farm" Specific commodities	"Total farm" "Total farm" with classification by main commodity specialization
Output variable	Sales revenue Value of production	Sales revenue Cost of products sold

We proceeded to analyze the technical efficiency of Russian farms after more than a decade of transition using two sources of data: the 2003 BASIS survey and the Goskomstat national dataset. A schematic comparison of the two datasets is given in Table 14.1. The 2003 BASIS survey covered some 500 farms, whereas the Goskomstat national dataset had over 20,000 farms. The 2003 BASIS survey collected cross-section data at a single point in time (2002), whereas the Goskomstat national dataset had panel data spanning the period between 1995 and 2002. On the other hand, the Goskomstat national dataset contained data for corporate farms only, whereas the 2003 BASIS survey provided detailed information for farms of different organizational forms, both corporate and individual (subdivided in turn into peasant farms and commercially oriented household plots). Finally, the detailed survey data on inputs and outputs in the 2003 BASIS survey were disaggregated by specific commodities, whereas the Goskomstat national dataset contained "total farm" information only (with a possibility of classifying the farms by main commodity specialization).

The differences in datasets made it possible to apply DEA and SFA methods at different levels of aggregation and over time. The particularly rich vector of farm inputs in the BASIS survey allowed "total farm" analysis as well as detailed commodity-by-commodity analysis of technical efficiency of production of grain, sunflower, beef, etc. The option of classifying the corporate farms in the Goskomstat database by their main commodity specialization (i.e., farms that produce mainly grain, farms that produce mainly vegetables, farms that produce mainly pigs, etc.) naturally led to technical efficiency analysis of farms of different specializations though not at the fine commodity-by-commodity detail of the BASIS survey. We thus analyze three levels of aggregation in this study: "total farm" models and commodity-specific models at the two extremes with "total farm" models with farms classified by main commodity specialization in between. Table 14.2 presents a schema of the various combinations implemented in our technical efficiency analysis and indicates the names of the researchers responsible for each combination.

Table 14.2. Schematic classification of technical efficiency calculations

Models	Dataset	Organizational type	DEA	SFA
"Total farm"	2003 BASIS survey	Corporate farms	Lerman	Lerman
		Individual farms	Lerman	Lerman
"Total farm" with different commodity specialization	Goskomstat national database	Corporate farms (two individual years)	Uzun	—
Commodity models	2003 BASIS survey	Corporate farms	Grazhdaninova	Brock, Grazhdaninova

"Total Farm" Analysis Using the 2003 BASIS Survey

The "total farm" approach using the 2003 BASIS survey made it possible to evaluate technical inefficiency of corporate farms on their own and also to analyze separately corporate farms, independent peasant farms, and household plots in a pooled sample. The analysis led to a comparison of the technical inefficiency of these distinct organizational types—something not often done in the literature. Of the three organizational types analyzed, household plots appear to have the highest mean technical efficiency using either SFA or DEA methods. This is quite interesting given that one cause of inefficiency may be poor management or a lack of a sense of ownership in the results of the farm.

"Total farm" technical efficiency was analyzed using both DEA and SFA methods (for the description of the software used see Coelli et al., 1998). The technical efficiency of corporate farms was estimated on a sample of 119 observations with all non-missing values. The value of agricultural production (aggregating crops and livestock) was used as the output variables; the inputs included agricultural land in use (hectares), number of workers engaged in agricultural activities, input costs (in rubles), and value of farm fixed assets.

Table 14.3. Technical efficiency scores for large farms in the 2003 BASIS survey

	Mean (*n*=119)	Rostov (*n*=54)	Ivanovo/Nizhnii Novgorod (*n*=65)
DEA CRS	0.18	0.23*	0.14
DEA VRS	0.28	0.34*	0.23
SFA without regional factor	0.67	—	—
SFA with regional factor	0.62	0.70*	0.56

*Difference statistically significant at 5%.

Fig. 14.3. Distribution of SFA efficiency scores for corporate farms

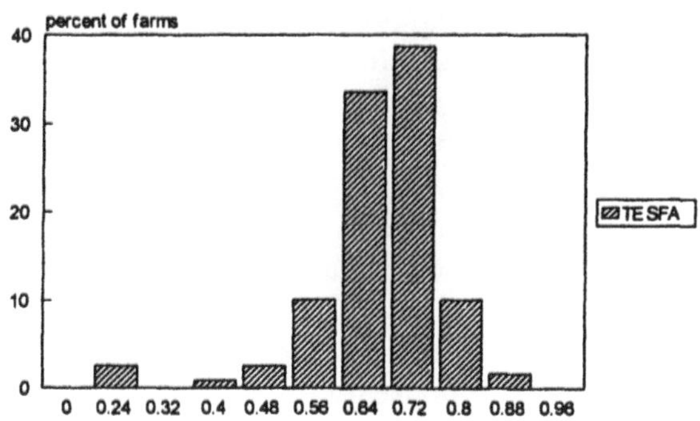

Source: 2003 BASIS survey

Fig. 14.4. Distribution of DEA efficiency scores for corporate farms

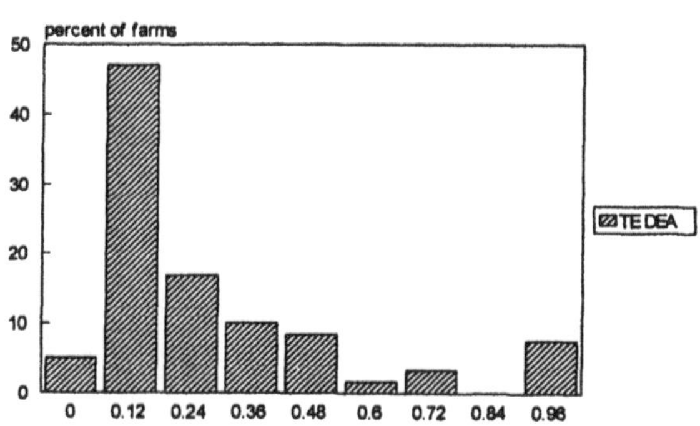

Source: 2003 BASIS survey

DEA (in both constant and variable return to scale specifications) suggests a large amount of technical inefficiency among corporate farms with mean TE scores in the 0.2-0.3 range. The SFA results yield a much higher mean efficiency in the 0.6-0.7 range (Table 14.3). The difference is visualized quite dramatically in Figures 14.3 and 14.4, which show the distributions of the efficiency measures produced by the two methods. While the SFA scores are bunched relatively close to the efficiency frontier (where TE = 1), the DEA scores are strongly shifted toward the lower tail of the distribution, characterizing a very high degree of inefficiency.

The literature suggests the differing results for SFA and DEA could be due to the different treatment of measurement error in the two methods (see *Technical Efficiency Framework* above). The SFA results obtained with the 2003 BASIS survey sample are similar to mean technical efficiency found on large Russian farms a decade earlier (Brock, 1996/1997), which would support the idea of a largely unrestructured corporate farm management (e.g., Liefert, 2001) even after a decade of reforms in other sectors.

Regional effects were explored by separating out the agriculturally rich southern province of Rostov. In all instances, the TE scores for Rostov were higher than for the agriculturally less endowed provinces of Ivanovo and Nizhnii Novgorod (Table 14.3; the differences were statistically significant). The regional factor was included as a Z-variable in SFA, allowing simultaneous one-step estimation of the TE scores with regional effects. In DEA the regional variable was included as a dummy in second-step regression.

For the analysis of the three organizational forms—large corporate farms, peasant farms, and household plots—in a pooled sample we had to use sales revenue, and not the value of farm production, as the output variable. This was inevitable because the data on the value of production in small individual farms turned out to be totally unreliable in the survey.[1] The vector of input variables included five inputs: agricultural land, number of agricultural workers, input costs (in million rubles), machinery (number of pieces adjusted for quality), and livestock (standard head). The selection of these variables maximized the number of valid observations: DEA, *n*=504 (non-logged) and SFA, *n*=499 (logged) out of 567 farms in the database.

Table 14.4. Technical efficiency scores for farms of different organizational forms in the 2003 BASIS survey

	Mean (*n*=499)	Corporate farms (*n*=121)	Peasant farms (*n*=211)	Household plots (*n*=167)
SFA*	0.745	0.736	0.699	0.811
DEA VRS	0.357	0.339	0.276	0.472

*With region and organizational form as Z factors.

[1] Uzun also had to use sales revenue in his analysis of all "large" corporate farms in Russia (see next section): this is the only output variable in the national Goskomstat databases.

The technical efficiency scores are presented in Table 14.4. Both the SFA and the DEA scores are significantly different across all three organizational forms (although for DEA some simultaneous pairwise tests fail to distinguish significant differences between corporate and peasant farms). The efficiency ranking by SFA scores and by DEA scores with variable returns to scale (DEA VRS) is Household plots > Corporate farms > Peasant farms. The small household plots thus appear to be the most productive by all methods, while peasant farms are least efficient.

Fig. 14.5. Distribution of SFA efficiency scores for corporate and individual farms

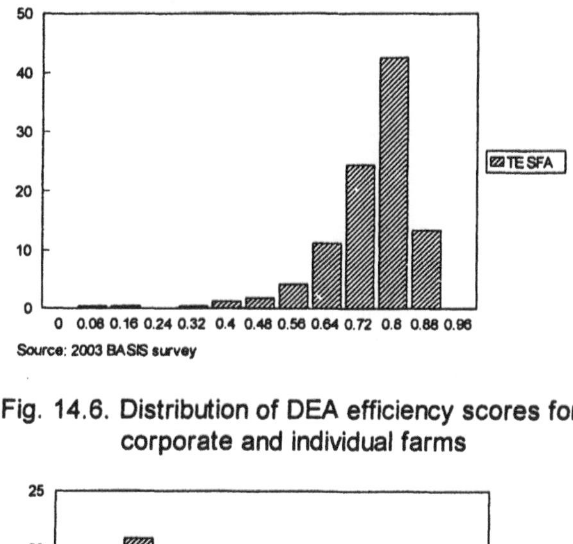

Source: 2003 BASIS survey

Fig. 14.6. Distribution of DEA efficiency scores for corporate and individual farms

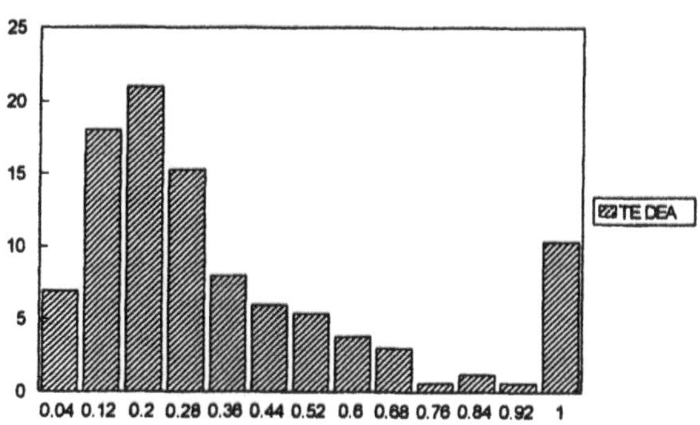

Source: 2003 BASIS survey

The frequency distributions of SFA and DEA efficiency scores for all 500 farms (large corporate farms and small individual farms) are presented in the two histograms in Figures 14.5 and 14.6. The SFA scores show the classical bunching near 1 (although the mode at 0.80 is relatively far from 1 by "Western standards"). The DEA scores, on the other hand, show a pathological reverse pattern with the mode at 0.20. The distributions for the pooled sample of corporate and individual farms are on the whole similar to those observed for corporate farms only (see Figures 14.3 and 14.4).

"Total Farm" Analysis Using the National Database

In addition to estimating technical efficiency from the data of the 2003 BASIS survey, we had the opportunity to calculate farm efficiency from the Goskomstat national database of all corporate farms in Russia (Table 4.15).

Table 14.5. TE scores of corporate farms with different commodity specializations 1995, 2002 (estimated by the DEA method from the Goskomstat national database)

Farm specialization	Number of farms	TE (mean)	St. dev.	% of farms		
				TE > 0.9	TE > 0.7	TE< 0.3
	(1)	(2)	(3)	(4)	(5)	(6)
Grain and technical crops						
1995	1,249	0.39	0.19	4	7	37
2002	1,296	0.24	0.21	4	5	77
Potatoes and vegetables						
1995	448	0.44	0.25	9	17	37
2002	298	0.47	0.30	15	23	36
Pigs						
1995	365	0.59	0.24	15	24	20
2002	198	0.64	0.26	23	32	15
Poultry						
1995	770	0.51	0.25	11	21	22
2002	541	0.51	0.28	16	25	28

Common wisdom suggests that farm productivity may vary depending on specialization. We accordingly classified the corporate farms by a simple sequential technique. First, the corporate farms were divided into three groups depending on the dominant component of their agricultural sales (crop farms, livestock farms, and other farms). The crop farms were in turn classified into three groups: those with predominance of grain and technical crops in their sales ("grain farms"); those with predominance of potatoes and vegetables ("vegetable farms"); and other crop farms. The livestock farms were divided into four groups by the main commodity contributing to their sales: cattle farms, pig farms, poultry farms, and other livestock farms. The number of farms in each specialization category is shown in the first column in Table 14.5. The TE scores were calculated for farms in each specialization category separately using input-oriented DEA with variable returns to scale.[2] 1995 and 2002 were compared to get a picture of efficiency changes over time.

Fig. 14.7. DEA efficiency scores of grain farms

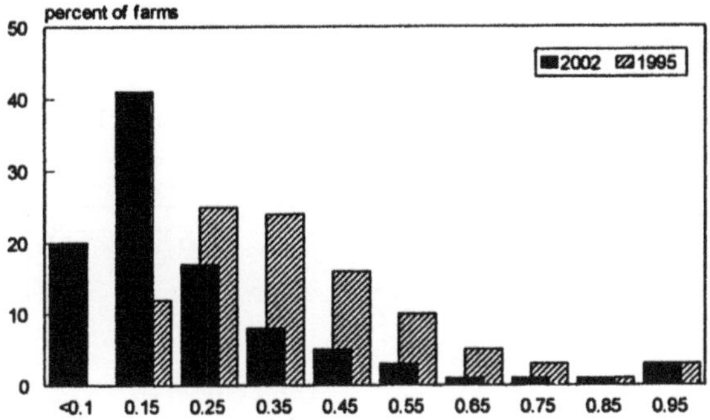

Source: Goskomstat national database

Fig. 14.8. DEA efficiency scores of vegetable farms

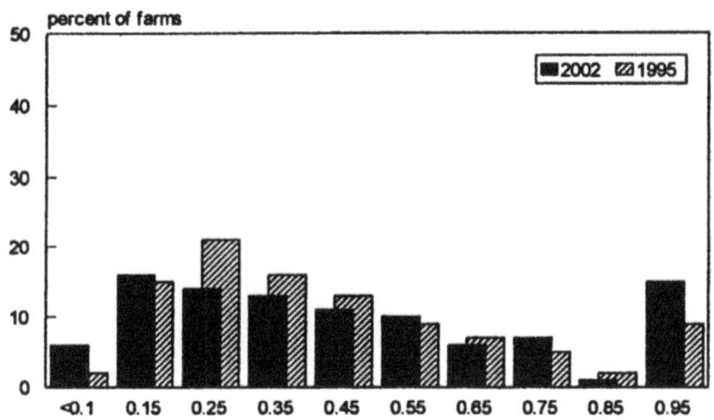

Source: Goskomstat national database

In all models, the output variable was the cost of products sold, including crop sales, livestock sales, and nonagricultural sales. Arable land, number of workers, and balance-sheet value of productive assets were used as the inputs.

Table 14.5 shows the TE scores for farms of different specializations in 1995 and 2002. In addition to the mean scores, the table also presents in-

[2] The calculations were carried out using the EMS software (http://www.wiso/ini-dortmund/de/LSFG/schttl/ems).

formation about the most efficient and the least efficient farms. The most efficient farms are those with TE > 0.9 (column 4), or less restrictively those with TE > 0.7 (column 5); these farms are close to the efficiency frontier, where technical efficiency reaches its maximum value 1. The least efficient farms lie far from the production frontier and have TE < 0.3 (column 6). The full distributions of corporate farms by TE scores in different specialization categories are shown in Figures 14.7-14.10.

Fig. 14.9. DEA efficiency scores of pig farms

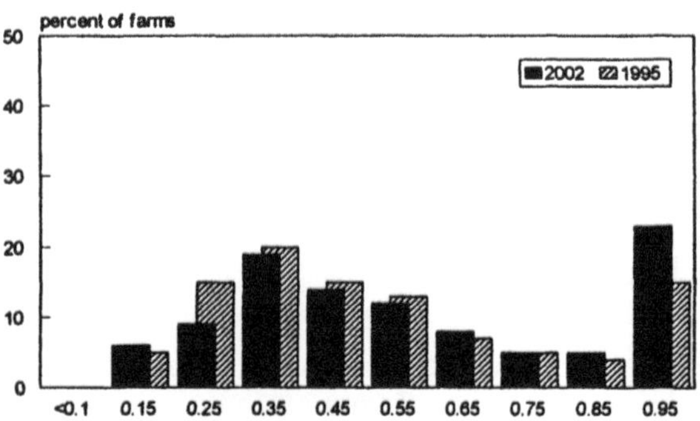

Source: Goskomstat national database

Fig. 14.10. DEA efficiency scores of poultry farms

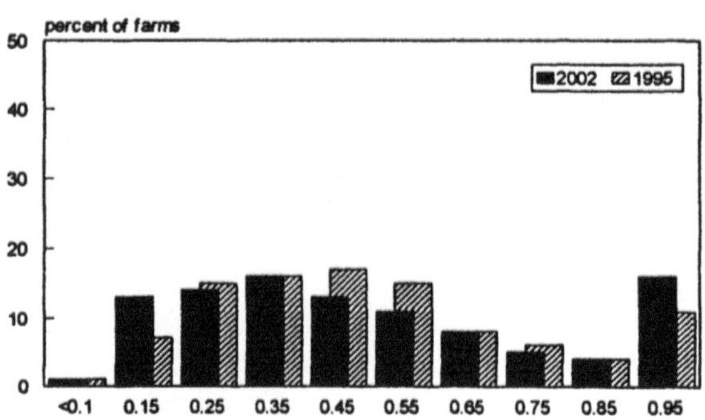

Source: Goskomstat national database

We see from Table 14.5 and from the distributions in Figures 14.7-14.10 that corporate farms show a high variability by technical efficiency. For most vegetable and poultry farms (about 60%) the TE scores are below 0.5. Pig farms are relatively more efficient, with TE scores below 0.5 for only one-third of the farms. Inefficiency is quite high for farms that produce grain and technical crops (Figure 14.7). For these farms the distribution of TE scores has a high peak at TE = 0.1-0.2 and nearly 80% of grain farms have TE < 0.3. The bulk of grain farms are thus quite inefficient, and furthermore their inefficiency increased markedly between 1995 and 2002 (see column 6 in Table 14.5).

The other distributions in Figures 14.8-14.10 follow a clear bimodal pattern. A substantial proportion (up to 20%) of vegetable, poultry, and pig farms lie close to the efficiency frontier with TE > 0.9 (see Table 14.5 and Figures 14.8-14.10). However, a wide chasm separates these highly efficient "leaders" from the bulk of the farms, for which the second peak occurs far from the efficiency frontier at TE around 0.2-0.4. The TE distribution in Russia sharply differs from the standard distributions observed in established market economies, where most farms are bunched in a strong peak near the efficiency frontier.

As noted previously, grain farms became much less efficient over time: their mean TE score dropped from 0.39 in 1995 to 0.24 in 2002. For other farms, the mean TE score did not change much, although this has been accompanied by increased polarization: both the percentage of most efficient farms (TE > 0.9) and the percentage of most inefficient farms (TE < 0.3) increased between 1995 and 2003 (see Table 14.5).

Commodity Models Using the 2003 BASIS Survey

The wide variety of microeconomic data gathered in the 2003 BASIS survey presented an opportunity for analyzing detailed commodity-specific models by SFA and DEA methods at a level of disaggregation that is quite unique. In this section we present the results obtained by the SFA method for two crops (grain and sunflower) and by the DEA method for a wider range of crop and livestock commodities covered in the survey.

The SFA method was applied to two specific commodities—grain and sunflower, which were chosen to get a highly disaggregated production function with the output and the main inputs measured in physical units (rather than monetary values).[3] The output is the quantity of each commodity produced by the corporate farms in the survey. The inputs include land, labor, fertilizer, fuel, seeds, and machinery—all expressed in physical units as allocated to each crop. The capital stock is adjusted for age (to reflect differences in quality) and the use of physical units for tractors and harvesters avoids problems with how farmers do depreciation accounting. Additional dummy variables were included for the weather, farm creation date

[3] The analysis was carried out using the standard Frontier 4.1 software (see Coelli et al., 1998).

(Soviet or post-Soviet), and location. The weather is not incorporated in the classical error term, but is explicitly included in the production function as reported by the farmers. The production function is about as close to a truly microeconomic production function of a specific commodity as one is likely to get for any farm. Importantly, because of the level of disaggregation, issues of output mix and capital stock mix that arise at a higher level of aggregation do not apply here. This detailed production function is then further enhanced by incorporating a range of Z-variables that might influence technical efficiency (distance from regional center, entitlement to subsidies, wage payment by task or by time, etc.).

Mean technical efficiency is found to be similar to the literature and to "total farm" findings reported in Tables 14.3 and 14.4, with grain production having a mean TE score of 0.80 and sunflower 0.65. Farms that are relatively more profitable and farms that did not exist in Soviet times are found to be more efficient than other farms in the sample. Distance from an oblast center does not impact on a farm's TE score, indicating that poor rural infrastructure may not be a constraint on farm performance with these two crops. How workers are paid also impacts on the efficiency of grain production: farms that pay workers by task and not a fixed salary are observed to be more efficient. It seems that, even if a farm cannot radically restructure, there may still be efficiency gains if the method of labor payment is changed alone, though more research beyond this small sample would be needed to support this conclusion. Clearly the TE index should be used as one of several measures of farm performance beyond those reported to tax authorities.

As expected, different regions had significantly different technical efficiency scores with the southern region of Rostov having slightly higher overall technical efficiency. This regional effect is consistent with the "total farm" result reported in Table 14.3. Farms that were profitable in the Soviet era had higher technical efficiency in the transition era possibly because this indicates better connections to suppliers or simply financial strength of the farm. Farms that specialized in a particular output or actively acquired land had higher technical efficiency as well—again suggesting this might proxy for better managed and simply actively managed farms. Finally, wage arrears (used as a Z-variable) negatively impacted on technical efficiency as hypothesized.

The technical efficiency of production for a wider range of commodities—both crop and livestock—was also estimated using input-oriented DEA models with variable returns to scale (VRS).[4] Input-oriented estima-

---

[4] The DEA program has been developed at AFE by Aleksandr Usol'tsev on the basis of standard linear programming algorithms published in the literature.

tion is more appropriate than the output-oriented alternative because one of the objectives of the study is to determine the efficiency of input use for the production of a given output and find ways to optimize input use. The impact of external factors on technical efficiency was then estimated by second-step regression analysis as the use of Z-variables in a one step procedure is not available in DEA.

Table 14.6 presents the TE scores for specific commodities. The high mean technical efficiency scores and the high frequency of "best practice" technologies in the sample limit the potential impact that can be expected from the adoption of "best practice" technologies by the inefficient farms. Thus, the adoption of "best practice" technologies will increase the production of beef by 22%-36% (Model 2 and Model 1, respectively). Results here suggest a disaggregate analysis of the sample reveals much more technical efficiency than a "whole farm" approach using the BASIS or Goskomstat data, which might seem contradictory. However, differences in coverage (120 farms in three oblasts in the sample-based analysis versus 1,200 farms in 75 oblasts in the national-level analysis), differences in the definitions of both input and output variables used, and the possible effect of "gross" versus "pure" TE mentioned above suggest that such differences should not be surprising. Indeed, the differences can help us focus on where the farm might improve its operations with much more attention perhaps given to how the farm management operates across commodities with issues such as the mix of output and equipment decided on the total farm level.

Table 14.6. Technical efficiency scores (TE) of corporate farms for selected commodities (estimated by the DEA method)

	TE (mean)	St. dev.	TE > 0.9, % of farms	TE > 0.7, % of farms	TE< 0.3, % of farms
Grain	0.77	0.23	39	64	2
Sunflower	0.71	0.24	35	49	2
Beef 1*	0.64	0.29	31	44	17
Beef 2*	0.78	0.26	54	64	6
Milk 1*	0.65	0.24	25	37	2
Milk 2*	0.82	0.22	55	69	1
Beef +milk 1 **	0.78	0.20	42	59	0
Beef +milk 2 **	0.88	0.16	65	80	0
Pork 1*	0.75	0.28	50	63	6
Pork 2*	0.88	0.18	67	79	0

*See notes on Beef/Milk Models 1 and 2 in Chapter 11, Table 11.2.

**Two-output DEA models (beef and milk). Without fattening operations, the inputs are shared by the two outputs.

The results of Model 1 for livestock products (in these models, animal feed is expressed in feed units) shed some light on the differences between farms near the efficiency frontier (those with TE > 0.9) and the highly inefficient farms with TE < 0.4. For the high-efficiency farms, the share of purchased feed is relatively high. It is 19% for the beef model, 14% for the milk model, and 10% for the pork model. For the low-efficiency farms, the respective purchased feed shares are 2.6%, 2.0%, and 3.4%. The mean share of purchased feed for all corporate farms in the sample is 7%. Purchased feed is mainly high-quality concentrated feed, whereas feed from own production is basically hay, pasture grasses, or low-quality concentrated feed. The farms at the efficiency frontier, using a high proportion of purchased feed in the ration, probably optimize the ration by analyzing the market price of feed, the cost of on-farm production, and the value of the end product. The inefficient farms, using a small proportion of purchased feed in their ration, probably follow the strategy of cost minimization for purchased inputs because of financial constraints.

Table 14.7. Factors that influence technical efficiency*

Factor	Grain	Sunflow-er	Beef 1	Beef 2	Milk 1
Ivanovo regional factor	−0.134		−0.337	−0.296	−0.279
Distance from oblast capital				0.243	
Augmentation of farm holdings	0.122			0.321	0.281
Wage arrears	−0.017				−0.022
Pre −reform status (profitable)		0.298			0.214
Crop specialization			−0.496	−0.613	−0.777
Controlling packet of shares	−0.175				
Surplus labor					

Factor	Milk 2	Beef+ milk 1	Beef+ milk 2	Pork 1	Pork 2
Ivanovo regional factor	−0.294			−0.492	−0.284
Distance from oblast capital		0.134			
Augmentation of farm holdings	0.298		0.167	0.267	
Wage arrears	−0.017	−0.018			−0.025
Pre −reform status (profitable)	0.185		0.171		
Crop specialization	−0.613	−0.533	−0.343	−1.179	−0.485
Controlling packet of shares					
Surplus labor			0.002		

* Only factors that are statistically significant at 10% level in one of the commodity models are shown.

The relationship of the DEA technical efficiency scores with various external factors, such as farm size, location, or financial conditions, may be determined in second-stage regression. Table 14.7 shows the estimation results for such a regression model. Many factors that a priori were expected to affect the technical efficiency proved to be not statistically significant and are not shown in Table 14.7. Thus, the farm size (measured by hectares of land used) does not have a statistically significant effect on technical efficiency. State subsidies and borrowing of any kind do not affect technical efficiency (however, very few farms in the sample provided information on these variables). The managerial qualification variable expressing knowledge of tax laws, lease payments, and land allocation procedures is not statistically significant in the model. The land utilization ratio (i.e., the share of land actually used in agricultural production) does not affect technical efficiency. It is quite possible that keeping agricultural land in the farm's possession without cultivating it is the most efficient strategy under the present circumstances because of complex alienation procedures and high transaction costs.

On the other hand, most models show a positive association between enlargement of holdings and technical efficiency. Farm enlargement (a yes/no variable that indicates if the farm has added new agricultural land to its holdings) is statistically significant in both crop and livestock production models. Farm enlargement not only represents expansion of the sown area, but it is also an indicator of management quality.

As suggested by a priori considerations, wage arrears have a negative effect on technical efficiency (in half the commodity models). If this factor is accepted as a proxy for financial health, we conclude that financially ailing farms are less efficient than the rest. This factor is also closely connected with management quality. The absence of wage arrears and good financial health reflect highly qualified management.

The impact of management structure is also reflected by the farms' pre-reform status. Farms that were profitable in the pre-reform period demonstrate higher technical efficiency in some models. The existence of a controlling packet of shares, which a priori led us to expect greater efficiency due to more effective control of the majority owner over management, is not statistically significant in most models, perhaps because of the very small number of farms with a controlling packet in the sample.

Livestock producers that add crops to their product mix are less efficient in all commodities. The regional factor has a statistically significant effect: farms in Ivanovo Oblast are the least efficient and Rostov farms are the most efficient in the sample. Distance from the oblast capital in general does not have a statistically significant effect on technical efficiency. The existence of surplus labor does not affect the technical efficiency of corporate farms either.

Conclusions and Further Research

Russian farms, like farms in other countries, reveal a variety of technical inefficiency results across commodities and many types of inputs. Technical efficiency scores can be used in conjunction with standard performance measures (e.g., reported profit, which unfortunately is prone to being manipulated for tax purposes) to fully understand how Russian farms are doing. Because Russian farms are generally not very specialized, our unique commodity-by-commodity results deserve special attention. Assigning various inputs to the production that they actually are used for without getting lost in aggregate measures needs to be done over time and in other regions to more fully understand how Russian agriculture is performing. While farms may be able to achieve best practice on the current frontier, that frontier is stagnant and not shifting out (technological progress) which is a major inhibitor to growth.

Enough technical inefficiency has been found in this study to stimulate further research in other regions and comparison papers where Russian farms as well as similar farms from other countries might be included in the same sample. The latter research would support a trend in the comparative economics literature towards looking at cross-country regressions to better understand institutional issues and other cross-country differences on performance. Russian farm data combined with, say, U.S. farm data in a single sample carefully controlling for location would seem to be one avenue to go down to better understand Russian farm performance. Adding another transition economy such as Ukraine would enrich the analysis further and lead to some interesting policy recommendations. Throughout this comparative analysis, our results using household plots and small farms should be incorporated to test for the influence of farm size and sense of ownership on technical efficiency across regions and countries. We know Russian household plots are relatively productive and have found limited evidence for them being relatively more efficient, but much more research would be needed to be conclusive about an optimal size for a Russian farm by crop or livestock type.

The cross-sectional nature of the survey data rules out panel analysis of technical efficiency over time. Further research might use the Goskomstat database to carefully construct a panel microeconomic dataset including the farms used in the project. While it is clear that agriculture did not become suddenly highly efficient after the freeing of prices in the early 1990s, technical inefficiency may be declining slowly over time, or conversely it may be increasing in the aftermath of the 1998 financial crisis. Only time-series analysis can answer this question. Finally, the impact of other variables on technical efficiency such as location would be much better understood if the impact is examined over time as well as space.

References

Aigner, D.J., Lovell, C.A.K., and Schmidt, P. (1977). "Formulation and Estimation of Stochastic Frontier Production Function Models." *Journal of Econometrics* 6(1): 21-37.

Audibert, M. (1997). "Technical Inefficiency Effects Among Paddy Farmers in the Villages of the 'Office du Niger', Mali, West Africa." *Journal of Productivity Analysis* 8: 379-394.

Bauer, P. (1990). "Recent Developments in the Econometric Estimation of Frontiers." *Journal of Econometrics* 46: 39-56.

Brock, G. (1996/1997). "Are Russian Farms Efficient?" *Journal of International and Comparative Economics* 5(1): 45-66.

Clayton, E. (1980). "Productivity in Soviet Agriculture." *Slavic Review* 39(3): 446-458.

Clayton, E. (1984). "Factor Endowment, Variable Proportions, and Agricultural Modernization: Regional Production Functions for Soviet Agriculture." in Robert C. Stuart, ed., *The Soviet Rural Economy*. Totowa, NJ: Rowman.

Coelli, T, Rao, D.S.P., and Battese, G. (1998). *An Introduction to Efficiency and Productivity Analysis*. Boston: Kluwer.

Danilin, V.I., Materov, I. S., Rosefielde, S., and Lovell, C. A. K. (1985). "Measuring Enterprise Efficiency in the Soviet Union: A Stochastic Frontier Analysis." *Economica* 52(206): 225-233.

Fare, R., Grosskopf, S., and Lovell, C.A.K. (1985). *The Measurement of Efficiency of Production*. Boston: Kluwer.

Forsund, F., Lovell, C. A. K., and Schmidt, P. (1980). "A Survey of Frontier Production Functions and of their Relationship to Efficiency Measurement." *Journal of Econometrics* 13: 5-25.

Gray, K. (1981). "Soviet Agricultural Prices, Rent and Land Cadastres." *Journal of Comparative Economics* 5(1): 43-59.

Grazhdaninova, M. and Brock, G. (2004). "Grain and Sunflower Crops on Russian Farms in 2001: How Efficient is Crop Production?" *Post-Communist Economies* 16(3): 297-305.

Johnson, D. G. and Brooks, K. (1983). *Prospects for Soviet Agriculture in the 1980s*. Bloomington: Indiana University Press.

Jondrow, J., Lovell, C. A. K., Materov, I. S., and Schmidt, P. (1982). "On the Estimation of Technical Inefficiency in the Stochastic Frontier Production Function Model." *Journal of Econometrics* 19: 233-238.

Lerman, Z. and Csaki, C. (2000). *Ukraine: Review of Farm Restructuring Experiences*. World Bank Technical Paper 459, Washington, DC: World Bank.

Liefert, W. (2001). "Agricultural Reform: Major Commodity Restructuring but Little Institutional Change." in *Russia's Uncertain Economic Future*, Joint Economic Committee, U.S. Congress. Washington, DC (December), pp. 253-281.

Wang, H. and Schmidt, P. (2002). "One-step and Two-step Estimation of the Effects of Exogenous Variables on Technical Efficiency Levels." *Journal of Productivity Analysis* 18(2): 129-144.

Chapter 15

Summary

Bruce Gardner

The preceding chapters constitute a wide-ranging compendium of information about Russian agriculture and agricultural policies since 1991. The depth of data and analysis pertaining to Russian agricultural input markets is unprecedented in prior publications. This chapter attempts to identify the main points of fact, explanation, and policy implications that the research project has developed. These points are focused on agricultural inputs and markets for them. The input-market focus is in some respects a limitation of the BASIS project and of this book, but an important conclusion of the project is that input market considerations are in fact central to the overall situation of Russian agricultural production and rural incomes at present, and to the prospects for rural economic growth in the future.

The Economic Situation of Russian Agriculture in 2000-2005 as Related to Input Markets

Discussion of Russian agriculture since 1991 typically begins with the decline of agricultural production. By 1998, gross agricultural output (GAO) was estimated at 56% of its 1990 level (Table 7.29).[1] The economic significance of that decline depends crucially on associated events in the input markets. Agriculture had been heavily supported by State direction and subsidies through the 1980s, and it was only to be expected that when that support evaporated in the 1990s, output would decline. The central development on the input side that makes the decline of agricultural output an

[1] Unless an explicit external reference is given, all references in this chapter refer to contents of this volume.

indicator of a serious economic problem is that employment of labor in agriculture did not decline nearly as much, and not at all in the years of transition through 1995 (Table 7.10). This in itself means agriculture generated a decreasing material basis for the standard of living in rural areas. That such a result persists, albeit with some improvement since 1998, after 15 years of transition to a market economy, goes strongly against the hopes and expectations of 1991. What happened?

The factual foundations for answering that question begin with events in input markets. Chapter 7 provides detail on rural employment, with estimates of labor devoted to agricultural and nonagricultural production. Chapter 8 develops estimates of machinery stocks and the use of purchased inputs including fertilizers, fuel, seeds, and other goods and services, and Chapters 2, 3, and 4 give data on agricultural land use in corporate, peasant, and household plot farm enterprises. Between 1990-1991 and 2002 agricultural labor declined from 9.7 to 7.7 million full-time equivalent workers (Chapter 7), and the planted area of land from 117 to 84 million hectares. Other inputs are incompletely measured but the data of Figures 8.3 and 8.4 indicate even larger percentage declines in tractors, combines, fuel, and fertilizer use over this period. Using roughly estimated factor share weights, an index of all these inputs declined by 51% between 1990-1991 and 2002.[2] With output declining 36% over this period (calculated as the change in natural logs), this implies an increase in total factor productivity of 15%.

The decline in output occurred almost entirely in the first five years after 1991. The same is true of the decline in purchased inputs, while the estimated labor input remained the same in 1995 as it had been in 1990 (Table 7.29). Thus total factor productivity declined during the initial years of transition, and labor productivity as calculated in Table 7.29 declined sharply (40% between 1990 and 1995); while the losses in total factor productivity, but not labor productivity (see again Table 7.29), were more than recovered between 1995 and 2002.

---

[2] The weights used are labor, 0.45, land, 0.20, fuel and fertilizer, 0.2, and machinery, 0.15. The decline in the index is not highly sensitive to the weights because all inputs declined substantially. The main differences among the inputs are that land and labor declined in the neighborhood of 25%-30% over the period while the purchased inputs declined 70%-90%. Note that the proper calculation of percentage changes between the earlier and later years is not straightforward. The percentage changes for purchased inputs in Chapter 8 use the initial period as a base, e.g., when fertilizer falls from 80 kg/ha to 40 kg/ha, this is a decline of 50%. It is equally valid to use the final period as a base, so the decline from 80 kg/ha to 40 kg/ha is 100%. The typical approach in recent index number calculation is to use the change in the natural log or cross-weights that use the average of the initial and final period as a base (giving 69% or 67%, respectively as the percentage decline from 80 kg/ha to 40 kg/ha). The calculations here use the change in natural logs.

Since Russian agriculture was considered inefficient prior to the transition, as discussed in Chapter 2, the lack of improvement in productivity immediately after 1991 and rather slow recovery to the present has been an unpleasant outcome and an unexpected surprise, at least to some economists. Much of the detailed analysis in the chapters of this book can be viewed as an attempt to understand better what underlies this outcome, and to suggest what might be done to achieve sustained productivity growth in Russian agriculture.

Constraints in Input Markets

A major issue in untangling the causes of changes in Russian agricultural production is identifying the extent to which problems are attributable to the availability of inputs, the cost of inputs, or the management of the conversion of inputs to output on farms. Assuming that the story as outlined in the preceding section is roughly correct, we can say that the output decline of the whole period 1991 to 2002 is attributable perhaps entirely, in the production economics context, to a reduction in inputs used. Efficiency in the management of the inputs used, while not increasing in the way one might expect as market incentives replaced the collective farm system, has apparently increased somewhat.[3] This does not of course mean that management inefficiency is absent. One of the most persistent results in the chapters of Part III from DEA and SFA analysis is that considerable inefficiency remains. What the TFP calculations suggest is that inefficiency was even more rampant pre-1991.

Given that increased management inefficiency did not occur after 1991, the output declines are attributable to reduced use of inputs. The question then is whether these reductions were a matter of economic choice by farm managers, or external constraints on input availability. Experience with oth-

[3] It is not possible to be confident about an increase in total factor productivity over 1991-2002 because of the uncertainties in input quantity data, which is unavailable for some inputs and difficult to measure for others, notably machinery and other capital goods, and the impossibility of accurate measures of factor shares to weight the inputs. However, sensitivity analysis suggests that the finding of some increase in TFP is robust. For example, measured TFP growth would have been less if the shares of labor and land had been higher and the shares of purchased inputs, especially fertilizer, lower. If we increase the share of labor from the 0.45 used above to 0.5, and decrease the share of fuel and fertilizer from 0.2 to 0.15, the estimated increase in TFP over the period is reduced from 15% to 10%. One would need implausibly high weights on land and/or labor, and low weights on purchased inputs, to estimate a decline in TFP. The more plausible sensitivity experiments would be to use lower labor and/or land weights and higher purchased input weights, which would raise estimated TFP growth above the 15% given in the text.

er countries as recounted in the literature on economic development, as well as some anecdotal evidence about Russia, suggested that input market supply constraints were a real possibility. The fact that the great bulk of Russian fertilizer has been exported while its use in Russia has been so greatly reduced suggests a supply-side problem. The idea that farms may be credit constrained as former state-supplied credit has been withdrawn is also plausible. Yet the predominance of evidence as developed in the BASIS project indicates that such constraints are not the main reasons for declines in input use.

A key indicator that suggests reasons for farmers' input-use choices to have changed a lot since 1991 is the terms of trade between agricultural output prices and input prices as shown in Figure 8.1. Output prices relative to input prices by 1994 were one fifth their level of 1990 (160% lower using the change in natural log as an indicator). Profit-maximizing farmers are bound to use substantially less purchased inputs in these circumstances. In this situation it makes a lot of difference for policy recommendations whether the adjustments were in fact ones that optimizing farmers would make, or whether they lacked a good economic rationale from the managerial viewpoint. If the latter, then steps to improve farm management (for example via restructuring of farms, or placing bankrupt farms under new management, or improving incentives for earning farm profits through strengthened property rights or incentive contract with managers) are a plausible means of improvement in the performance of the sector.

Assessment of Management Performance

Unfortunately, the large changes in quantities and proportions of inputs used make it difficult to apply standard tests of performance based on indicators of efficiency/inefficiency. The difficulties of measuring total factor productivity have been discussed above. The chapters in Part III work with two alternative approaches to measuring efficiency that do not require as much aggregation of disparate inputs: estimates of value of marginal product as compared to factor prices, and nonparametric data analysis of shortfalls of production from the production frontier—the maximum that could be obtained with given inputs.

Russian and Western studies of efficiency trends in Russian agriculture in the 1990s produce widely differing results. Some researchers find that efficiency increased during the transition, while others conclude that efficiency decreased.[4] We can only say with certainty that the huge improvements in efficiency originally expected at the beginning of reforms in 1991-1992 have not materialized. This study applied the data of the BASIS survey conducted in 2002-2003 to examine two forms of efficiency: allocative efficiency and technical efficiency. Allocative efficiency measures show how far the farm is from the point of maximum profitability given the existing market prices for inputs and products. Thus, allocative efficiency determines whether the factors of production are used in proportions that ensure maximum output at given market prices. Technical efficiency measures show how efficiently the farm uses the available inputs to produce a given output. In other words, technical efficiency determines whether the farm achieves maximum output using a given bundle of factors of production.

The basic approach to estimating allocative efficiency is through the *value of marginal product (VMP)*. The value of marginal product is calculated from econometrically estimated production functions. The use of an agricultural input is allocatively efficient if the value of marginal product is equal to its price. Technical efficiency is usually estimated by one of two approaches: parametric (*stochastic frontier analysis, SFA*) or non-parametric (*data envelopment analysis, DEA*). Both methods calculate an efficiency index, which measures the distance of the observed firm (corporate farm in our case) from a point on the production frontier. Firms lying on the production frontier are absolutely efficient, and the "inefficiency" of the remaining firms increases with the distance from the production frontie

Estimation of Value of Marginal Products Compared to Input Prices

Chapter 10 reviews the efficiency criterion that optimal level of use of each input occurs when the value of the marginal product (VMP) generated by the input just equals the cost of the input, and summarizes the results of Chapters 11-13 pertaining to input use on large corporate farms.

One result of the analysis that may be surprising is that the labor input is typically found to have a value of marginal product in excess of the wage rate. The seven different estimates reported in Table 10.2 show VMPs that range from 3% less to 32% more than the relevant cost of labor. Given the standard errors of the estimated production function coefficients underlying the VMPs and the difficulties of determining the appropriate cost of labor for comparison, if one were to take efficiency (VMP equal to factor price) as the null hypothesis that has to be rejected at some low probability of er-

---

[4] The first view is advanced in Lerman et al. (2003) and Voigt and Uvarovsky (2001); the second view in Osborne and Trueblood (2006) and Sedik et al. (1999). The references are at the end of Chapter 10.

ror, one would be likely to retain the null hypothesis (efficiency). But given that six of the seven estimates indicate a VMP that exceeds labor cost, we are inclined to take the conclusion seriously that, at least, there is no case for an excess supply of labor in Russian agriculture. Note that the finding of VMP greater than labor cost is also consistent with the response to BASIS questionnaires that many farms were experiencing a labor shortage.

For land the VMP analysis generates the opposite finding. Six of the seven estimates reported in Table 10.2 generate a VMP of land that falls short of the price (rental value) of land. This is not unexpected. Indeed several production functions were not used in the VMP analysis because the estimated coefficients in the production function were not significantly different from zero. This suggests that the low VMPs reported, from those estimated production functions in which land was a significant variable, actually overstate the contribution of land. The result is also consistent with the BASIS questionnaire results, which did not find respondents pointing to a lack of land availability, and with the observation of substantial areas of farmland remaining idle in some places.

Purchased inputs present a mixed picture. Fertilizer shortages have been much discussed and the VMP estimates indicate that indeed optimal resource allocation would imply more fertilizer use given product and factor prices. Why so much Russian fertilizer is exported in these circumstances is discussed in Chapter 9.[5] But the generic category of purchased inputs is estimated to be generally over-used as reported in Table 10.2.

Nonparametric Efficiency Measures

The difficulties of estimating a production function using a cross-section of heterogeneous observations of farms, many of which are likely not maximizing profits, led the BASIS researchers to devote substantial efforts to methods of analysis that do not require statistical estimation. Chapter 14 undertakes, for different data sets, the two most widely used methods of nonparametric efficiency analysis, Data Envelopment Analysis (DEA) and Stochastic Frontier Analysis (SFA). Both are methods of systematically measuring how far individual farms are from the performance in converting inputs to outputs of the farms whose input and output data indicate the best performance in the sample. Thus the spirit of the analysis is quite different. The estimated production functions assume all the farms are identical in the parameters that convert inputs to output, while DEA and SFA

[5] See also W. Liefert, B. Gardner, and E. Serova. "Allocative Efficiency in Russian Agriculture." *American Journal of Agricultural Economics* 85: 1228-1233 (2003).

permit all the farms to be different in that respect (but that all have a common frontier, or maximum achievable output from given inputs). DEA is open to the criticism that the best-performing farms in any cross-sectional sample may well have achieved that eminence via randomly high production, or even measurement error, so that shortfalls from that frontier have no implication of economic inefficiency in resource allocation. SFA is an attempt to adjust for random error in production.

The findings of both the DEA and SFA research on several data bases show a strikingly large spread between the best performers and a large number of farms whose performance is considerably worse. The DEA analysis indicates many farms obtaining only 20% to 30% of the output for given inputs that the best-performing farms attain. SFA results indicate strikingly less inefficiency, as discussed in Chapter 14, roughly to the same extent as earlier studies had found for pre-1991 collective farms. It seems likely that the limitations of DEA in setting a benchmark for efficiency that is due to random or unique characteristics of a few farms is causing a substantial problem of interpretation of DEA findings.

Causes of Inefficiency and Implications

Chapter 14 carries out an econometric investigation intended to uncover the causes of farm-to-farm variations in performance. Regional factors are a significant contributor, with Rostov farms being most efficient in the BASIS sample. This result does not tell us the extent to which physical characteristics such as soils or weather are responsible, or differences in policies or institutions in the different regions. Variables intended to get at measurable indicators of managerial differences, such as a concentrated controlling interest in the farm. There is evidence that financial problems (indicated by arrears in wage payments) are associated with inefficiency and that farm enlargement is associated with efficiency. This tells us that the efficiency measures really do get at the relative performance of farms (as opposed to being statistical artifacts) but still do not provide a sense of what it is that enables better-performing farms to perform better.

In conjunction with the VMP analysis summarized in Chapter 10, the following are given as possible explanations for allocative inefficiency on a sector-wide basis: (1) deficiencies of farm management decision-making; (2) market power of farm managers (monopsony power) over their workers; (3) a lack of working capital needed to purchase inputs; and (4) rationing of input quantities by suppliers. This last fact appears pertinent to fertilizer, where producers who prefer to export their output given the structure of prices available to them, even though the value of the marginal product in agriculture exceeds the price farms pay. The suggestion is that this price paid is in effect a subsidized price, at the expense of fertilizer producers, and that consequently suppliers limit their sales at those prices. A lack of working capital is plausible in helping to explain why farms to not offer more to buy fertilizer, but the lack of a general finding that purchased inputs have VMPs that exceed the input prices does not indicate a severe credit constraint. Nonetheless, the association of financial stringency with inefficiency in the DEA/SFA analysis suggests that finance is a significant problem for many farms—but not necessarily that there is a market failure in the provision of credit.

Index

Contributors

Vladimir Bogdanovskii
Candidate of economic sciences, VNIETUSKh—Institute for Economics, Labor, and Management in Agriculture, Moscow, Russia.
E-mail: bogdanovsky@bk.ru

Gregory Brock
Associate professor of economics, School of Economic Development, Georgia Southern University, Statesboro, Georgia, USA.
E-mail: gbrock@georgiasouthern.edu

Bruce Gardner
Professor of agricultural economics, College of Agriculture and Natural Resources, University of Maryland, College Park, MD, USA.
E-mail: bgardner@arec.umd.edu

Margarita Grazhdaninova
Candidate of economic sciences, AFE—Analytical Centre for Agri-Food Economics, Moscow, Russia. E-mail: margo@iet.ru

David Epshtein
Doctor of economics, professor, chief researcher at the Northwestern Institute of Agricultural Economics, St. Petersburg.
E-mail: epshtein@DE1150.spb.edu

Robert Jolly
Professor in the Department of Economics, Iowa State University, Ames, IA, USA. E-mail: rjolly@iastate.edu

Irina Khramova
Candidate of economic sciences, AFE—Analytical Centre for Agri-Food Economics, Moscow, Russia. E-mail: irakhramova@yahoo.com

Zvi Lerman
Professor in the Department of Agricultural Economics and Management, The Hebrew University of Jerusalem, Israel. E-mail: lerman@agri.huji.ac.il

William Liefert
PhD in economics, senior agricultural economist with Economic Research Service, U.S. Department of Agriculture, Washington, DC, USA.
E-mail : wliefert@ers.usda.gov

Leonard Rolfes, Jr.
Deputy director of RDI—Rural Development Institute, Seattle, Washington, USA. E-mail: leonardr@rdiland.org

Dmitri Rylko
Candidate of economic sciences, general director of IKAR—Institute for Agricultural Market Studies, Head of IMEMO Agribusiness Center, Moscow, Russia. E-mail: D.Rylko@ikar.ru

Sergei Sazonov
Doctor of technical sciences, professor, director of the Independent Economic Analytical Center of Peasant Farms, Tambov, Russia.
E-mail: KFX@pub.tmb.ru

Damira Sazonova
Candidate of economic sciences, chairman of the board of the Independent Economic Analytical Center of Peasant Farms, Tambov, Russia.
E-mail: KFX@pub.tmb.ru

Evgeniya Serova
Doctor of economics, professor of agricultural economics at the Moscow Higher School of Economics, President of AFE—Analytical Centre for Agri-Food Economics and Higher School of Economics, Moscow, Russia.
E-mail: serova@iet.ru

Olga Shick
Candidate of economic sciences, project manager at AFE—Analytical Centre for Agri-Food Economics, Moscow, Russia. E-mail: shick@iet.ru

Natalya Shagaida
Doctor of economics, lead researcher at VIAPI—Nikonov Institute of Agrarian Problems and Informatics, Moscow, Russia.
E-mail: nshagaida@raf.org.ru

Aleksandr Subbotin
MA in economics, AFE—Analytical Centre for Agri-Food Economics, Moscow, Russia. E-mail: danilevski@mail.ru

Vasilii Uzun
Doctor of economics, professor, holder of Russia's Distinguished Scientist award, Head of Department of Agricultural Reform at VIAPI—Nikonov Institute of Agrarian Problems and Informatics, Moscow, Russia.
E-mail: vuzun@raf.org.ru

Olga Yastrebova
Candidate of economic sciences, associate professor at Lomonosov State University, general director of Ecorys-NEI Research and Consulting, Moscow, Russia. E-mail: yastrebova@ecorys.ru

www.ingramcontent.com/pod-product-compliance
Ingram Content Group UK Ltd.
Pitfield, Milton Keynes, MK11 3LW, UK
UKHW012034150926
475560UK00001B/12

9 780739 120095